Boundaries of the Mind
The Individual in the Fragile Sciences
Cognition

Where does the mind begin and end? Most philosophers and cognitive scientists take the view that the mind is bounded by the skull or skin of the individual. Rob Wilson, in this provocative and challenging new book, provides the foundation for the view that the mind extends beyond the boundary of the individual.

The approach adopted offers a unique blend of traditional philosophical analysis, cognitive science, and the history of psychology and the human sciences. There are highly accessible discussions of the origin of psychology, nativism about the mind, contemporary views of computation, mental representation, consciousness, the metaphysics of mind, the idea of group minds, and how to think about the individual in the cognitive, biological, and social sciences, what Wilson refers to as the fragile sciences. A companion volume *Genes and the Agents of Life* explores this general theme in the biological sciences.

Written with verve and clarity, this ambitious book will appeal to a broad swath of professionals and students in philosophy, psychology, cognitive science, and the history of the behavioral and human sciences.

Robert A. Wilson was born in Broken Hill, Australia, and lives in Edmonton, Canada. He is the author or editor of five other books, including the award-winning *The MIT Encyclopedia of the Cognitive Sciences* (1999).

Boundaries of the Mind
The Individual in the Fragile Sciences

Cognition

ROBERT A. WILSON
University of Alberta

PUBLISHED BY THE PRESS SYNDICATE OF THE UNIVERSITY OF CAMBRIDGE
The Pitt Building, Trumpington Street, Cambridge, United Kingdom

CAMBRIDGE UNIVERSITY PRESS
The Edinburgh Building, Cambridge CB2 2RU, UK
40 West 20th Street, New York, NY 10011-4211, USA
477 Williamstown Road, Port Melbourne, VIC 3207, Australia
Ruiz de Alarcón 13, 28014 Madrid, Spain
Dock House, The Waterfront, Cape Town 8001, South Africa

http://www.cambridge.org

© Robert A. Wilson 2004

This book is in copyright. Subject to statutory exception
and to the provisions of relevant collective licensing agreements,
no reproduction of any part may take place without
the written permission of Cambridge University Press.

First published 2004

Printed in the United States of America

Typeface ITC New Baskerville 10/13 pt.   *System* LATEX 2_ε [TB]

A catalog record for this book is available from the British Library.

Library of Congress Cataloging in Publication Data
Wilson, Robert A., 1964–
Boundaries of the mind : the individual in the fragile sciences / Robert A. Wilson.
p.   cm.
Includes bibliographical references and index.
ISBN 0-521-83645-X – ISBN 0-521-54494-7 (pbk.)
1. Philosophy of mind.   2. Mind and body.   I. Title.
BD418.3.W535   2004
128'.2–dc22   2003065268

ISBN 0 521 83645 x hardback
ISBN 0 521 54494 7 paperback

For Selina

Contents in Brief

PART ONE: DISCIPLINING THE INDIVIDUAL AND THE MIND

1	The Individual in the Fragile Sciences	page 3
2	Individuals, Psychology, and the Mind	27
3	Nativism on My Mind	50

PART TWO: INDIVIDUALISM AND EXTERNALISM IN THE PHILOSOPHY OF MIND AND THE COGNITIVE SCIENCES

4	Individualism: Philosophical Foundations	77
5	Metaphysics, Mind, and Science: Two Views of Realization	100
6	Context-Sensitive Realizations	120
7	Representation, Computation, and Cognitive Science	144

PART THREE: THINKING THROUGH AND BEYOND THE BODY

8	The Embedded Mind and Cognition	183
9	Expanding Consciousness	214
10	Intentionality and Phenomenology	242

PART FOUR: THE COGNITIVE METAPHOR IN THE BIOLOGICAL AND SOCIAL SCIENCES

11	Group Minds in Historical Perspective	265
12	The Group Mind Hypothesis in Contemporary Biology and Social Science	286

Notes	309
References	335
Index	355

Contents

List of Tables and Figures	*page* xiii
Acknowledgments	xv

PART ONE
DISCIPLINING THE INDIVIDUAL AND THE MIND

1	The Individual in the Fragile Sciences	3
	1 Individuals and the Mind	3
	2 Individuals and Science	4
	3 The Fragile Sciences	8
	4 Individualism in the Cognitive, Biological, and Social Sciences	9
	5 Inside the Thinking Individual	14
	6 The Beast Within	17
	7 Culture, Nature, and the Individual	19
	8 The Metaphysical Picture: Smallism	22
	9 A Path Through Boundaries of the Mind	24
2	Individuals, Psychology, and the Mind	27
	1 Psychology amongst the Fragile Sciences	27
	2 The Disciplining of Psychology	30
	3 From Physiology to Philosophy: Wundt and James	31
	4 Disciplining the Social Aspects of the Mind	36
	5 Wundt's Individuals	40
	6 Galton's Individuals	41
	7 Nativism and the Continuity Thesis	45
	8 Overlaying the Mind	48

3 Nativism on My Mind 50
   1 Nativist Threads 50
   2 From Chomsky to Fodor to Pinker: A Thumbnail 51
   3 Empiricist Alternatives to Nativism 54
   4 The Two-Dimensional Approach 56
   5 Making Do with Less? 60
   6 Satisfying Some Desiderata 65
   7 But Could Two Dimensions Be Enough? 67
   8 Nativism about Cognition and Biology 68
   9 Conceptual Analysis and Nativism 72

PART TWO
INDIVIDUALISM AND EXTERNALISM IN THE PHILOSOPHY
OF MIND AND THE COGNITIVE SCIENCES

4 Individualism: Philosophical Foundations 77
   1 Making Sense of the Individualism-Externalism Debate 77
   2 Individualism, Taxonomy, and Metaphysical Determination 79
   3 Getting to Twin Earth: What's in the Head? 82
   4 The Social Aspect to Having a Mind 87
   5 Narrow and Wide Content 90
   6 Functionalism, Physicalism, and Individualism 93
   7 The Appeal to Causal Powers 96
   8 Metaphysics and the Fragile Sciences 98

5 Metaphysics, Mind, and Science: Two Views
   of Realization 100
   1 The Metaphysics of Mind and the Fragile Sciences 100
   2 Realization within the Philosophy of Mind 101
   3 A Sketch of Two Views of Realization 102
   4 The Standard View (I): Realizers as Metaphysically Sufficient 103
   5 The Standard View (II): Realizers as Physically Constitutive 104
   6 Smallism, the Standard View, and the Fragile Sciences 105
   7 Context-Sensitive Realization and Metaphysical Sufficiency 107
   8 Physical Constitutivity and Wide Realizations 111
   9 Wide Realizations in the Biological and Social Sciences 114
   10 Two Views Reconsidered 117

6 Context-Sensitive Realizations 120
   1 Adjusting One's Metaphysics 120
   2 Microphysical Determinism, Relations, and Smallism 121
   3 Dispositions and Science 125
   4 Nonreductive Materialism 128
   5 The Modified Standard View: Causation and Realization 130

	6 Context Sensitivity within the Standard View	133
	7 Keeping Realism Afloat	137
	8 Pluralism about Realization	139
	9 Abandoning the Subject?	141
	10 Putting Our Metaphysics to Work	143
7	Representation, Computation, and Cognitive Science	144
	1 The Cognitive Science Gesture	144
	2 Individualism in Cognitive Science	145
	3 Mental Representation as Encoding	147
	4 The Debate Over Marr's Theory of Vision	150
	5 Segal and Egan on Computation and Representation	155
	6 Exploitative Representation and Wide Computationalism	162
	7 Narrow Content and Marr's Theory	172
	8 Locational versus Taxonomic Externalism	174
	9 Having It Both Ways?	178
	10 Beyond Computation	179

PART THREE
THINKING THROUGH AND BEYOND THE BODY

8	The Embedded Mind and Cognition	183
	1 Representation and Psychology	183
	2 Life and Mind: From Reaction to Thought	184
	3 The Embeddedness and Embodiment of Higher Cognition	187
	4 Memory	189
	5 Cognitive Development	198
	6 Folk Psychology and the Theory of Mind	206
	7 The Mind Beyond Itself	210
9	Expanding Consciousness	214
	1 The Return of the Conscious	214
	2 Processes of Awareness and Phenomenal States	215
	3 Expanding the Conscious Mind: Processes of Awareness	217
	4 Arguing for Expanded Consciousness	221
	5 Global Externalism and Phenomenal States	225
	6 TESEE and Sensory Experience	232
	7 Individualistic Residues	238
	8 Global Externalism and the TESEE View	240
10	Intentionality and Phenomenology	242
	1 The Relationship Between Intentionality and Phenomenology	242
	2 Dimensions of the Inseparability Thesis	244
	3 Deflating the Inseparability Thesis	246

	4 Phenomenal Intentionality	252
	5 Individualism and Phenomenal Intentionality	255
	6 How to Be a Good Phenomenologist	260

PART FOUR

THE COGNITIVE METAPHOR IN THE BIOLOGICAL AND SOCIAL SCIENCES

11	Group Minds in Historical Perspective	265
	1 Group Minds and the Cognitive Metaphor in the Biological and Social Sciences	265
	2 Two Traditions	267
	3 The Collective Psychology Tradition	269
	4 The Superorganism Tradition	274
	5 Group Minds and the Social Manifestation Thesis	280
	6 Collective Psychology, Superorganisms, and Socially Manifested Minds	282
	7 From the Past to the Present	284
12	The Group Mind Hypothesis in Contemporary Biology and Social Science	286
	1 Reviving the Group Mind	286
	2 On Having a Mind	288
	3 Minimal Minds, Consciousness, and Holism	293
	4 The Contemporary Defense of the Group Mind Hypothesis	295
	5 The Social Manifestation Thesis	299
	6 The Cognitive and the Social	302
	7 From Group Minds to Group Selection	303
	8 Groups, Minds, and Individuals	306

Notes	309
References	335
Index	355

List of Tables and Figures

TABLES

6.1	Realization in Psychology, Biology, and Chemistry	*page* 127
8.1	Locus of Control and Representational Type	187
8.2	Higher Cognition and Its Realizations	188

FIGURES

1.1	Nativism and Cognition	16
1.2	Nativism and Inheritance and Development	19
1.3	Culture and the Individual Mind	21
2.1	Psychology amongst the Fragile Sciences	29
3.1	Nativism and Cognition	59
3.2	Nativism and the Theory of Mind	61
3.3	Nativism, Genetics, and Development	71
5.1	(a) Constitutive Decomposition Involving Entity-bounded Realization and (b) Integrative Synthesis Involving Wide Realization	112
7.1	Standard Computationalism	165
7.2	Wide Computationalism	166
8.1	Rush Hour: A Problem	193
8.2	(a) and (b) Rush Hour: A Solution	194

Acknowledgments

I began thinking about some of the material in this book in the mid-1990s as work on three distinct projects progressed. Some of the metaphysical issues concerning the mind, particularly what I think of as the hard-nosed physicalist challenges to nonreductionist views posed by Jaegwon Kim and David Lewis, had been passed by too quickly in my first book, *Cartesian Psychology and Physical Minds*. These called for further reflection. I had also started developing, slowly and through trial-and-error teaching, some background in biology and the philosophy of biology. Finally, I was being forced to think about the diversity of views within cognitive science through my role as general editor for *The MIT Encyclopedia of the Cognitive Sciences*. It took a few years for these interests and issues to coalesce, and for the project of which this book is a part to be articulated.

The project itself was initiated while I held a fellowship at the Center for Advanced Study at the University of Illinois, Urbana-Champaign, in the spring of 1998. The center also played a broader supportive role for me throughout my time at Illinois, for which I am grateful. Rough versions of the first chapter and those in Part Two were drafted in the fall of 1999 while holding a fellowship courtesy of the Program for Liberal Arts and Sciences Study in a Second Discipline at Illinois. During this period, the Cognitive Science Group at the Beckman Institute at Illinois provided a stimulating intellectual home that regularly transgressed disciplinary boundaries of all kinds. I thank my colleagues there and in the Department of Philosophy – especially Gary Dell, Gary Ebbs, Steve Levinson, Patrick Maher, Greg Murphy, and Fred Schmitt – for fostering a constructive and welcoming academic environment in which a somewhat open-ended project could be undertaken.

Most of the writing for the project has been done while I have been at the University of Alberta these past two or three years, and with generous release time provided by grant SSHRC 410-2001-0061 from the Social Sciences and Humanities Research Council of Canada. As the project developed, it became clear that it could not be completed within a single book. Much of the last year has been spent making both this book and *Genes and the Agents of Life* (Cambridge 2005), walk the line that independent but related books must. I thank my departmental chair, Bernie Linsky, for his flexibility in assigning my teaching load, giving me the whole of calendar 2002 to concentrate on both books. My thanks also to my research assistants over the last two years – Ken Bond, Jennie Greenwood, Li Li, and Patrick McGivern – and to members of an upper-division and graduate course in the spring of 2003 who grappled with, and usefully critiqued, material from both books.

For feedback and encouragement along the way, I would like to thank Karen Bennett, Paul Bloom, Andy Clark, Lesley Cormack, Gary Ebbs, Frances Egan, John Heil, Terry Horgan, Frank Keil, the late David Lewis, Patrick Maher, Robert McKim, Ruth Millikan, Alex Rueger, Robert Smith, Andrea Scarantino, Gabe Segal, Larry Shapiro, Sydney Shoemaker, Bob Stalnaker, Kim Sterelny, Paul Teller, John Tienson, J. D. Trout, Peter Vallentyne, Steve Wagner, David Sloan Wilson, and Steve Yablo. Greg Murphy and Larry Shapiro provided detailed written comments on the penultimate manuscript, for which I am grateful, and Michael Wade did the same for the material that comprises Part Four. The editorial and production team at Cambridge – Stephanie Achard, Jennifer Carey, Shari Chappell, and Carolyn Sherayko – have been a pleasure to work with throughout, and have improved the final manuscript considerably. I would also like to thank four reviewers for Cambridge University Press who provided feedback and guidance on the overall shape of the project, and Terence Moore at Cambridge for his editorial leadership.

Many of the chapters draw on and develop material that I have published elsewhere. I would like to acknowledge publishers for permission to include material drawn from the following publication sources:

Chapters 4 and 7: "Individualism," in Stephen P. Stich and Ted A. Warfield (eds.), *The Blackwell Guide to Philosophy of Mind*. Oxford: Blackwell Publishers, 2002, pp. 256–287.

Chapters 5 and 6: "Two Views of Realization," *Philosophical Studies*, 104 (2001), pp. 1–30. © 2001 Kluwer Academic Publishers.

Chapter 6: "Some Problems for 'Alternative Individualism,'" *Philosophy of Science*, 67 (2000), pp. 671–679. © 2000 Philosophy of Science Association. All rights reserved. The University of Chicago Press.

Chapters 7 and 8: "The Mind Beyond Itself," in Dan Sperber (ed.), *Metarepresentations: A Multidisciplinary Perspective.* Volume 10, Vancouver Studies in Cognitive Science. New York: Oxford University Press, 2000, pp. 31–52.

Chapter 8: "The Individual in Biology and Psychology," in Valerie Gray Hardcastle (ed.), *Where Biology Meets Psychology: Philosophical Essays.* Cambridge, MA: MIT Press, 1999, pp. 357–374.

Chapter 10: "Intentionality and Phenomenology," *Pacific Philosophical Quarterly*, 84 (2003), pp. 413–431. Oxford: Blackwell Publishers.

Chapter 12: "Group-Level Cognition," *Philosophy of Science*, 68 (2001), pp. S262–S273. © 2001 Philosophy of Science Association. All rights reserved. The University of Chicago Press.

Edmonton, Alberta
June 2003

Boundaries of the Mind
The Individual in the Fragile Sciences
Cognition

PART ONE

DISCIPLINING THE INDIVIDUAL AND THE MIND

1

The Individual in the Fragile Sciences

1 INDIVIDUALS AND THE MIND

Where does the mind begin and end? We think of the mind as tied to and delimited by individuals. Minds do not float free in the air or belong to larger, amorphous entities, such as groups, societies, or cultures. No, they are tightly coupled with individuals. Minds exist inside individuals, and the particular mind that any individual has constitutes an important part of what it is to be that individual. We may not know precisely when during ontogenetic development the mind begins to exist and when it ceases to exist. Indeed, we might think that there is no such precise time, and that to think otherwise is to fall into some sort of conceptual muddle. But that a particular mind's temporal boundaries are delimited by the life of the individual is reflected in both Western science and law.

Likewise, we might quibble about how far throughout the brain and central nervous system the mind extends spatially. But again, the boundary of the mind is no greater than the boundary of the individual. If it doesn't stop further in, in the brain, it at least stops at the skin.

There are ways in which these ideas about the mind may appear to be challenged by pervasive systems of thought beyond science. For example, on many religious views, at least something very like the mind is neither temporally nor spatially bounded by the body by which we usually identify an individual. Minds leave bodies when a person dies, and can find themselves in places beyond Earth, or in other bodies on Earth. This sort of view is even accommodated within analytic discussions of personal identity in philosophy. The soul of a prince may end up in the body of a pauper, or (in the modern version) a person's mind might be stored in a computer and downloaded in other matter, or teletransported Star Trek-style to another spatial location.

In fact, such widespread views reflect and reinforce, rather than challenge, the tie between minds and individuals. For in effect what they do is identify an individual in terms of his or her mind. It is that very same individual who sins on Earth and is thus punished in the afterlife, and the very same individual who emerges from the teletransporter as the one who stepped into it. If this were not true, then Hell would deliver only a Kafkaesque notion of justice and desert, and teletransportation would be suicide and the creation of a new life all in one.

This book is about the boundaries of the mind. The preceding two paragraphs aside, it has nothing directly to say about religious thought or about work on personal identity. Rather, its focus is on the idea of the individual as a boundary for the mind in the cognitive sciences and the philosophy of mind. There are various strands to the idea that the individual serves as a boundary for the mind in those sciences, and I want to tease them apart, to probe and examine them, and to question at least some of them. I take seriously the idea that there are important senses in which the individual is not a boundary for the mind, but do so from within the confines of the cognitive sciences. This will involve saying much about what I think the mind is. And it will be hard to say much of use on this topic without also entering into discussion of what individuals are. That, I suggest, takes us immediately beyond the mind and the cognitive sciences to a larger arena within science.

2 INDIVIDUALS AND SCIENCE

The concept of the individual is central to how we think about the mind, about living things, and about the social world. The sciences that concern each of these domains – the cognitive, biological, and social sciences – have developed independently. Human agents are often taken to be paradigms of individuals in each of these sciences, what I shall collectively refer to as the *fragile sciences*. Yet there exists little systematic discussion of the roles and conceptions of individuals across the cognitive, biological, and social sciences. *Boundaries of the Mind* focuses on the role that the individual has played and continues to play in guiding our thinking about the mind within cognitive science. It conceptualizes that science (or, better, cluster of sciences) as part of a broader range of sciences, the fragile sciences, that use individuals as a touchstone. It aims to contribute and draw on the fragile sciences.

In everyday life human agents are our paradigm example of individuals. We take for granted human agents in our everyday lives, and they feature centrally in our myths, histories, literature, and artistic representations. They are perhaps the most familiar part of our landscapes and our memories. Individuals are not some theoretical abstraction or posit, but perceived and known things. They are something that we can feel sure about, something basic or foundational, something beyond question.

Individuals have also been taken for granted within the various explanatory enterprises that form part of contemporary science. But here the sense of security and surety can begin to dissipate as we register some of the variation that exists in just how human agents are thought of within the domain of the human sciences. Within evolutionary biology, human agents are conceived as animals with a phylogenetic history and a particular range of ecological niches. Within anthropology, human agents are interpreters of meaning and creators of culture. Within cognitive science, they are the locus for computational programs and modules. Within economics, rational decisionmakers, optimizers of utility. The claim that individuals play a central role in these sciences is platitudinous. Yet recognition of the various roles that individuals have in these sciences, and of the presuppositions and implications of such roles, have been limited enough to warrant using the platitude as a focal theme for a broader discussion of the individual in the fragile sciences.

So my first point about individuals and science is that while we can readily accept human agents as paradigmatic individuals, how such individuals are conceptualized varies across different sciences. This should occasion no real surprise, given that these sciences have developed with considerable autonomy from one another, and the central role that "the individual" plays in each. There are, however, various kinds of project that might take this as a point of departure. For example, there are historical investigations of specific conceptions of the individual, detailed comparisons of distinct traditions and thinkers across these disciplines, and synthetic overviews that weave a narrative revealing affinities and ruptures within these fields of thought. While the kind of project that I am undertaking incorporates historical, comparative, and synthetic perspectives, the unifying structure to it lies in the interplay between the role and conception of individuals and the way in which the corresponding sciences have developed.

In both common sense and the sciences we have a firm grip on what individuals are: They are individual human beings like you and me, and by extension, individual thinkers, organisms, and agents. What warrants reflection, however, are the various ways in which these thinkers, organisms, and agents have been construed, respectively, in psychology, biology, and the social sciences. For these construals are pivotal to many of the most important and controversial topics in those sciences and the sort of sciences they become.

This brings me to a second point. How one conceives of individuals and the role that one ascribes to individuals structure and constrain how any "individual-focused" science is theorized and practiced. We can illustrate both of these points with an example from the biological sciences, that of the role of individuals in the theory of natural selection.

In the traditional Darwinian theory of natural selection, the individual organism plays the central role as the agent on which natural selection operates. Organisms are the individuals that bear phenotypic traits, that vary in their fitness within a population, and that, as a result, are selected for over evolutionary time. Organisms are the bearers of adaptations, such as thick coats in cold climates, or porous leaves in humid climates. They are the agents of selection. On Darwin's own view, units larger than the individual, such as the group, were (more or less) unnecessary, and units smaller than the individual, such as the gene, unknown.

By contrast, in the postsynthetic view of evolution by natural selection that is often glossed in terms of the concept of the selfish gene, individuals play a very different role. On this view, it is genes rather than individuals that are the agents of selection, and that come to play many of the roles (and have many of the features) that are ascribed to organisms on the traditional Darwinian view. On this view, individuals are not much more than ways in which genes get to propagate themselves. In terms that Richard Dawkins uses, they are the *vehicles* in which the real agents of selection, genes, the *replicators* in the story of life, are lodged. Genes are the bearers of adaptations, and the units between which variations in fitness provide the basis for the process of natural selection. Furthermore, not only is the individual organism no longer the agent of selection, but as Dawkins has also argued, it is only an arbitrary boundary for phenotypes, which should be seen as extending into the world beyond the organism.[1]

Each of these conceptions of the role of the individual in the theory of natural selection carries with it implications for a number of issues to which that theory is central. I shall mention just two here.

The first is what is usually called the problem of altruism. Altruistic phenotypes and behaviors are those that decrease the relative fitness of the organisms that bear them. On the traditional Darwinian view of natural selection, focused as it is on organisms and their reproductive success, the existence of altruism represents a puzzle. If organisms are the agent on which natural selection operates, then natural selection will select those organisms that have a relatively higher level of fitness. Thus, organisms adapted to increase the relative fitness of *other* organisms, that is, decrease their own relative fitness, could not evolve through this process, except as incidental by-products. The problem of altruism, in this view, is how to explain the existence of organism-level altruism.

On the gene-centered view of natural selection, by contrast, this problem does not exist. Or, rather, it is solved by showing how altruism, so conceived, is a result of the process of natural selection operating on genes and maximizing their reproductive success. For copies of the same genes can exist in different organisms, and so altruistic behavior would be predicted by a gene-centered view of natural selection where organisms share significant proportions of their genes, such as when they are kin.

A second issue for which the individual- and gene-centered views of natural selection have implications is how important higher-level selection is in shaping the tree of life. Such selection operates on agents larger than the individual, anything from temporary dyads, to demes, to species, to whole clades. Both the traditional Darwinian and the gene-centered view are skeptical about the need to posit higher-level selection, and posit higher-level selection only when explanation at their preferred level is not empirically adequate. On the traditional Darwinian view, these higher-level agents are conceptualized very much *as* organisms in that they are seen as sharing many properties ordinarily ascribed to organisms. But on the gene-centered view, groups, species, and clades are simply larger pools of genes, different sized vehicles, if you like, but never truly the agents of selection.[2]

This example brings out another dimension to discussion of the individual in the cognitive, biological, and social sciences. Human agents are the paradigmatic individuals in these sciences, but they are neither the only entity that serves as an individual nor always the most central such entity. Entities both smaller (for example, information-processing modules) and larger (for example, whole species) than our paradigm individuals are sometimes conceptualized either as individuals or as having many of the properties possessed by paradigmatic individuals, and thus as being like individuals. Sometimes it is these entities that are central to the relevant sciences, with our paradigm individuals receding to the background. We can raise the same questions about the conception and role of such entities as we have about our paradigm individuals. As I will argue in the final chapters, this may be true even when our focus is on cognition, and where the relevant individuals are groups of some kind.

3 THE FRAGILE SCIENCES

I have introduced "the fragile sciences" as shorthand for the cognitive, biological, and social sciences, and here I want to say something about the neologism itself. Given that human agents are paradigmatic individuals in these sciences, one might think that "human sciences" is descriptively more informative. But like "behavioral sciences" or "special sciences," this is a term whose connotations are more misleading than helpful, and whose extension differs from that of the range of sciences that I have in mind.

The first respect in which this is true is that "the human sciences" denotes simply those sciences that attempt to understand human nature in one or more of its dimensions: biological, psychological, behavioral, social. The term thus suggests continuities between humanistic studies of human nature that precede the disciplinization of the psychological and social sciences in the nineteenth century and subsequent, disciplinary-based research that I shall not discuss at all. Roger Smith's *The Norton History of the Human Sciences* is an excellent work with this focus, but my concern is largely with conceptions of the individual that postdate the formation of psychology and the various social sciences as distinct disciplines.

A second but related reason for not speaking of the human sciences is that even if human agents are paradigmatic individuals in the cognitive, biological, and social sciences, their conceptualization here seldom even attempts to capture the essence of humanity or to grapple with the loftier goals of earlier inquiries, such as "man's place in nature." The conceptualization of the individual has become more partial and less encompassing, but also more closely tied to models, techniques, and research strategies in particular sciences. It is these ties, rather than the study of human agency or human nature, that interest me.

A third reason to coin a term rather than make do with an existing one, such as "the human sciences," is that the greater part of the biological sciences, as well as a sizable portion of the cognitive sciences, have at least as much to say about nonhuman as about human agency. The scientific study of intelligent capacities has sometimes been used as a brief characterization of cognitive science, to which artificial intelligence, the study of such capacities in machines, has been, historically and substantially, a major contributor. Human beings constitute one special object of study within zoology. If "the human sciences" is taken, as it often is, to refer to psychology plus the social sciences, then we need a term that refers to these plus the biological sciences.

But why *fragile* sciences? Two brief reasons. The first is that it both parodies and transcends traditional divisions between the sciences: between the natural and social sciences, hard and soft science, and the physical and human sciences. As a sometimes philosopher of science, I have found these dichotomies too often driving views of the nature of science, and as inevitably privileging the natural, the hard, and the physical over the social, the soft, and the human. "Fragile sciences" serves as a partial counter to these tendencies pervading not only philosophy but also education, popular culture, and science itself. The second is the cluster of ideas that fragility calls to mind. Fragile things can be easily broken, are often delicate and admirable in their own right, and their labeling as such carries with it its own warning, which we sometimes make explicit: Handle with care! But they are also both strong and weak at the same time, and their fragility lies both in their underlying physical bases and in how it is that we treat them. No doubt, "fragile sciences" triggers other meanings, and should two reasons not be reason enough, let that be a third. Welcome to the fragile sciences!

4 INDIVIDUALISM IN THE COGNITIVE, BIOLOGICAL, AND SOCIAL SCIENCES

One way in which the role of the individual has been made prominent in the fragile sciences is via the defense of one or another form of individualism. I have examined individualism in psychology in detail elsewhere, concluding with the suggestion that the relationship between the various individualistic theses in psychology, biology, and the social sciences deserved substantive exploration. While that exploration does not exhaust the content of either *Boundaries of the Mind* or the broader project of which it is a part, the relationship between these individualisms constitutes a reference point for discussion of individuals in the fragile sciences. I shall begin to discharge the promissory note that there is something to be gained by considering "the individual" across the fragile sciences by sketching a common, simple framework in terms of which these forms of individualism can all be understood.[3]

In psychology and the cognitive sciences, more generally, individualism is the thesis that psychological states should be construed without reference to anything beyond the boundary of the individual who has those states. This is the thesis that Jerry Fodor has called *methodological solipsism*, and it requires that one abstract away from an individual's environment in taxonomizing or individuating her psychological states. A more precise and common expression of individualism in psychology says that psychological states should be taxonomized so as to supervene on the intrinsic, physical states of the individuals who instantiate those states. (A property, A, supervenes on another, B, just if no two entities can differ with respect to A without also differing with respect to B.) Those who deny individualism are *externalists*. In Parts Two and Three I shall develop several varieties of externalism about the mind. Individualism about mental states requires, and will there receive, more detailed articulation. I briefly note three things about it here.[4]

First, individualism is a normative thesis about how one ought to do psychology: It proscribes certain views of our psychological nature – those that are not individualistic. For this reason, I consider individualism in psychology as a putative constraint on the sciences of cognition.

Second, this constraint itself is claimed to derive either from general canons governing science and explanation or from entrenched assumptions about the nature of mental states themselves. It is not construed as an *a priori* constraint on how one does psychology, but one derivative from existing explanatory practices that have met with considerable success in the past.[5]

Third, combining the previous two points, approaches to the cognitive sciences that are not individualistic are both methodologically and metaphysically misguided. They go methodologically awry in that the most perspicuous examples of explanatorily insightful research paradigms for cognition or for science more generally have been individualistic in the corresponding sense. Included here are computational approaches to cognition and the taxonomy of entities in science more generally by their causal powers. Given the successes that such individualistic approaches to the mind have had, those rejecting individualism are left in methodological limbo.

Externalist approaches to the mind go metaphysically awry in two corresponding ways: Either they relinquish our only real insight into the nature of mental causation – cognition is computational – or they imply that the mind is not governed by principles that apply more generally to the physical world – such as the supervenience of a thing's properties on its intrinsic, causal powers.

These three features of individualism in psychology are shared by individualistic theses in both the biological and social sciences. There are various forms that individualism can take in the biological sciences. For now I elaborate on an example that I have already mentioned, that concerning the agents of selection.

In evolutionary biology there has been a sustained and continuing debate over the agents of selection: At what level or levels in the biological hierarchy – from the very small, such as cells and their constituents, to the very large, such as whole species or larger taxonomic units, such as genera or families – does the chief force of evolutionary change, natural selection, operate? Three putative agents of selection have been most frequently proposed and defended: the individual organism, with individual selection being what I have already referred to as the traditional Darwinian view; the gene, a unit inspired by the rise of population genetics through the evolutionary synthesis, with genic selection associated most often with George Williams and Richard Dawkins; and the group, with discussion of the process of group selection experiencing a contemporary revival, due largely to the work of David Sloan Wilson and Elliott Sober. While much of the debate over the agents of selection concerns the relative strength and thus importance of each of these selective forces – individual, genic, and group – there is a strand to the debate that has concerned *the* level at which selection operates, and thus, which is less pluralistic than is suggested by such a construal of that debate. It is this strand that I want to focus on here.[6]

In the context of this debate over the agents of selection, individualism is the view that the organism is the largest unit on which natural selection operates. Thus, proponents of genic selection who claim that natural selection can always be adequately represented as operating on genes or small genetic fragments are individualists about the agents of selection, as are those who adopt the traditional Darwinian view that allows for only individual-level selection. Like individualism in psychology, individualism in evolutionary biology is a putative normative constraint that derives from existing explanatory practice, and whose violation, according to its proponents, involves both methodological and metaphysical mistakes. Let me explain.

Individualism about the agents of selection implies that individual organisms act as a boundary beyond which evolutionary biologists need not venture when attempting to theorize in considering the nature of what it is that competes and is subject to natural selection, and thus evolutionary change. By focusing on the individual and what lies within it, one can best understand the dynamics of adaptive change within populations of organisms, whether it be via population genetic models, through the deployment of evolutionary game theory, or by means of the discovery of the forms that individualistic selection takes. This constraint on how to think about the operation of natural selection builds on the specific explanatory successes of models of kin selection, reciprocal altruism, and other processes that articulate strategies that individuals might adopt in order to maximize their reproductive success.

Finally, those who go beyond the individual in postulating selective processes that flout individualism create both methodological and metaphysical problems that individualistic approaches avoid. Methodologically, since the full range of observed behaviors can be explained in evolutionary terms by models of evolution that are individualistic about the agents of selection (for example, via kin selection and reciprocal altruism), those who reject individualism are abandoning real explanatory achievement. As we saw was the case with the rejection of individualism in psychology, precisely because the constraint of individualism in evolutionary biology rests on a history of explanatory success, those advocating a nonindividualistic view of the agents of selection are faced (and not unreasonably) with the question of what their perspective buys that cannot be purchased with an established currency. The corresponding metaphysical problem lies in the idea that natural selection – somehow – transcends the level of the individual and goes to work directly on groups. A common response to this sort of nonindividualistic suggestion is either that the appropriate model of group selection really boils down to a variant on an existing individualistic model of selection, or that it requires assumptions that rarely, if ever, hold in the actual world. In the former case, we don't, in fact, depart from individualism; in the latter, we are committed to views about what there is that do not correspond to how the world actually is.

Consider now the social sciences. In the social sciences, methodological individualism is a cluster of views that give some sort of priority to individuals and their properties, particularly their psychological properties, in accounting for social phenomena. As with mental phenomena, although there is a sense in which we have an intuitive grasp of what social phenomena are, there are questions of how those phenomena fit into the natural order of things, and of how we should go about studying them. Methodological individualism has sometimes been expressed as the view that all social phenomena must be explained, ultimately, in terms of the intentional states of individual agents. Like individualism in psychology, it has also sometimes been formulated as a supervenience thesis, the thesis that social phenomena supervene on the psychology of individuals.[7]

These two formulations, what we might think of as the explanatory and ontological formulations, are usefully seen as linked by viewing methodological individualism as a thesis about the taxonomy or individuation of social phenomena, as I have suggested that individualism in psychology is most usefully viewed. Methodological individualism is the view that social phenomena should be taxonomized so as to supervene on the intrinsic, physical properties of the individuals who participate in or sustain those phenomena. Thus, our explanations of social phenomena should be restricted to positing states and processes that respect this normative, individualistic constraint. This is what is thought to give the intentional states of individuals both explanatory and ontological priority over social phenomena, just as, given individualism in psychology, the intrinsic, physical states of individuals have explanatory and ontological priority over mental phenomena.

Fueling the idea that methodological individualism is a constraint on the social sciences is also a claim about a tradition of explanatory successes, although unlike the case of the cognitive and biological sciences, this claim about an established track record for individualistic approaches to the social sciences has a significantly slimmer evidential basis. This is in large part because relatively uncontroversial explanatory success stories are rare in the social sciences; it is also because the best-known defenses of methodological individualism in the social sciences, such as those of Karl Popper and John Watkins, predate most of those successes that we can identify. Historically, it is more accurate to see the rise of methodological individualism in the social sciences as a gamble, a promissory note about what approaches to understanding society would prove most fruitful, though contemporary advocates of methodological individualism do, I think, share the "past success" view of their putative constraint on the social sciences, listing to their credit, for example, a variety of applications of rational choice theory in economics and schema theory in cognitive anthropology.[8]

The parallel to the third feature of individualism in psychology that I drew – the claim that violation of the constraint brings with it related methodological and metaphysical problems – is, I think, relatively clear. Individualists have claimed that approaches to and theories in the social sciences that are non-individualistic, such as Emile Durkheim's account of suicide, or Karl Marx's theory of capitalism, to take two well-known examples, have proven to be methodological dead ends. As with claims about individualistic success stories in the biological sciences, these claims about explanatory failure are contentious. In any case, with respect to the social sciences more emphasis has been given to the metaphysical pitfalls of a denial of methodological individualism: social science that is non-individualistic reifies social objects and categories, such as class interest (sociology), markets (economics), and cultures (anthropology). It posits an inflated social ontology of entities that have some type of *sui generis* existence, and provides no promise of how to integrate the developing social sciences with the natural sciences.[9]

There are questions about these three forms of individualism that this brief comparative treatment of them leaves open, and we will return to explore in detail those concerning cognition in Part Two. My main point here has been to sketch a common framework in terms of which these distinct individualistic theses can be viewed. The existence of such a framework suggests that problems, challenges, solutions, and responses that have arisen with respect to a given debate over individualism may also exist or be lurking in the background of another debate over individuals and their role in explanation in the fragile sciences. This common framework will thus function like a trampoline for some individualism hopping across the fragile sciences.

Thus far I have attempted to show that three individualistic theses in the cognitive, biological, and social sciences that have seldom been discussed together in fact share an important set of affinities. The natural and further suggestion is that the debates into which an individualistic or externalist position in any of these sciences launch one can be mutually informative. That in turn presupposes that each of these individualistic perspectives has important ramifications for central debates in each of these fields. It is in part by way of supporting this claim that I turn now to consider innateness hypotheses in the fragile sciences. I shall propose a two-dimensional analysis of a range of nativist debates both that shows what they share despite their distinct subject matter, and that reveals the link between nativism and individualism. I begin with the cognitive sciences, where nativism has received its most extensive treatment.

5 INSIDE THE THINKING INDIVIDUAL

Nativists about the mind hold that the mind itself is an important source of, and imposes a structure on, the nature of human knowledge and human mental representation. As we will see in more detail in Chapter 3 when we discuss nativism about the mind at length, contemporary nativism about cognition in general has been inspired by nativism about language more specifically. Language acquisition, and cognitive development more generally, are viewed as proceeding maturationally, rather than through some type of environmentally driven process, such as learning. And these maturational processes are viewed by nativists as requiring specialized, internal cognitive machinery, rather than general purpose, one-size-fits-all mechanisms.

In the contemporary cognitive sciences, the chief antinativist cluster of views are *empiricist*, holding that it is our sensory or perceptual apparatus, not something distinctly cognitive or mental, that structures human knowledge and mental representation. Both nativists and empiricists recognize that there is a given, built-in organismic contribution to cognition. But since empiricists locate this contribution within perception, "upstream" of the mind, what enters the mind proper is a function of experience in the world. This experience may vary with one's environment and shapes an organism's particular cognitive structures. Cognitive structures are thus not built into the design of the organism's mind. In this sense, its cognitive architecture is not innate but contingent on the nature of its environment.

There are two distinct dimensions to contemporary debates over nativism about the mind. The first concerns the nature of an individual's internal structure, while the second concerns the causal role that an individual's external environment plays in the acquisition and development of a given cognitive process. These dimensions can be represented as two theses about cognition as follows. Letting X stand in for some particular cognitive ability, phenomenon, or behavior, these theses are:

The Internal Richness Thesis: Structures and processes internal to the individual that are important to the acquisition and development of X are *rich*;

and

The External Minimalism Thesis: Structures and processes external to an individual play at best a secondary causal role in the acquisition and development of X.

A cognitive structure is rich just if it is specialized, localized, internally complex, and causally powerful. Intuitively, the internal richness thesis says that there is a wealth of structure built into the individual that explains why it possesses a particular cognitive ability or manifests some specific behavior.

Consider a strong candidate for an innate human cognitive capacity, such as the capacity to acquire language. Nativists about language acquisition hold that language learners are equipped with a "language acquisition device" that is rich in the sense I have specified. While they acknowledge environmental input as essential, it clearly plays a secondary causal role in language acquisition. By contrast, consider the ability to do calculus, something usually acquired through classroom instruction. Here whether one acquires the ability to do calculus depends in large part on the structure of one's environment, in particular, on whether it contains a teacher who can transmit the relevant knowledge. Since calculus is a relatively recent human innovation, it is unlikely that the brain contains structures specifically for calculus.

Paradigmatic nativists in the cognitive sciences, such as Chomsky and Fodor, accept both the internal richness and the external minimalism thesis about a range of cognitive capacities. Paradigmatic antinativists, such as behaviorists and early connectionists, reject both of these theses. One implication of the two-dimensional view of nativism is that we can understand such views as diametrically opposed to one another, as Figure 1.1 makes clear: In virtue of accepting both theses about cognition, Fodor and Chomsky are what I shall call *strong nativists* about cognition, while behaviorists and at least early connectionists, by virtue of rejecting both theses, are *strong antinativists*. As Figure 1.1 makes clear, these strong positions do not exhaust the range of options in the debate over nativism about the mind, a point to which I shall return in Chapter 3.

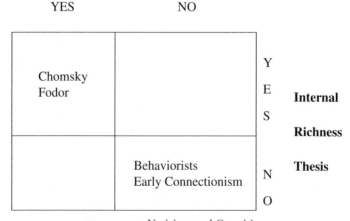

FIGURE 1.1. Nativism and Cognition

Nativist views, particularly what I have been calling strong nativism, provide at least prima facie support for individualistic views of the mind. Consider the external minimalism thesis. If the environment plays a secondary role in the acquisition and development of a given cognitive ability – and in the extreme case is best viewed as little more than a trigger for that ability – then there is a clear sense in which the environment can be bracketed out when we formulate taxonomies and explanations for mental phenomena. This strand to strong nativism suggests a "methodologically solipsistic" research strategy in psychology. And if one accepts the internal richness thesis, then there is a sense in which cognitive structures for specific tasks are programmed within the organism, and so again it would seem that bracketing out the environment for the purposes of psychological taxonomy and explanation would be appropriate. In short, if our innate cognitive machinery is rich in structure, and our psychological taxonomies aim to carve mental reality at its joints, then those taxonomies should be individualistic.

There is more to be said not only about nativism about the mind, but also about these putative relationships between nativism and individualism, and we will return to such issues in due course. This introduction of the two-dimensional view of nativism is, however, intended primarily to exemplify the trampoline hopping across the fragile sciences that I mentioned at the end of the previous section. Let's hop, then, from cognition to biology.

6 THE BEAST WITHIN

Debates over nativism also have a well-worn history within the biological sciences. The two-dimensional view of the contemporary debate over nativism within the cognitive sciences helps to illuminate these debates.

One such debate concerns the role and conception of inheritance and organismic development. How should we understand inheritance and organismic development? Gross bodily traits in human beings, such as number of arms and legs, eye color, and position of the mouth, are typically considered to be innate, and this ascription of innateness fits with the two-dimensional analysis of nativism. The environment plays a secondary causal role in the development of such traits, with the primary role being played by rich structures internal to the organism. By contrast, for traits that are typically viewed as acquired, such as height or number of scars borne by age five, neither the external minimalism nor the internal richness thesis is true. As in the cognitive case, the two-dimensional view provides a way of demarcating paradigmatically innate from paradigmatically acquired bodily traits.

The parallel with nativism about the mind runs deeper, however, in that the two-dimensional view also specifies strong nativist and strong antinativist views of inheritance and development. Within classical and molecular genetics the prevalent view of the inheritance and development of phenotypic traits gives genes a preeminent role to play, holding that genes both code for and direct organismic development. Here genes play much the role that modules do in nativist accounts of cognitive development: They are internally rich structures that play the principal role in guiding development. Thus, while an organism's environment plays a necessary role in inheritance and development, it is not the primary causal agent in those processes. This view of the nature and role of genes in inheritance and development represents a strong nativist view within the biological sciences.

Corresponding to the strong empiricist views about cognition is a strong antinativist view of inheritance and development associated most directly with Richard Lewontin, sometimes referred to as *interactionism* or *constructivism* about development. To be clear, no one denies the existence of intracellular units, genes that play an important role in inheritance. What has been questioned, however, is (a) whether genes play either a unique, universal, or asymmetrical role in the processes of inheritance and development with respect to other cellular (and extracellular) components; and (b) whether the concept of a gene, and the roles that it plays in our theories of inheritance and development, can be understood in an environment-independent way. To call (a) into question is to imply that genes are not internally rich structures in the sense that I have specified; to call (b) into question is to challenge the external minimalist strand to strong nativism.[10]

Figure 1.2 shows both the opposition between these "strong" views and the open space that is represented by less radical positions about nativism and biology. As with nativism about the mind, here a strong nativist position, particularly acceptance of the external minimalism thesis, makes an individualistic position in biology attractive. Even if claims about genes as "master molecules" directing hereditary and developmental traffic are relaxed in favor of a more encompassing view of the nature of those processes, so long as we remain within the organism, the environment of that organism can, in effect, be bracketed off from our taxonomic and explanatory practices.

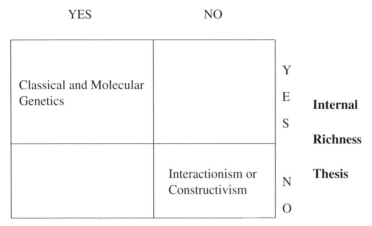

FIGURE 1.2. Nativism and Inheritance and Development

I shall say more about this characterization of nativism in biology in Chapter 3. As we will see there has been some recent skepticism within biology over whether the notion of innateness has outlived its usefulness.[11]

7 CULTURE, NATURE, AND THE INDIVIDUAL

In the social sciences, perhaps the most hotly contested arena in which the nativism debate has been played out concerns culture, particularly within anthropology. In the first half of the twentieth century the concept of culture came to define the core domain that anthropologists studied, just as biologists studied the domain of the living and psychologists the domain of the mind. The anthropologist Adam Kuper has provided an elegant history of the conception of culture within anthropology and beyond it (within "cultural studies," for example). It is only one strand to the recent part of that history that is my concern here. Paralleling the cases of cognition and biology, we can capture two opposed views of the relationship between culture and what individuals are innately endowed with by using the two-dimensional view of nativism.[12]

The prevalent view of culture within anthropology is strongly anti-nativist in rejecting both the external minimalism and internal richness theses about culture. Culture is not a monolithic structure that all humans share, but something that varies across different groups of people, who socially construct their cultures through particular traditions, rituals, and customs. Cultures, in all their rich detail, rather than some abstraction, "culture," are what shape up individual patterns of behavior and what anthropologists study. Individuals come to acquire the ways of thinking and behaving particular to their culture through various forms of social transmission, and the particular structure to their environments determines what cultural traits they come to possess. While a certain internal complexity, particularly mental complexity, is needed for groups of organisms to develop a culture, there are no built-in, specialized, mental structures necessary for culture.

This view of culture and its acquisition by individuals has been challenged by recent evolutionarily inspired views, typified by John Tooby and Leda Cosmides. Sociobiologists such as Edward O. Wilson had earlier argued for the in-principle extension of the ethologists' treatment of behaviors as part of an organism's phenotype to the full range of human social behaviors. Tooby and Cosmides have advocated a similar view of human cultures, emphasizing the particular cognitive adaptations that, they claim, generate human cultures. On their view, there are a variety of cultural universals that derive from internally rich cognitive programs that are built into our minds. The structures of an individual's environment explain only the particular cultural variants that he or she acquires, not what lies at the heart of culture. This is a view of human cultures that parallels the linguistic nativist's view of natural languages. Indeed, this strong nativist view of culture derives from strong nativist views of language and the mind. The cognitive scientist Steven Pinker has popularized these views and the connections between them in a series of books.[13]

Clearly then, part of what is at issue in this debate over culture and "what's within" is the conception of the place of the mind within anthropology. Few (if any) strong antinativists would deny that the mind is entirely irrelevant to culture. Rather, they propose that we understand the mind through culture. The most influential expression of this view is the anthropologist Clifford Geertz's interpretationism, which holds that culture lies in the actions and artifacts of agents in the observable, physical world rather than in any inner theater of the mind. The task of a cultural anthropology is to interpret these actions and artifacts where this involves uncovering their meaning, what they symbolize, and how they function in a given society. In "Thick description," the opening essay to Geertz's *The Interpretation of Cultures*, having said that "[c]ulture is public because meaning is," Geertz continues:

The generalized attack on privacy theories of meaning is, since early Husserl and late Wittgenstein, so much a part of modern thought that it need not be developed once more here. What is necessary is to see to it that the news of it reaches anthropology; and in particular that it is made clear that to say that culture consists of socially established structures of meaning in terms of which people do such things as signal conspiracies and join them or perceive insults and answer them, is no more to say that it is a psychological phenomenon, a characteristic of someone's mind, personality, cognitive structure, or whatever than to say that Tantrism, genetics, the progressive form of the verb, the classification of wines, the Common Law, or the notion of a 'conditional curse'... is.[14]

"Socially established structures of meaning" are the meeting point of mind and culture, the trading zone in which psychology and anthropology exchange their goods.

The fragment of the broader debate over culture and the mind that I have recounted as one between strong nativists and antinativists about culture (see Figure 1.3) suggests three points. The first is that views about the nature of culture in anthropology are intimately entwined with views about the nature of the mind. The second is that the two-dimensional analysis of this debate here implies that, as in the cognitive and biological sciences, there are positions that depart from both strong nativism and strong antinativism that are worth exploring. The third is the possibility that "taking culture seriously" may require thinking beyond the boundary of the individual not only in how we think of culture itself, but also in how we think of the mind.

In the cases of nativism about cognition and biology, I have suggested that strong nativist views promote a corresponding form of individualism

External Minimalism Thesis

YES	NO	
View of Culture from Evolutionary Psychology		YES
	Interpretationism	NO

Internal Richness Thesis

FIGURE 1.3. Culture and the Individual Mind

in leading one to focus exclusively on what lies within the boundary of the individual organism in investigating psychological or developmental processes. Perhaps unsurprisingly, none of the general views about culture that I have demarcated are individualistic in this sense: It is not in dispute that one must look beyond the head of individuals in order to describe and explain culture. But those who have accepted strong nativist views of culture have, I think, come close to expressing a methodologically individualistic view of culture as either reducible or explainable in terms, ultimately, of the thoughts and actions of individuals. In traditional sociobiology, culture is simply the sum of individual behaviors; in evolutionary psychology, it is sometimes construed as the sum of individual cognitive programs. Thus again, strong nativist views of culture fit naturally with the corresponding form of individualism, in this case methodological individualism.

8 THE METAPHYSICAL PICTURE: SMALLISM

I have thus far outlined the general framework shared by individualists across the fragile sciences, tried to show the affinities that exist between a cluster of nativist views in those sciences, and connected these nativist views to corresponding forms of individualism. I want to round out this introductory chapter by suggesting that there is a metaphysical picture that makes individualism and nativism and the methodologies they both invite seem quite compelling. The metaphysics here is a twentieth-century generalization of seventeenth-century corpuscularianism, a view that played a central role in the scientific views of some of the leading scientists of that period, including Robert Boyle and Sir Isaac Newton.

This modern day generalization of corpuscularianism constitutes, in my view, a form of metaphysical discrimination that has left its mark on a range of sciences. In keeping with these politically correct times, I shall refer to this form of discrimination as *smallism*, discrimination in favor of the small, and so, against the not-so-small. Small things and their properties are seen to be ontologically prior to the larger things that they constitute, and this metaphysics drives both explanatory ideal and methodological perspective. The explanatory ideal is to discover the basic causal powers of particular small things, and the methodological perspective is that of some form of reductionism.[15]

Corpuscularianism is the view that matter is ultimately made up of simple corpuscles or atoms, which themselves possess a relatively small number of intrinsic properties. It is these corpuscles and their associated properties, the *primary qualities*, that explain all observable physical and chemical phenomena. While corpuscularianism itself reached its height at the end of the seventeenth century and is no longer taken seriously as a hypothesis about the structure of the physical world, some of its features have structured smallist views in the psychological, biological, and social sciences. To see this, consider corpuscularianism itself.

This view was developed in seventeenth-century Europe chiefly in the works of Galileo Galilei, Pierre Gassendi, and Robert Boyle. The sort of atomism that it promoted in metaphysics and science can be found throughout the works of many thinkers of the period, including those of Thomas Hobbes and John Locke. While the list of primary qualities has varied across these authors, an aggregated list includes the following properties: solidity, extension, figure, mobility, motion or rest, number, bulk, texture, motion, size, and situation. Most relevant for my purposes here are two features of corpuscularianism: The idea that primary qualities are inherent in corpuscles, and the claim that all of an entity's properties could be explained in terms of corpuscles and their primary qualities.[16]

Primary qualities are inherent in the sense that the things that have them do so in and of themselves, without reference to any other thing whatsoever. More generally and precisely, a quality Q is inherent to a thing X just if X would have Q even if X were the only thing to exist in the world at all. The idea that the real and ultimate properties that a thing has are inherent in this sense plays an important role in the compositional nature of corpuscularian metaphysics and explanation. Corpuscles are (physical) parts of observable things, and parts of the parts of observable things. It is properties of the parts of observable things, and ultimately of the corpuscles themselves, that cause, and thus explain, the observable properties of macroscopic objects. Given the inherentness of primary qualities, this implies that the properties of the things that their subjects constitute will be a function of these inherent properties, plus the relations that hold between the things that have them.

It should be clear how this sort of corpuscular metaphysics makes individualism an inviting position to hold. If the ultimate properties that are the causal movers and shakers in general are inherent in individuals, then systematic explanations ought, in the long run, to make some sort of appeal to those properties. We need, then, to constrain our higher-level, fragile sciences in a way that will allow them to hook up with the sciences that discover the primary qualities of the most basic material things. Since these are inherent properties, the constraint should be one that limits the higher-level sciences to positing inherent properties. This is just what the individualistic constraints that I have outlined in this chapter attempt to do: An entity's properties must supervene on that entity's intrinsic, physical properties, where the most significant entity is a paradigmatic individual. So the fragile sciences should focus on what's inside individuals, not what lies beyond their boundaries. This is true whether we focus just on our paradigmatic individuals human agents, or whether we think of units both larger and smaller than our paradigms as individuals.

This emphasis on what is inside is the common point between individualistic and nativist views across the fragile sciences. What's inside is construed as being rich in structure and fixed despite any variation in organism-environment interactions. Individuals are not just concentrations of especially important properties, but have those properties by virtue of something about those individuals themselves: They have them inherently.

One general problem for smallism is shared by individualistic (and I think nativist) views in the fragile sciences. Many of the kinds of things that there are in the world – modules, organisms, and species for example – are relationally individuated. Thus, what they are cannot be understood solely in terms of what they are constituted by. Moreover, regardless of how the entities themselves are individuated, many of their most salient properties – their functionality, fitness, and adaptedness, for example – are relational properties. As such, since these properties do not inhere in the entities that have them, they cannot be fully understood by focusing exclusively on what falls inside the boundaries of those entities. This metaphysical point itself carries the methodological imperative to look beyond the boundary of the individual in exploring the ways in which individuals and their components function causally in the world.

The prima facie problem that relational properties and kinds pose for smallist and individualist views requires further articulation, and it invites some obvious responses. We will see several forms that this problem takes, and explore some smallist responses, in Part Two.

9 A PATH THROUGH *BOUNDARIES OF THE MIND*

This book has four parts. In the remainder of Part One, I aim to further round out the start made in this chapter in articulating the idea of the individual as a boundary for the mind. In the next chapter, I shall introduce the idea that both the mind and its scientific study, as well as individuals, have been "disciplined" in Michel Foucault's sense. Neither the mind nor its study are givens, but are, in some sense, constructed through genealogical processes. Thus, in this chapter, I want to convey some sense of psychology's recent history (even if not of its "long past"), and the conception of minds and individuals within it. In Chapter 3, I shall return to put some flesh on the bones of my discussion of nativism. Rather than using the two-dimensional view of nativism illustratively, as I have in the current chapter, in Chapter 3 I will aim to defend the two-dimensional view of nativism about the mind, and to explore some of the claims about its significance left undeveloped here.

Parts Two and Three form the core of the book, and focus exclusively on individualism and externalism about the mind. The three chapters in Part Two are foundational, with Chapter 4 discussing individualism in the philosophy of mind in detail, and Chapters 5 and 6 moving to discuss the metaphysical notion of realization. The idea that mental states are realized by physical states of the brain is commonplace in the philosophy of mind. Although realization is a technical notion within philosophy, it is linked to the concepts of mechanisms and dispositions, each prevalent within the cognitive sciences themselves. One of the key ideas in these chapters is that the standard way of thinking about realization in the philosophy of mind carries with it a smallist bias, and that recognizing this opens the way to an externalist view of the mind. Chapter 7 begins to develop such a view with reference to computational cognitive science.

So, Part Two contributes to the increasingly elaborate debate between individualists and externalists about the mind conducted within the philosophy of mind and computational cognitive science. In Part Three, the externalist view of the mind I defend is extended beyond these domains to areas of psychology that typically are less explicitly computational (Chapter 8) and to consciousness (Chapter 9). Chapter 8 both points to existing work on memory, cognitive development, and folk psychology that can be viewed from an externalist point of view and suggests how externalism may be extended further. Chapter 9 introduces what I call the TESEE conception of consciousness: consciousness as *t*emporally *e*xtended, *s*caffolded, and *e*mbodied and *e*mbedded. In this pair of chapters my aim is to continue with the articulation of externalism undertaken in Chapter 7 by reference to existing empirical work on the mind.[17]

In the final chapter in Part Three, I return to more "purely philosophical" work in exploring the relationship between intentionality and phenomenology. Much of the motivation for externalist views in the philosophy of mind has derived from reflection on the nature of intentionality, and at least some of the resistance to externalism stems from views of consciousness and phenomenology. Chapter 10, in part, addresses this resistance through a critical examination of two recent proposals by philosophers for thinking about the relationship between intentionality and phenomenology.

This chapter began with the idea that individuals and minds were strongly connected in several ways. One of these connections manifests itself in the innocent-sounding claim that individuals have minds, but that groups, societies, and cultures do not. In the two chapters in Part Four, I explore the claim, common enough a hundred years ago in the nascent social sciences, that groups can be said to have minds. This group mind hypothesis was important both at the interface of social psychology and sociology, and in early systematic work on animal and plant communities, including that on insect colonies as superorganisms. Chapter 11 discusses these ideas in their historical context. It has, however, a philosophical moral, one that, I shall argue in Chapter 12, is highly relevant to recent revivals of the group mind hypothesis in both the biological and social sciences. The biologist David Sloan Wilson has defended the group mind hypothesis in the context of his broader defense of group selection in evolutionary biology. The anthropologist Mary Douglas has suggested that there is more to Ludwig Fleck's notion of a *Denkkollektiv* or "thought collective" than contemporary social scientists have recognized. As I shall argue, there is something right about both of these claims – just what is the topic of Part Four.[18]

2

Individuals, Psychology, and the Mind

1 PSYCHOLOGY AMONGST THE FRAGILE SCIENCES

In Chapter 1, I identified psychology as one of the fragile sciences. It is the science of the mind. Historically, psychology developed as the institutional home for the scientific study of the mind in a variety of dimensions, including cognitive, biological, and social dimensions. Along with these distinct aspects to the mind, psychology has also encompassed many different organizing theories or paradigms. Consider just those that have structured research within the cognitive dimension to psychology. These range from introspectionism, Gestalt psychology, and behaviorism in the first half of the recent history of psychology, to computationally based paradigms, such as the "rules and representations" approach that began the cognitive revolution in the 1950s, and connectionism and dynamic systems theory that have challenged that approach more recently.

The contemporary discipline of psychology – sprawling, heterogeneous, and perhaps unitary in only the most attenuated of senses – is the result of this ecumenical and paradigm-shifting history. Departments of psychology in major research institutions are characteristically organized into divisions or subfields whose descriptive adjectives – such as cognitive, developmental, social, perceptual, decision making, clinical, quantitative, and personality – correspond to clusters of research interests, publication venues, and professional organizations. These divisions frequently carry with them distinct courses, degree requirements, hiring procedures, graduate admissions policies, and laboratory and seminar meetings. In moderate to large psychology departments, it is common for researchers to work exclusively within their division, and even to have little or no professional contact with the majority of their departmental colleagues in other divisions throughout their careers.

As well as being characterized as the science of the mind, "the science of mental life," as William James put it more than one hundred years ago, psychology is also commonly viewed as being wedged between biology and the social sciences. This is so not only in representations of hierarchies of sciences familiar from the logical positivist tradition within the philosophy of science, but in characterizations of the methodologies appropriate for the scientific study of mental life, which are drawn from both the biological and the social sciences.

There is something right about both of these common views of psychology and the mind: The study of the mind is psychology's prerogative, even if psychology is a discipline sealed from neither the biological nor the social sciences. As our brief discussion of culture and nativism in Chapter 1 illustrated, the study of the mind can be directly relevant to, indeed, a part of, the social sciences. Consider areas of the social sciences that are concerned primarily with the operation of institutions or large-scale social changes rather than the minds of individuals. These operations and changes are mediated by the actions of individuals. Accounts of them typically propose or presuppose some principles governing the minds of those individuals, whether this be rational choice theory (in economics) or schema theory (in anthropology).

Aspects of the mind and mental life are also claimed by the biological and medical sciences as their operational domain. Pathological development of the mind, such as that leading to mental retardation or a variety of syndromes typically named after a person instrumental in their isolation and demarcation (for example, Williams, Asperger, and Down's syndromes), is often studied by geneticists and by medical doctors. Institutionally, psychiatry is a specialization within medicine, rather than a field of psychology. This is motivated in part by the idea that mental disorder should be assimilated to bodily disease. Moreover, techniques developed within the biological sciences, such as single-cell recording or positron-emission tomography (PET), have found widespread use in the last twenty years in studying distinctively mental abilities and traits.

Despite contemporary psychology being a house of many mansions, at its core is a focus on distinctly mental processes of individuals, human agents. Minds belong to individuals, and the processes that constitute minds occur inside the boundary of those individuals. Core topics within the study of cognition, such as memory, language, or learning, are construed in this individualistic way. But such a view permeates psychology

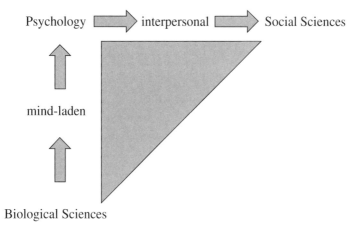

FIGURE 2.1. Psychology amongst the Fragile Sciences

more generally. How does what is inside an individual, particularly what is inside her head, enable her to act on her environment, interact with others, and dynamically adapt her behavior? Without distinctly mental processes, we are in the realm of mere biology. And if we move on to encompass an exploration of the relationship between individuals, of processes and actions that extend beyond the boundary of the individual, we move from psychology to the social sciences. We might encapsulate this idea by saying that psychology is both *mind laden* (rather than mind free) and *individual bound* (rather than interpersonal) and depict the place of psychology within the fragile sciences through Figure 2.1: Since psychology leaks into the rest of the fragile sciences, we should properly locate some of its sub- (or related) disciplines, such as neurophysiology and psychophysics somewhere along the axis linking psychology to the biological sciences, and others, such as social psychology and parts of clinical psychology, along the axis linking psychology to the social sciences.

Neither psychology as a discipline nor mind-laden individuals as the subject matter of that discipline are givens. The discipline has not always existed, and there is a history or a genealogy to the formation of the discipline that has involved construing individuals and minds in different ways at different times. This chapter takes the historical contingency of psychology seriously, and attempts to draw several genealogical lines from its early history to current presumptions about the boundaries of the mind. I begin in the next section with some brief, general comments about the disciplining of psychology before moving on in the following sections to discuss how individuals and the mind were construed as psychology was disciplined.

2 THE DISCIPLINING OF PSYCHOLOGY

In *The Order of Things*, Michel Foucault provided a sweeping analysis of the emergence of the disciplines of biology, economics, and philology in the eighteenth century. Foucault's analysis was aimed, in part, at illustrating how both subject areas and individuals were disciplined in what I am calling the fragile sciences. This theme of the way in which the formation of areas of discourse and knowledge is intertwined with the construction of particular types or kinds of individuals became central to his more widely read works, such as *Discipline and Punish* and *The History of Sexuality*.[1]

Disciplines have a contingent rather than necessary existence. They come into existence through a complex process with conceptual, social, political, technical, and methodological aspects, and construct, rather than simply discover, their subject area in the process. When their subject area concerns the individual, as it does for the most part in the fragile sciences, that means constructing individuals in certain ways: identifying some of their properties rather than others as important for this particular discipline, emphasizing some categories and kinds and not others as those needed to classify individuals. Not just discourse and knowledge is disciplined; so too are individuals.

Thus, the disciplining of psychology is intimately connected to the disciplining of the individuals it studies. In part, this is because both processes are subject to the same historical contingencies, but also because the individual and what lies inside it has become central to the very conception of the psychological. Certain construals of the individual, and ways of studying the mind of the individual, are entrenched within the discipline of psychology, and those construals are good for extracting some types of knowledge, and bad for others.

Ebbinghaus's famous remark that psychology has a short history but a long past, made as the distinct disciplinary pathways of psychology and philosophy were being sculptured in the late nineteenth century, has a certain charm to it. Philosophers and psychologists alike regard it with some fondness – in my view, with too much fondness. Although philosophers have offered various accounts of the mind and its place in nature since at least the Greeks over 2,400 years ago, it was only in the nineteenth century that the conceptual, methodological, and institutional prerequisites for a distinct science of psychology coalesced and the disciplining of psychology began. No point in time, of course, demarcates the beginning of the scientific discipline of psychology from either prescientific psychological thought or scientific nonpsychological investigation. Yet two landmark events have been claimed as marking the beginning of psychology in the origin myths that contemporary psychologists often tell themselves, a myth that I will rely on in what follows.[2]

The first is Wilhelm Wundt's foundation of the first experimental laboratory devoted to the study of psychological phenomena in Leipzig in 1879. Wundt's designation of his laboratory space as "psychological" was preceded by a number of significant steps down the path to the formation of a discipline. Included here are several decades of experimental work in perceptual physiology and what Gustav Fechner had called *psychophysics*, work headed by Hermann von Helmholtz, Eward Hering, and Fechner, as well as Wundt's own publication of the widely used and cited *Foundations of Physiological Psychology* in 1874. The second is the publication of William James's *Principles of Psychology* in 1890, the book that James began by identifying psychology as the "science of mental life, both of its phenomena and their conditions." James's *Principles* sought to integrate the rich, philosophical tradition of work on the mind that ascribed a central place to inward reflection on one's conscious mental life with the developing physiology of the nervous system. While both Wundt and James recognized the debt that psychology, conceived as a distinct science, owed to physiology and psychophysics, both also characterized its distinctiveness in terms of the nature of the phenomena that it investigated. These were distinctly mental, and such phenomena could not be exhaustively understood in terms of existing physiological and psychophysical theory and practice.[3]

As the origin myth implies, Wundt and James clearly did play important roles in the disciplining of psychology from physiology and philosophy. In the next section I shall spell out these roles more precisely before turning to the question of the disciplining of psychology from the social sciences. Individuals come to be conceptualized differently in different parts of the fragile sciences, and this has left a legacy concerning how psychologists conceptualize individuals, minds, and the relationships between them.

3 FROM PHYSIOLOGY AND PHILOSOPHY: WUNDT AND JAMES

Physiology is the science that studies the structure and function of bodily systems, including the nervous system. Like psychology, it did not always exist. It was disciplined and emerged from anatomy as a distinctly experimental science in the first half of the nineteenth century, both in Germany under Johannes Müller and more particularly his students, such as Emil Du Bois-Reymond and Helmholtz in the 1840s, and in France initially under François Margendie and then later Claude Bernard in the 1850s. Previous to this period, physiology had been secondary to anatomy, and as such had focused on the identification and operation of physically localized biological structures, paradigmatically bones, muscles, and organs. Physiology then was concerned principally with the internal organization of the organism, rather than its dynamic functioning. As a systematic attempt to explain not just biological structures but also biological functions in physical terms, the beginnings of experimental physiology were instrumental for those of psychology in an indirect and in a direct way.

Indirectly, the focus on function itself allowed physiologists to abstract from particular, observable structures and explore the putative systems performing those functions of which those structures formed a part. By experimentally intervening in the operation of these systems, they could reveal some of the complexity to the dynamics of physiological systems without antecedently having to identify and characterize all of the component parts of those systems. Or, to put it in terms that will feature centrally in my discussion of realization in Part Two, experimental physiology constituted a form of *integrative synthesis*. Here anatomically identified organs were located within some broader, functional system, with integration replacing localization as the corresponding technique of investigation. Thus, experimental physiology carried with it the general idea that there could be a properly scientific investigation of an organism that was, in a certain sense, autonomous of detailed anatomy and further decompositional analysis.[4]

More directly, since some of the functions had a mental aspect to them in that they involved the reactions and registrations of conscious beings, experimental physiology could be directly applied to understanding the mind, or at least parts of it. This was not least because the mind itself was increasingly conceived as being realized in the brain and nervous system, and this system was subject to increasing functional analysis throughout the nineteenth century, beginning with the positing of distinct mental functions in different parts of the brain by phrenologists in the early part of the century. The first widely accepted evidence for the localization of a specific mental ability – that of speech – in the cerebral cortex was Paul Broca's description of patients with damage in their left frontal lobes in 1861, patients who suffered from what is now known as "Broca's aphasia."

I have said that both Wundt and James viewed psychology in terms of its concern with a distinctly mental life. This characterization of psychology not only distinguished it from physiology, as I noted there, but also allied it with traditional philosophical discussions of the mind that can be found throughout the history of Western philosophy. This is one sense in which psychology emerged between physiology and philosophy, between the empirical and experimental exploration of the material world and the *a priori* investigation of the mental world. It is perhaps worth reflecting further on the gulf that existed between these two worlds, and so on what sort of bridgework was done in the formation of psychology as a discipline.

The dichotomy between mind and body has pervaded both common sense and scientific thinking for the last 400 years. The study of the mind was traditionally the province of philosophers, either as that branch of "metaphysics" concerned with nonmaterial (but earthly) reality, the mind, or as a branch of "morals" concerned with human nature and the aptitudes contained within human agents. As the material world became the subject of various disciplinary projects through the scientific revolution in the seventeenth century, questions concerning the nonmaterial part of reality became more pressing. What was the reach of the corpuscularian, mechanical philosophy? Where did the immaterial parts of reality fit into this worldview? How did these two worlds, that of the mind and that of matter, connect?

Body and mind, and the relationship between them, constituted the most obvious locus for the discussion of such questions. Bodies were material. Individual human bodies became the subject of scientific investigation not only through the development of the practice of medical dissection, but also with corpuscular theory as an overarching metaphysics for all material things, including bodies. Everything material, including bodies, was made up, ultimately, of corpuscles, and could be understood in principle in terms of their properties. If such a view pointed to the way in which bodily activity could be understood, then the question of whether this form of materialism could be extended to the mind itself became a real one. René Descartes took his mechanical view of the world to extend partly to what we might think of as the mind – to sensation and reflexive movements, for example – but claimed that the most important aspects of mind, typically reified as Reason and the Will, could not be understood through this kind of science. This general view remained widely shared in the seventeenth and eighteenth centuries by both rationalists and empiricists, both within and beyond the corpuscularian framework.

This was a legacy in part of dualism about mind and body, and in part of a concern with how far the knowledge new in the seventeenth century could extend.

In Chapter 1, I claimed that this corpuscularian metaphysics has underwritten a quite general, pervasive attitude in the fragile sciences in the twentieth century – that of smallism, discrimination in favor of the small and so against the not-so-small. The application of corpuscular thinking to the body constituted a way to extend a smallist metaphysics into the realm of individuals, as did the positing of the "mental atoms" of sensation in the hands of British empiricists and their nineteenth-century positivist successors. If dualism about mind and body preserved at least aspects of the mind as exceptions to the reach of corpuscularianism, the physiological basis to psychology whittled away this exceptionalism. Experimental psychology emerged in the nineteenth century in the grip of the smallist visions that corpuscularianism had inspired.

Here, then, is the picture of the late nineteenth-century emergence of the discipline of psychology that I am sketching. As physiology consolidated as a distinctive science, systematic accounts of sensory phenomena, such as haptic and auditory discrimination, were developed, and seemingly more and more of the mind admitted of a physiological or physical explanation. Combining the general point from physiology – that an organism's functions could be experimentally investigated independent of a detailed knowledge of that organism's physical anatomy – with the legacy of dualism – that not all of the mind could be understood via the dominant scientific paradigms – created space for an autonomous experimental science of the mind that went beyond "mere" philosophy. This was the science of psychology.

The method of this new science was, like that of physiology, experimental. Complementary to the idea that psychology was the science of mental life was the development of an experimental methodology that was also introspective in nature. The combination of introspection and the experimental method was the crux of Wundt's contribution to the emergence of psychology. Wundt was critical of the reliance on what we might think of as unconstrained (that is, nonexperimental) introspection in studying the mind (what he called *innere Wahrnehmung* [inner perception]), distinguishing this from the objective, controlled self-observation (*Selbstbeobachtung*) around which he built his conception of psychology. Both the emergence of the discipline of psychology from those of physiology and philosophy and its continuing connections with those disciplines, and thus the fluidity of disciplinary boundaries at the founding of psychology, are reflected in the careers of Wundt and James.⁵

Wundt began teaching physiology at Heidelberg in the late 1850s after having taken a doctorate there in medicine. He held professorships in both anthropology and medical psychology at Heidelberg in the mid-1860s, and then in the mid-1870s moved to Leipzig where he held the position of professor of philosophy. Wundt remained influenced by the philosophical work of Immanuel Kant and Arthur Schopenhauer throughout his life, and published more purely philosophical works in logic and ethics as well as those for which he is better remembered. Although Wundt's *Foundations* was the best known and the most widely read of his works during his lifetime, his conception of psychology is better represented by the earlier publication of a pair of books: his *Contributions to the Theory of Sensory Perception* in 1862, which concentrated on perceptual phenomena, and *Lectures on Human and Animal Psychology* in 1863, a more wide-ranging and longer work that encompassed culture, society, emotion, and aesthetics. As both of these works indicate, Wundt's publication of his ten-volume "folk psychology" between 1900 and 1920 did not represent a shift from his earlier conception of psychology as a methodologically introspective, experimental discipline, as is sometimes claimed, but, rather, reflects Wundt's initial conception of a discipline with two faces: one experimental and concerned with in-the-head mental processes, the other observational and focused on complicated mental processes that involved external cultural representations.⁶

James had taken a medical degree at Harvard in 1869 and began his teaching career there in the early 1870s in anatomy and physiology, moving on to teach physiological psychology in 1875, and finally giving his first lectures in philosophy in 1879. In conjunction with the course in physiological psychology, James had established a minor psychology laboratory prior to Wundt in the late-1870s, but unlike Wundt, James did not use it as a base for recruiting and training graduate students. (James disliked doing laboratory work.) James's professorial appointments at Harvard were in physiology in 1876, philosophy in 1885, and then in psychology in 1889, shortly before the publication of his *Principles* in 1890. Like Wundt, James formed part of a philosophical tradition, in James's case, the relatively recent and still developing tradition of American pragmatism, whose best known exponents, apart from James himself, were Charles Sanders Pierce and John Dewey. And again like Wundt, after forging a connection between physiology and psychology, James further explored the mind in a nonexperimental context. In James's case, the focus was on psychical and supernatural phenomena, religious belief, and pedagogy.

Psychology may have emerged from physiology and philosophy, but for both Wundt and James it was never simply the extension of physiological methodology to mental functions, nor merely the application of scientific techniques to resolve traditional philosophical questions about the mind. It is not just that both Wundt and James had a more encompassing conception of what the discipline of psychology was and to which other fields it was allied than do the majority of contemporary psychologists, but that they exploited the fluidity of the discipline in recognizing problems with and limits to specific methodologies and assumptions. Perhaps this is simply one way in which they both remained philosophers in establishing the discipline of psychology.

4 DISCIPLINING THE SOCIAL ASPECTS OF THE MIND

If the object of study of psychology distinguished it from physiology as an antecedently (but recently) established discipline within the university system, and its experimental methodology distinguished it from philosophy, it was psychology's concern with mental processes that occurred within the individual that distinguished it from the later-developing field of sociology. Social psychology is concerned largely with social aspects of individual cognition, while sociologists are concerned with social systems that are constituted by and influence the actions of individuals. This basic division remains in place today, with the field of social cognition, a recent incarnation of how psychologists working within a cognitive and often computational framework approach the social aspect to individual cognition. Perhaps the most integrative perspectives within the two disciplines are to be found in the revival of symbolic interactionism within both recent social psychology and sociology via a reappreciation of the work of George Herbert Mead and Irving Goffman.[7]

August Comte has sometimes been represented as the founder of both sociology and social psychology, and in characterizing sociology as a later-developing discipline than psychology, I might be accused of overlooking this or of ignoring the contribution of Comte to both disciplines. Comte's views themselves are interesting, especially in the context of thinking about the later demarcation of sociology from psychology in terms of what occurs within the boundary of an individual.

Comte coined the term *sociologie* as one of two species of "psychic" science. In doing so, Comte conceived of two aspects to the mind, leading to two distinct paths to its study: the "static aspect," leading to the study of the organic, bodily basis for the mind, and the "dynamic aspect," leading to sociology. According to Comte, this latter study "will simply amount to tracing the course actually followed by the human mind in action, through the examination of the methods really employed to obtain the exact knowledge that it has already acquired." That is, sociology, what Comte also called "social physics," would be a science based on the methods of natural science, these being those that have led to exact knowledge. It would study complex relations between organisms in the way in which physiology was beginning to study complex relations within organisms.[8]

This left no room for what would become the distinctive science of psychology. Indeed, writing in the second quarter of the nineteenth century, Comte was skeptical of the idea that there could be a distinctive psychological science, particularly one that made use of introspection, on the grounds that this constituted a metaphysical rather than a positive stage in the development of the science. Somewhat later, in his *System of Positive Polity*, Comte foresaw the necessity of a science that integrated physiology and sociology. Calling this science *la morale*, Comte claimed that it would be "able to reduce to a system the special knowledge of man's individual nature, by duly combining the two points of view, the Biologic, and the Sociologic, in which that nature must be necessarily regarded." But Comte died before this sort of suggestive comment was developed more fully. Writing before psychology had been disciplined, and given Comte's own positivism about scientific inquiry, it is not clear what shape *la morale* was to take.[9]

The interface between psychology and sociology *qua* disciplines in the late nineteenth century was the nascent field of social psychology, with both disciplines able to treat it from their own perspective. The first textbook with "social psychology" in the title appeared in 1908, and was written by the sociologist Edward A. Ross. Ross used the metaphors of planes and currents of mental states to characterize social psychology, claiming that "[s]ocial psychology... studies the psychic planes and currents that come into existence among men in consequence of their association." Ross was concerned to explain the "planes of uniformity" that one finds across individuals with respect to beliefs and actions as well as the dynamics of the currents that sweep up individuals when they act as members of a group at a particular time, such as during a lynching. As a form of psychology, social psychology focuses on these planes and currents as they act on individuals. This contrasts with sociology, which focuses on human groups and structures themselves. What demarcates social psychology from psychology more generally for Ross is that social psychology is concerned to identify causes of individual human behavior that derive from outside the individual; these Ross calls *suggestions*, and his focus is on social factors that influence the human psychological disposition to suggestibility. Note that for Ross, social psychology is a discipline that theorizes exclusively about human beings.[10]

There is an interesting contrast between this perspective on social psychology and that of the psychologist William McDougall, whose *An Introduction to Social Psychology* also appeared in 1908. McDougall is chiefly concerned to show that there is an area of psychology – social psychology – that treats an individual's psychology as more encompassing than simply that individual's consciousness or mental life, and that serves as the foundation for the social sciences. Put the other way around, McDougall seeks to show, against sociologists such as Durkheim and Comte, that the social sciences are not independent of psychology, since there is a conception of psychology that is broader than the experimental introspectionism that had been extracted from Wundt's approach to scientific psychology. This aspect of psychology studies an individual's innate instincts, and McDougall identifies its appeal to evolutionary theory as the ground for the expectation of psychological continuities between humans and (other) animals. The first three-quarters of McDougall's book (Section I) is devoted to his instinct theory; only the last quarter (Section II) turns to how some of these instincts operate to form the basis of human societies. Amongst McDougall's list of instincts are flight, repulsion, curiosity, pugnacity, self-abasement, and self-assertion – which play little role in his account of human social organization – and the parental, reproductive, and gregarious instincts – which play a more significant role here.

Here I want to register three points of contrast between Ross and McDougall in terms of how they conceptualize the social aspect of psychology. First, while McDougall is particularly concerned to explain human social life on the basis of a subset of his instincts, on his conception instinct theory should also explain features of nonhuman social life. Thus, the scope of social psychology is broader than it is under Ross's conception. Second, as a psychologist, McDougall begins with a rich conception of what is within the individual and offers a much thinner characterization of the social phenomena that his instincts purport to explain, while Ross, as a sociologist, has just the opposite emphasis or bias. Third, McDougall is wary not only of social science that ignores psychology but that which builds on what he regards as a misleading psychology. Included here would likely be Ross's own appeal to suggestibility as a basic feature of the human mind, it being little more than an attribution to individuals of a psychological disposition that parallels a social phenomenon, that of individuals whose actions are influenced by the suggestion of others.

In the previous section, I argued, in effect, that psychology overcame a putative subsumptive challenge from both physiology and philosophy by demarcating itself in terms, respectively, of its endorsement of experimental methodology and its articulation of a distinct level of organismic abstraction. One might wonder, by contrast, whether the "within the individual" criterion really functioned (and functions) to demarcate social psychology from sociology. In fact, sociology never posed a threat of subsumption to psychology; rather, from the outset, sociology itself was concerned to establish its own autonomy from other disciplines, including psychology.

Perhaps best known here is Emile Durkheim's statement that "a social fact can be explained only by another social fact... Sociology is, then, not an auxiliary of any other science; it is in itself a distinct and autonomous science." This aspect to Durkheim's thought is provokingly developed in a classic discussion of the nature of representation in psychology. Writing after the foundations had been laid for the discipline of psychology, Durkheim recognized an individual level of representation that governs individual behavior, but claimed that in addition there are *collective representations* that are not in any sense reducible to or explainable by reference solely to what goes on in the heads of individuals. Just as individual representations are no mere "echo" or epiphenomena of cerebral activity, claimed Durkheim, so too are collective representations not inherent in or intrinsic to individual minds. Durkheim sought to distinguish such collective representations from anything interior to individuals, and in so doing was the first in a long line of social scientists to create or maintain a space for social sciences, which relied little, if at all, on psychology. We will return to discuss this idea of collective representations in Part Four.[11]

The foregoing historical reconstruction of the partial origins of psychology as a distinct discipline, brief as it is, takes us beyond the obvious fact that psychology concerns first and foremost individual human agents to the more interesting issue of the ways in which human agency has been construed in psychology. For philosophers, physiologists, and sociologists not only generalized and theorized about the individual: Each construed the individual in a particular way. Philosophers concerned with the mind thought of the individual as a receiver of sense impressions from the world (in the empiricist tradition) or as the source of innate ideas (in the rationalist tradition). Physiologists viewed individuals as complicated sets of biological systems that were integrated into a functioning whole. And sociologists saw the individual as a Janus-faced figure at the intersection of the purely biological or physical world and the social world. Each of these disciplines, of necessity, abstracted away from individuals *in situ*.

In the next two sections, and following the work of the historian of psychology Kurt Danziger, I shall focus on the abstractions made within psychology itself and the ways in which they have shaped the sort of science that psychology has become, particularly the ways in which psychology was developed as an experimental science. In particular, I shall discuss the conception of individuals in the psychological paradigms initiated by Wundt, and then turn to that within the Galtonian paradigm for psychology. Both abstract individuals from their social and physical contexts in an attempt to make individuals scientifically and mathematically tractable, but do so in very different ways.[12]

5 WUNDT'S INDIVIDUALS

For introspection to yield a repeatable, quantifiable result, not any old inward-looking report under physical stimulation would do, and unregimented introspection yielded reports that varied in accord with the theoretical proclivities of the laboratories in which they were produced. Within Wundtian experimental psychology, one solution was to ensure that the subject was someone with a sort of expertise in being able to attend to and report on the contents of his or her consciousness correctly under experimental conditions. It is for this reason that Wundt himself or one of his assistants typically played the role of subject in his experiments; as Danziger says, "the experimental subject was the scientific observer, and the experimenter was really a kind of experimental assistant." The reliability of one's findings was measured in the first instance by intrasubjective agreement over repeated trials, with replication of this reliability across subjects being initially of less significance. As Danziger puts it, in the Wundtian paradigm the individual subject was an "exemplar of the generalized human mind," and intersubjective replication of an experimental finding served simply to confirm an antecedently established finding.[13]

As we have seen, Wundt himself thought that you could investigate only elementary psychological processes through this method, with complex processes being studied in the broader social contexts in which they were embedded, and through methods that linked psychology to the humanities and social sciences. Whatever Wundt's own intent, his view amounted to a sort of bipartite view of the psychological, with the experimental subject investigated individualistically and the individual's broader psychological features, such as his personality, his temporally extended consciousness, his sociality, divorced from this paradigm and studied largely beyond the laboratory. It was a relatively small step from such a bipartite view of the psychological to full-scale psychological apartheid, whereby only the individualistic construal of the individual survived as a part of psychology proper.

Thus, even though the particular identity of the individual subject was important to the success of the Wundtian experiment, this was a thin identity, not one to which the person more fully considered was relevant. This assumption about individuals – that specific aspects of their mental processing can be experimentally investigated in isolation from the broader psychological and social profiles of those individuals – has been generalized in contemporary psychology and survived through the behaviorist and cognitive revolutions of the twentieth century.

The Wundtian paradigm thus both relied on a thin construal of the subject and recognized limitations of the scope of experimental psychology. These two aspects of the Wundtian construction of the individual were corollaries of his attempt to study the "generalized human mind" scientifically, and both promoted an individualistic view of the discipline of psychology. The individualism within the experimental part of the Wundtian paradigm – its bracketing off of the individual's broader psychological and social life in order to focus on understanding elementary psychological processes – might be thought to be a necessary corollary of any experimental investigation of psychological processes that occur within the head of an individual and that are not simply idiosyncratic to that individual. The very point of experimentation, after all, is to isolate the specific factor responsible for a certain, observable effect: Bracketing off or controlling certain factors is necessary in order to investigate others. But while this general point about experimental control is true, this leads to individualism about psychology only on the assumption that embedded aspects to an individual's mental states can always be (or even are ideally) bracketed out through experimental control and leave something of genuine psychological interest. In Parts Two and Three, I shall argue that an individual's mind may be more deeply socially and physically embedded than this implies.

6 GALTON'S INDIVIDUALS

Sir Francis Galton, a first cousin of Charles Darwin, was an idiosyncratic pioneer in a variety of areas of what was becoming psychology in the late nineteenth century. Consider three of his better known initiatives, all of which have left their trace on contemporary psychology. Galton developed techniques of composite portraiture and applied them to the study of judgments of beauty; he introduced statistical techniques for generalizing about groups of subjects; and he initiated the use of twin studies in studying the mind. In addition, Galton systematically explored specific mental faculties (such as that for mental imagery), and placed the scientific examination of mental qualities in an evolutionary and anthropological context. Unlike Wundt and James, Galton conducted most of his psychological investigations as a gentleman scholar, rather than as a university professor. Galton's influence on the practice of psychology, however, is at least as great as that of Wundt and James. This legacy is manifest in much that is at the core of experimental methodology within psychology, and was established in part through Galton's influence on "applied" areas of psychology, such as education and what we might think of as public mental health.

The Galtonian paradigm arose to prominence with and through the development of mental testing and has had more far-reaching effects on the conduct of psychology than perhaps any other single approach. Galton began by combining his belief in the heritability of traits of all sorts with his adaptation of Adolphe Quetelet's insights about statistical measurement in a population, originally with *Hereditary Genius* and later in *Inquiries into Human Faculty and its Development*. At his notorious "anthropometric laboratory" at the International Health Exhibition in London in 1884, Galton sought to measure human psychological abilities by having many individuals perform certain tasks, and then applying statistical analysis to the population of results obtained. These abilities were quite general, characterlike intellectual dispositions that were close to ordinary folk psychological ways of describing the stable mental life of individuals.[14]

Galton's interest in such abilities derived from two beliefs he held. The first was that mental characteristics should be treated just like bodily characteristics and other features of the natural world. The second was that since mental traits were heritable they could either be selected for or against in a population. Hence, the link between the Galtonian paradigm and the eugenics movement. Galton's interest was not so much in what cluster of abilities his subjects possessed, but, rather, in the distribution of abilities in a population of subjects, and thus by extrapolation, in the population more generally. Such a distribution was used to establish norms that could be used to make judgments about particular individuals, where what was being judged were the innate biological propensities of those individuals. Thus, individuals remain the subject matter of psychological attribution, but their role as subjects of experimentation is very different than in the Wundtian paradigm. As Danziger puts it:

> To make interesting and useful statements about individuals it was not necessary to subject them to intensive experimental or clinical exploration. It was only necessary to compare their performance with that of others, to assign them a place in some aggregate of individual performances. Individuals were now characterized not by anything actually observed to be going on in their minds or organisms but by their deviation from the statistical norm established for the population with which they had been aggregated.

Thus, the individual experimental subject simply provides data used to construct an aggregate or collective subject about which generalizations are formed, and individuals are then judged relative to this collective subject.[15]

In contrast to the regimentation of the individual that characterized the Wundtian experimental paradigm, the Galtonian paradigm of psychology encouraged an almost playful interaction between investigator and subject. This was not, however, because the Galtonian paradigm was any less individualistic but because it developed a battery of measurements whose reliability did not depend on control of the precise conditions under which they were taken. Indeed, the Galtonian and experimental paradigms share the idea that psychological propensities inhere in subjects and could be investigated without regard to the social backgrounds of those subjects. As with the Wundtian paradigm, the mental abilities of individuals within the Galtonian paradigm are not simply in the individual but are so in such a way as to imply that they can be adequately investigated in abstraction from that individual's environment.

Consider the basic process leading to psychological judgment and generalization in the Galtonian paradigm. First, we begin with tasks that can be understood with minimal explanation, attempted by anyone regardless of their particular background, and completed through the agency of an individual person. Second, we devise a performance scale that we can use to grade individual performance on these tasks, ideally one that can be calibrated around a standard distribution curve. Third, we measure the performance of a large number of individuals on these tasks. Fourth, we statistically analyze the performance of this population of individuals and use this to construct a collective subject. Fifth, we locate, rank, or taxonomize particular individuals relative to this collective subject.

While I think that an individualistic perspective permeates all five of these steps, I want to focus on Galton's tasks and the testing situation invoked at the first step. As Danziger notes, performances on these tasks "defined characteristics of independent, socially isolated individuals and these characteristics were designated 'abilities.'"[16] The governing assumption of the testing situation itself, however, is that the psychological abilities of interest are those that can be assessed by probing an individual in abstraction not only from her real life, social environment, but from any substantial social environment. For Galton, this assumption was underwritten by the search for heritable, biological factors governing cognition. But the assumption itself is independent of Galton's particular motivation, and I want to ask why one might think that it is true.

One reason, one that I have found sometimes offered by psychologists themselves in conversation, is that this is just what psychological abilities are: They are dispositions that individuals carry around with them from situation to situation. Psychological abilities are intrinsic dispositions, and as such they do not rely in any substantive way on specific social or other extraindividual circumstances. This is an individualistic view of dispositions, and in Part Two, I shall introduce a framework in which there is a place for an alternative understanding of dispositions, including psychological dispositions. On this view, dispositions may be irreducibly embedded in that they are not simply fixed or determined by the intrinsic, physical properties of the individuals that have them. This view lays a metaphysical foundation for developing an externalist psychology that departs from both the Wundtian and Galtonian paradigms.

One might be both more metaphysically reserved about the nature of psychological dispositions than encapsulated in the above line of argument, and more pluralistic about the place of heritable mental traits in the study of the mind than was Galton himself. Suppose then that there are just some psychological abilities that are intrinsic to individuals, for whatever etiological reason. For these, at least, surely the Galtonian paradigm of mental testing provides a way to meaningfully measure the mind.

This seems to me to be likely to be true, although it stops short of establishing the claim that such psychological abilities are, in some way, more important than embedded abilities, or that they provide the key to understanding such abilities and the individuals who have them. To view the Galtonian paradigm from both a less individualistic and a more pluralistic perspective is, I suspect, to assign it a more restricted role within future psychology than it has had historically.

7 NATIVISM AND THE CONTINUITY THESIS

I would like to close the substantive part of this chapter by returning to Ebbinghaus's remark that psychology has a short history but a long past. This chapter has focused on the short history and ignored the long past. A substantive reason for this focus is my view that the disciplining of the individual within psychology in the particular ways that continue to pervade psychology are inextricably entwined with the disciplining of psychology as an academic field in the late nineteenth century. But there are also ways in which psychology's long past can distort, and not just dilute, our view of the individual, and I want to return to the debate over nativism that we discussed in Chapter 1 in order to illustrate this claim.

The debate over nativism is typically considered as exemplifying the sort of truth expressed by Ebbinghaus's remark, with the long history reaching back via the positions of classic rationalists and empiricists in the seventeenth and eighteenth centuries all the way to Plato. This view of the debate involves accepting what I shall call the *continuity thesis*, the claim that contemporary positions in this debate are continuous with thinkers of times long gone, particularly early modern "rationalists" and "empiricists." The continuity thesis is close to ubiquitous amongst cognitive scientists who glance back at the history of philosophy in locating their views.[17]

While the broad sweep that the continuity thesis invites can be useful, I think that we should be more cautious about its embrace than most have been. There are important differences between the traditional and contemporary discussions. Using the two-dimensional analysis of the nativism debate from Chapter 1 as a lens, I will focus on three related features of the classical positions and debate that make for important differences with the contemporary literature. These are the concern with justification and thus normativity, the (near) absence of a psychology, and the commitment to a dualist view of the mind.

First, the traditional rationalist-empiricist debate is largely concerned with investigating the foundations, scope, and limits of human knowledge. That is, it is a debate within epistemology in general and over the nature of knowledge and justification in particular. This normative dimension is peripheral to contemporary discussions of nativism in the cognitive sciences. Broadly speaking, early modern empiricists, such as John Locke, George Berkeley, and David Hume, hold that all of our knowledge derives from our sensory, experiential, or empirical interaction with the world, either in the sense that all knowledge is acquired, ultimately, from the senses, or in the sense that the ultimate justifications that there are for the knowledge we have are grounded in the senses. Early modern rationalists, such as René Descartes, Gottfried Leibniz, and Benedict Spinoza, by contrast, hold the negation of either or both of these views, that is, either that there is some knowledge that does not derive from experience, or there is some knowledge that cannot be justified by appeal to experience.

These claims about acquisition and justification were intimately connected in all of these writers in the classical rationalist-empiricist debate. Their primary concern, however, was with the epistemological question of whether certain knowledge claims were justified, and the nature of that justification. For them, the threat of skepticism, the view that there is no knowledge, loomed large. Claiming that certain ideas were or were not acquired in a certain way established a position on this epistemological issue. For example, to claim that the notion of causation could not be acquired through sense experience, as David Hume famously argued, was a way of undermining whatever justification there was for relying on that notion in our reasoning and theorizing about the world. To claim, as Locke did at the outset of Book II of *An Essay Concerning Human Understanding*, that all ideas derive from experience, is to imply that one cannot justify any ideas by appeal to their innateness. These links between the acquisition and justification of ideas are largely absent in contemporary debates over nativism – except when authors are attempting explicitly to support the continuity thesis.

Suppose that we put this difference to one side and focus instead on what might be thought to be a common preoccupation with the acquisition of knowledge and ideas. This brings me to a second reason for resisting the continuity thesis. While at a superficial level there is a shared interest in "where ideas come from," both the external minimalism thesis and (more obviously) the internal richness thesis are claims about the processes governing cognition. This concern with how cognition works is all but absent in both classical rationalists and empiricists, who disagree over the products of cognition, particularly over whether those products, especially ideas, were acquired. This could be put (no doubt, too strongly) by saying that there is no psychology in the classical tradition. Let me say more about what I mean by this and the contrast I have in mind.

Since at least our paradigms of knowledge – knowledge of our immediate environments, of common physical objects, of scientific kinds – seem obviously to be acquired through sense experience, empiricism has some prima facie intuitive appeal. Why not suppose that all our knowledge is like those paradigms? Rationalism, by contrast, requires further motivation: minimally, a list of knowables that represent a putative challenge to the empiricist's global claim about the foundations of knowledge. Early modern rationalists included knowledge of God, substance, and abstract ideas (such as that of a triangle, as opposed to ideas of particular triangles), claiming that such knowledge and the corresponding ideas do not derive from sense experience but are part of the mind's innate endowment that is merely elicited by environmental interaction. But there is no particular claim about the cognitive processes and structures that underlie such innate ideas; indeed, this type of nativism is compatible with a consistent skepticism or agnosticism about our ability to know about the structure of the mind at all. Hence, while early modern rationalists share with contemporary nativists something similar to the external minimalism thesis, they hold such a thesis with respect to ideas, rather than the processes underlying those ideas; moreover, they have nothing like an internal richness thesis on the table.

Likewise, while early modern empiricists deny something like the external minimalism thesis about the contents of the mind, they too have little to say about the structure of the mind, and lack any view corresponding to a denial of the internal richness thesis. Given the currency of the continuity thesis, particularly amongst those who see empiricism as constituting a sustained tradition, this claim requires further discussion.

Some contemporary empiricists view our cognitive architecture as structured by a relatively small number of general principles and mechanisms that govern all of our perceptual and cognitive abilities, with these mechanisms being deployed in different ways in light of particular experiential inputs. Thus, in connectionist models, backpropagation constitutes a general type of learning algorithm that is tailored by means of particular learning sets to guide performance in various perceptual, cognitive, and linguistic tasks. Early modern empiricists might be thought to share this same general view of the structure of the mind: It is governed by a few global processes, such as association and similarity, and so constitutes a domain-general device. Common as this view is, it paints a misleading picture of early modern empiricism.

Common to all early modern empiricists is some version of the distinction between, to use Locke's terms, simple ideas and complex ideas. Global processes are introduced not to explain how original or simple ideas are acquired (and thus justified), but how we move beyond these to derived or complex ideas. The account of how we acquire our simple ideas is extremely thin, discussion being focused instead on the simple ideas themselves and how they are taxonomized. Locke, for example, discusses simple ideas in terms of the ways in which they are acquired, that is, by one sense, multiple senses, reflection, and both sensation and reflection. But the focus here is on the ideas themselves, particularly on what introspection can reveal about them, rather than on the processes generating them. Furthermore, although Locke lists what he calls combination, association, and abstraction as the three chief ways in which complex ideas are derived from simple ideas, there is little attention to how these processes operate. The bulk of the remaining twenty or so chapters in Book II are devoted, again, to the ideas themselves, not to the processes putatively generating them.[18]

A third reason for denying the continuity thesis, or viewing it as of limited significance, is that both rationalists and empiricists were, by and large, dualists, either about the nature of substance or about the nature of the properties that substances had. That is, since neither rationalists nor empiricists were physicalists about the mind, their views of how the mind was structured were views about the contents of minds, the products of mental substance that could be examined through introspection. This dualism also explains, I think, why within the traditional debate between classical rationalists and empiricists there is not only no argument over the internal richness thesis but little by way of a psychological story at all cast in terms of the mechanisms governing the processes that generate our ideas. Very few cognitive scientists are professed dualists about cognition.

In short, the continuity thesis minimizes the normative dimension to the traditional debate between rationalists and empiricists, suggests a concern with the nature of psychological processes where it does not exist, and overlooks the significance of the difference between "psychology" as conducted within dualist and physicalist frameworks.

8 OVERLAYING THE MIND

I have been interested in this chapter in identifying some of the abstractions from what we might think of as full individuality that are made within particular paradigms of psychology, and have done so via a very partial treatment of aspects of the early history of psychology. I have focused on cognitive abilities (rather than, say, on personality traits), and on abstractions in which an individual's social, ecological, and physical "contexts" are thought of as something into which an individual can be inserted once we complete our characterization of that individual's psychology. As the subject of psychological states, individuals have often been construed individualistically in the specific sense that factors outside of the boundary of the individual organism are bracketed off or treated in some secondary way. This has meant that features of individuals that seem intuitively to be important to how those individuals function psychologically – such as their location in a particular set of cultural practices, their thoughts having particular meaning, or their belonging to groups with specific characteristics – have been viewed as peripheral to psychology per se. As the psychologist Jerome Bruner has put it, they have been viewed as *overlays to*, rather than constituents of, an individual's psychology.[19]

In drawing attention to this dimension of psychology, I hope at least to have raised the question of whether psychology need be individualistic in this sense, and whether some of the ways in which individuals have been viewed within psychology are as innocuous as they might initially seem. The externalist psychology developed in Parts Two and Three provides a way, or a variety of ways, of developing cultural, social, and semantic aspects to cognition within one's psychology, rather than laid over on top of it. This has implications, in turn, for how we conceptualize the relationship between psychology and some of the disciplines with which it shares a border zone.

3

Nativism on My Mind

1 NATIVIST THREADS

The issue of nativism was used in Chapter 1 to illustrate two points. First, adopting a shared framework for understanding various individualistic theses across the cognitive, biological, and social sciences can be mutually informative. Second, these individualistic perspectives have ramifications for central debates across the fragile sciences. To make these points I introduced a two-dimensional analysis of a range of nativist debates: about the mind, biology, and culture. Here I develop and defend that analysis in more detail with a particular focus on nativism about cognition and the mind.

I begin in section 2 by tracing the most influential nativist lineage in the cognitive sciences. This lineage begins with the Chomskyan revolution in linguistics, a revolution that is generalized to other aspects of cognition by philosophers and developmental and cognitive psychologists and then further extended and given a Darwinian twist by evolutionary psychologists. I then turn in section 3 to the chief alternative to such nativist views, often characterized as "empiricist," and briefly explain why behaviorist and connectionist views of cognition are paradigms of such alternatives. In sections 4–6, the heart of the chapter, I reintroduce, develop, and defend the two-dimensional account of the debate over nativism about the mind. There has been a recent revival of attempts to analyze nativism about cognition in fewer and in more dimensions than two, and part of my defense of the two-dimensional analysis will involve showing its superiority to its contemporary competitors. Finally, in sections 7–9, I return to issues raised in considering nativism debates across the fragile sciences: whether we need more than two dimensions in order to understand some specific nativist views, and how to build on the application of the two-dimensional view to the biological sciences sketched in Chapter 1.[1]

2 FROM CHOMSKY TO FODOR TO PINKER: A THUMBNAIL

Nativist views of the mind came to prominence with the rise of the cognitive sciences through the revolution in linguistics led by Noam Chomsky. The phonological and syntactic regularities to be found in spoken and heard sentences in particular natural languages required explanation, as did the apparent ease with which those languages were acquired, and the flexibility of children as language learners. Chomsky pointed to an underlying cognitive complexity, one that arose from the mind itself, rather than the "stimulus" that led to the acquisition of spoken language, as lying at the root of a common explanation for these facts about natural languages. As tragic cases of extreme linguistic deprivation showed, linguistic stimulation of some kind was necessary for the acquisition of language. But the "poverty of the stimulus" suggested that the contribution of the mind to natural language was innate rather than learned or in some other way a reflection of complexity in the environment. This innate contribution, universal or generative grammar, is itself structured, and encompasses an individual's phonological, syntactic, and semantic knowledge.

Many of the aspects to this Chomskyan paradigm for language and their implications for cognition more generally are expressed succinctly in the following passage from Chomsky's work:

> [H]uman cognitive systems, when seriously investigated, prove to be no less marvelous and intricate than the physical structures that develop in the life of the organism. Why, then, should we not study the acquisition of a cognitive structure such as language more or less as we study some complex bodily organ? . . . Even knowing little of substance about linguistic universals, we can be quite sure that the possible variety of languages is sharply limited. . . . The language each person acquires is a rich and complex construction hopelessly underdetermined by the fragmentary evidence available [to the child]. Nevertheless individuals in a speech community have developed essentially the same language. This fact can be explained only on the assumption that these individuals employ highly restrictive principles that guide the construction of grammar.[2]

As the first half of this quotation says, the part of the mind responsible for language should be viewed as complex bodily organs, such as kidneys or hearts. Yet as suggested by the remainder of the quotation, what the language organ produces is itself structurally complex. Linguistic complexity is to be explained by an appeal to complexity in the mind of the organism. Thus, the language organ must have a structure that can be expressed as a series of principles generating the complexity in the natural language it produces.

Although Chomsky focuses on language and how facts about its complexity, acquisition, and universality should be explained, as the reference above to "human cognitive systems" suggests, Chomsky sees his views about language as having broader implications for how we should study cognition and behavior more generally. The basic properties of cognitive systems are innate in the mind, part of the human biological endowment, on a par with whatever determines the specific, internally directed course of embryological development, or of sexual maturation in later years.[3]

Jerry Fodor has been foremost in extending a number of aspects of Chomskyan nativism to the mind more generally. In particular, he has argued for two views that can be seen in this light. First Fodor has argued that there is a general language of thought that is innate, a sort of universal grammar for thinking, whose atomic components (concepts) are built into the structure of the mind. Second, Fodor views the cognitive architecture of the mind as including a range of mental organs, each operating in some particular domain.[4]

The first of these claims has been defended as part of Fodor's computational theory of mind, and includes a radical form of nativism about concepts. As Fodor says,

such cognitive theories as are currently available presuppose an internal language in which the computational processes they postulate are carried out.... [T]he same models imply that that language is extremely rich (i.e., that it is capable of expressing any concept that the organism can learn or entertain) and that its representational power is, to all intents and purposes, innately determined.[5]

Fodor is skeptical of all extant accounts of concept acquisition, and he views virtually all lexical concepts as innate. As Laurence and Margolis note in a recent discussion, this radical concept nativism has won very few converts in the cognitive sciences over the almost thirty-year period during which Fodor has advocated it.[6]

By contrast, Fodor's development of the idea of a mental organ has been enthusiastically endorsed by perceptual, cognitive, and developmental psychologists. Indeed, this modularity thesis has been extended beyond the bounds that Fodor originally set for it – sensory perception plus language – to distinctly "central processes," such as reasoning and inference, an extension that Fodor remains critical of. The most enthusiastic application of the modularity thesis has been made by evolutionary psychologists, who view the mind as composed of thousands of innately specified modules that operate across the full range of cognitive domains, not just those that are perceptual.[7]

Evolutionary approaches to understanding the mind start with Darwin and Spencer themselves, but it is only via the notion of modularity that such approaches have entered into the mainstream of thinking about cognition. Evolutionary psychologists not only blend an evolutionary perspective on cognition together with a strong modular view of the mind, but do so using the tools of experimental psychology that are central to the cognitive sciences. Foundational here has been the work of Leda Cosmides and John Tooby on social reasoning.[8]

Cosmides sought to understand the varying (and often dismal) performance on a standard, experimental reasoning task that involved subjects deciding which of four double-sided cards to turn over in order to determine whether a given, simple rule had been adhered to or broken. The task was initially devised in the 1960s by the psychologist Peter Wason to explore Karl Popper's ideas about the role of falsification in science, and the literature that this paradigm had generated over a twenty-year period was vast. Cosmides's general idea was that we could best make sense of basic performance on the task, where less than 10% of subjects succeed, and that on modified versions of the task, by thinking of reasoning not as a general ability but as a domain-specific faculty that evolved to solve social problems. Cosmides's specific hypothesis is that the aspect of this ability drawn on in the Wason tasks and its variants is the detection of those who break social rules. In short, what happens in such tasks is that a social reasoning module, equipped with a cheater-detection algorithm, is pressed into service beyond the reasoning domain in which it evolved.

The basic metaphor that evolutionary psychologists have used for the structure of the mind is that of the mind as a Swiss army knife, containing many specialized pieces. These pieces, mental modules, can be pressed into service for a range of tasks, but each evolved in the history of the species (and beyond) to fulfill some particular cognitive task within some specific domain. These tasks were set as evolutionarily significant problems, and they were solved through the evolution of specific mental modules. Amongst the many modules posited by evolutionary psychologists, apart from those for social reasoning, are those that govern criteria for mate selection and standards of beauty, eye direction detection and the theory of mind, and principles of generalization. As Cosmides and Tooby make clear in their thinking about the adapted mind more generally, and as the psychologist Steve Pinker has emphasized, evolutionary psychology revolutionizes how we should think about the social sciences and the relationship between the mind and culture.[9]

3 EMPIRICIST ALTERNATIVES TO NATIVISM

Although nativist views of the mind have been influential within the cognitive sciences, they are far from having won ubiquitous support. The chief antinativist views about cognition are *empiricist*. Nativists and empiricists share a commitment to the idea that there is at least some given, built-in organismic contribution to cognition, but differ on the nature of this contribution. In general terms, empiricists locate this contribution in our perceptual or sensory apparatus, not in the mind itself, and hold that the only genuinely cognitive structures and representations that are innate are general purpose, of the one-size-fits-all variety. What ends up in the mind, both content and structure, is a function of an organism's perceptual and sensory experience of the world. Cognitive structures and representations are thus not hard wired in the design of the organism but are contingent in some deep way on the nature of the interactions between an organism and its environment across the sensory and perceptual interface.

In the dialectic between nativism and empiricism in the twentieth century study of cognition, behaviorism looms large. Classic behaviorism dominated psychology from roughly 1920–60, and the cognitive revolution in both linguistics and psychology was very much directed at overthrowing this paradigm, beginning with Noam Chomsky's famous critique of B.F. Skinner's *Verbal Behavior*. And critiques of and recent alternatives to the nativist views sketched in the preceding section are often characterized as neo-behaviorist. There are many strands that run through behaviorist views in psychology: the emphasis on observation and empirical data, the corresponding skepticism about "theoretical" posits, the priority given to laboratory studies of model organisms and the putative generalizability of results from such organisms to human beings. However, it is not my intention here to provide a detailed survey of behaviorism, but both to show the opposition between empiricism and nativism and to explain why behaviorism is a paradigm of empiricism about cognition.[10]

The two aspects of behaviorism most relevant for these purposes are the primacy it accords perception and sensation and the central role that learning plays as a mechanism for the acquisition of complex behaviors. For behaviorists, sensory and perceptual abilities provide the developmental basis for the acquisition of all other psychological and behavioral capacities, and organisms move from this sensory base to what might be called cognition proper through one or another process of learning. Learning involves a change in the capacities of an organism or system in response to its interactions with the environment, where these capacities are cognitive in some broad and intuitive sense. (Building up your muscles at the gym satisfies the chief clause, but not the qualification.) This process may take various forms, including *associative* learning, where two or more presented stimuli become linked together by the organism; *latent* learning, where information about an environment (for example, spatial layout) is gathered at some time but not manifest or deployed until some subsequent occasion; or *skill* learning, where what is learned is a particular skill, such as bicycle riding, reading, or the ability to play chess. For behaviorists, the most important of these is associative learning, since it is a general process that, in principle, allows any perceptible stimuli to be connected by the organism and leads to the modification of its behavior. It is the means by which an organism, supposedly beginning just with sensory primitives and general psychological mechanisms, acquires the variety of knowledge that it ends up possessing; it is how environmental complexity leads to or produces organismic complexity. The centrality of learning for behaviorists is reflected in their emphasis on techniques of conditioning and in their attempted development of a generalized learning theory that applies to any behavior that is cognitive in a broad and intuitive sense.

If the twentieth-century nativist-empiricist dialectic begins with the Chomskyan challenge to the empiricist paradigm of behaviorism, it has been continued by a newer empiricist alternative, that provided by connectionism. Connectionism was developed in the early 1980s as a computational modeling technique alternative to that embodied by "classical" or "traditional" artificial intelligence. Like behaviorism, its empiricism was manifest in its emphasis on explaining complex capacities in terms of the operation of learning strategies or rules on relatively simple inputs. Rather than viewing the innate cognitive contributions of the organism to be specific, rich, and "high-level" in structure, connectionists place an emphasis on environmental stimuli (the initial data set) and powerful, general computational algorithms that are reapplied to an initial data set to generate complex outputs.

One computational model that illustrates the challenge that connectionist models pose to nativist views of language is the connectionist model for the formation of past tense developed by David Rumelhart and James McClelland. What is significant about this model in the current context is that it does not have a preexisting linguistic rule that distinguishes regular from irregular cases, but eventually develops this distinction from repeated applications of an algorithm. The model was taken by its proponents to exemplify the idea that a sensitivity to linguistic structure could be developed through repeated interactions with an environment, rather than having to be "built in" to the language learner, as nativists had claimed. As Pinker notes, the "model irrevocably changed the study of human language. . . . it is no longer possible to treat the mere existence of linguistic productivity as evidence for rules in the head." While nobody views this model as depicting how we learn the past tense in natural language, the model has spawned over twenty descendants that have led proponents of "rules and representations" views of language acquisition to attenuate their views. These have also motivated the development of hybrid models that integrate the insights of such views with their connectionist competitors.[11]

4 THE TWO-DIMENSIONAL APPROACH

The basic idea of the two-dimensional approach to the debate between nativists and empiricists, recall, is that two simple, independent theses determine a two-by-two space in this debate. The two-dimensional approach does three things. First, it helps to identify strong forms of both nativist and empiricist views. Second, it points to views that are less extreme and, in some sense, intermediate between the strong versions of nativism and empiricism. Third, it provides a common framework for understanding debates over nativism elsewhere in the fragile sciences.

The first thesis, the internal richness thesis, concerns the nature of an individual's internal structure, while the second, the external minimalism thesis, concerns the causal role of an individual's external environment in the acquisition and development of a given cognitive process, structure, or phenomenon. Where X stands in for the relevant cognitive process, structure, or phenomenon, these two theses can be stated as follows:

Internal Richness Thesis: Structures and processes internal to the individual that are important to the acquisition and development of X are rich.
External Minimalism Thesis: Structures and processes external to an individual play at best a secondary causal role in the acquisition and development of X.

Richness here is a technical term, but it is intended to draw on the ordinary sense that the word has. Internal structures are rich, recall, just if they are antecedently specialized, localized, internally complex, and causally powerful. Proponents of the "modularity of mind" are paradigms of those who accept the internal richness thesis about cognition, while those who hold that cognitive processes are "domain general" or close to nonexistent (for example, behaviorists) are paradigms of those who reject that thesis. Likewise, those who conceptualize the environment as a "trigger" for cognition accept the external minimalism thesis, while those who view learning as the central mechanism for the acquisition of cognitive structure reject that thesis.

As these examples might suggest, strong nativist positions – exemplified by Chomsky and Fodor – accept both the internal richness and the external minimalism thesis. Although Chomsky and Fodor are strong nativists with respect to a range of mental processes and structures, Chomsky's view of generative grammar and Fodor's views of concepts exemplify their commitment to strong nativism as I have characterized it. Strong empiricist positions – exemplified by behaviorists and at least early connectionists – reject both the internal richness and the external minimalism thesis. For both behaviorists and early connectionists, an organism's cognitive structures are acquired through the process of learning. This process is not internally rich in the sense specified above, and ascribes a primary role to the structure of the environment in the process of acquisition. That both of these theses capture standard nativist and empiricist positions is reflected in Steve Laurence and Eric Margolis's recent characterization of these views:

> Though there is clearly a continuum of positions, empiricist models are on the side that attributes few innate ideas, principles, and mechanisms, and generally considers the innate material of the mind to be domain-neutral and relatively simple; empiricist models also tend to give special weight to the role of sensory systems, in the most extreme cases maintaining that all concepts are constructed from sensory constituents. By contrast, nativist models are on the side that views the mind as highly differentiated, composed of far more innate elements, including domain-specific systems of knowledge or principles of inference, and innate concepts of arbitrary levels of abstraction.[12]

Although these two theses are independent, strong nativists view them as mutually reinforcing. Here's why. Suppose that we accept the internal richness thesis. If our cognitive architecture is internally rich, there is a complexity to what we bring to any cognitive task that leaves less role for an appeal to the complexity of the environments in which we operate.

In the extreme case, environments are just triggers for the operation of innately complex structures. Conversely, suppose that we accept the external minimalism thesis. Since external structures play a minimal role in the acquisition and development of cognition, any complexity that cognition generates must be due to something else, and what lies within the cognizer is a strong candidate for that something else.

To the extent that the two-dimensional analysis accurately classifies paradigmatic nativist and empiricist positions, it passes one relatively undemanding test for the adequacy of an analysis of the debate. Yet the two-dimensional view is well suited to capture more of this logical geography, including positions less extreme than either of the paradigmatic nativist and empiricist views, and I consider this one of its chief virtues. There are various ways to reject paradigmatic nativist and empiricist views of cognition, and the two-dimensional view is not only "empirically adequate" but informative about this aspect to the logical geography of the debate over nativism and cognition.

Consider the option of accepting the external minimalism thesis but rejecting the internal richness thesis about cognition. Classic ethologists and proponents of general intelligence as a built-in, global cognitive ability are both advocates of such a position. In fact, we might locate much of cognitive psychology here, insofar as it gives little role to the environment in structuring cognition yet is cast largely in terms of domain-general capacities, such as memory and reasoning, emotions and imagination.

The converse alternative – of holding the internal richness thesis while rejecting the external minimalism thesis about cognition – has also been endorsed. For example, there are modularity theorists, such as Annette Karmiloff-Smith, who view modules as being assembled through environmental interaction, rather than endogenously prespecified or governed. For forms of cognitive activity in which there can be specialized, acquired expertise, such as chess playing or mathematical problem solving, such a view is particularly plausible. Proponents of socially distributed cognition, such as Edwin Hutchins, seem also to endorse this sort of view. On Hutchins' view of cognition, the structure of the social environment of the individual plays a primary role in generating the structure of both individual and group cognition. Yet individuals themselves are, in effect, internally rich devices for the performance of cognitive functions.[13]

Figure 3.1 provides a summary of how the two-dimensional view applies to this range of positions.

The two-dimensional view also sheds some light on how each of two neurally inspired views depart from strong nativist views of cognitive

External Minimalism Thesis

	YES	NO	
Y E S	Strong Nativism	Socially Distributed/ Acquired Cognition	**Internal**
			Richness
N O	General Cognitive Psychology	Strong Antinativism	**Thesis**

FIGURE 3.1. Nativism and Cognition

development, and how they are related to one another. The "constructive learning" view of neural and cognitive development that computational neuroscientists Steve Quartz and Terry Sejnowski articulate exemplifies a strong rejection of nativism. It ascribes a central role to the structure of the environment in the construction of representations (and so rejects the external minimalism thesis), while denying the need to posit antecedently richly structured cognitive modules (and so rejects the internal richness thesis). By contrast, neuronal selection theory, whereby neural development is explained as a selective process responsive to endogenous events, shares the internal richness thesis with strong nativist views, but appeals to underlying neural features, rather than cognitive modules, to explain the structure of cognition. This affinity between neural selection theory and strong nativist views, as well as their shared rejection of the external minimalism thesis, underpins, I believe, Quartz and Sejnowski's rejection of both.[14]

Another criterion for evaluating the two-dimensional view is to see how adequately it captures more specific debates over nativism amongst cognitive scientists. For example, we might turn to debates over the acquisition of concepts in general, of word meanings or other aspects of language, such as syntax, or those over specific postulated cognitive domains, such as the theory of mind.[15]

Consider first the last of these examples, that of the theory of mind within developmental psychology. The two-dimensional account classes the views of Alan Leslie and Simon Baron-Cohen, who hold that children's psychological attributions and understanding of the mind are governed by a domain-specific module that develops maturationally, as strong nativist positions. Strong antinativist positions, by contrast, are exemplified by Peter Hobson and Michael Tomasello. They view these abilities as deriving not from dedicated modules but from more general symbolizing capacities acquired through social experience.[16]

But again there are those who have maintained views less extreme vis-à-vis nativism. For example, "theory theorists," such as Alison Gopnik and Joseph Perner, take children to develop a theory of mind through the same mechanisms of theory construction that children apply elsewhere in acquiring knowledge of the world (and so reject internal richness). Yet Gopnik and Perner ascribe only a limited role to the environment in the generate-and-test process that governs their conceptual change, and so share the external minimalism thesis with strong nativists. By contrast, theorists such as Jay Garfield and Candida Peterson maintain a view that appropriates the mediational views of Vygotsky – and so reject the external minimalism thesis – but recognize a modular component to children's acquisition of knowledge of the mind, even though they do not think that there is a domain for the theory of mind per se, and so accept a form of the internal richness thesis.[17]

Figure 3.2 summarizes the two-dimensional view of the nativist aspect to the theory of mind debate. As one might expect, there are further aspects to the debate over the theory of mind that the two-dimensional view itself does not capture: For example, that between theory theorists and simulationists, that over the relationship between language and the theory of mind, and that concerning the significance of false belief task performance. This should remind us that although nativism is one issue in play in work on the theory of mind, it is not the only one.

5 MAKING DO WITH LESS?

Traditional philosophical analyses, cast in terms of the conditions putatively necessary and sufficient for the concept of innateness, have typically tried to understand nativism in one dimension, even if that dimension has some internal complexity to it. Several recent accounts of nativism attempt to make do with just one dimension, and I want to concentrate in this section on problems that such views face. The two views on which I shall concentrate, those of Muhammad Ali Khalidi and Richard Samuels, have been developed as alternatives to existing one-dimensional views, such as those that view innate structures as those present at birth, or

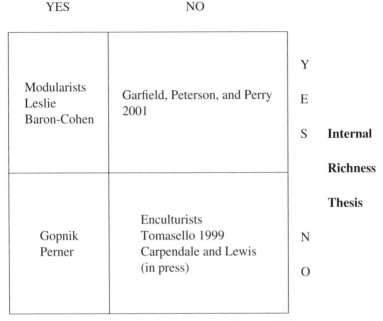

FIGURE 3.2. Nativism and the Theory of Mind

those that develop in the normal course of an organism's life. In what follows, I shall take for granted the criticisms that Khalidi and Samuels make of such views.[18]

Khalidi has defended the adequacy of a triggering view of innateness over what he calls the disease model and plasticity accounts of nativist views and argued that domain-specificity stands in an evidential rather than a constitutive relationship to nativism. In essence, Khalidi's account makes do with just one of the two dimensions that I have posited to the debate over nativism, that encapsulated by the external minimalism thesis. Khalidi begins by registering the variety of cognitive scientists – including Hirschfeld, Gelman, Landau, Cosmides, Tooby, Keil, Gopnik, and Meltzoff – whom he considers as viewing innateness and domain-specificity as constitutively linked. He then states his view that an "innate mental state or cognitive capacity is one that would be triggered by the environment," a view articulated earlier by the philosopher Stephen Stich. The remainder of Khalidi's paper criticizes views that attempt to show that there is some constitutive link between this notion of innateness and domain-specificity.[19]

There are two methodological problems with this view and how it is defended. The first is that it is rather philosophically high-handed in implying that many of those at the center of contemporary debates over nativism are conceptually confused about what nativism is. No doubt, there is some confusion. But defending a view of nativism that implies that a wide variety of researchers working on nativism are infected with this particular confusion is like insisting that lawyers are conceptually confused when they treat spoken agreements as contracts on the grounds that one has analyzed the concept of a contract in terms of signed, written agreements. The implication itself should be grounds for rethinking the grounds of the analysis.

The second problem is that the arguments that Khalidi provides can at best show that internal richness is neither identical to, nor reducible to, nor entailed by something like the external minimalism thesis. This point should be granted by a proponent of the two-dimensional view. Indeed, one might take this, in conjunction with how researchers themselves talk about nativism, as a reason for insisting that we need at least two independent dimensions to understand the debate over nativism. Moreover, since the arguments against incorporating something like the internal richness thesis into the account of nativism are all arguments that presuppose the priority of a one-dimensional triggering criterion, they beg the question against a proponent of the two-dimensional view.

Richard Samuels provides another one-dimensional analysis of innate cognitive structures, the *primitivist account*, that proceeds in two steps. According to Samuels, "a psychological structure is innate just in case it is a *psychological primitive*," and (step two) psychological primitives are structures posited by some correct scientific psychological theory, structures whose acquisition is not in turn explained by any such theory. (Putative counterexamples to the sufficiency of the account lead Samuels to modify it by making an additional appeal to the acquisition of the structure "in the normal course of events"; I ignore this here since nothing I say turns on this modification.)[20]

There are three related problems with this view: it (a) places a burden on the notion of a complete psychological theory heavier than that notion can bear; (b) presupposes a tidy alignment between psychological structures and the mechanisms that mediate their acquisition, and (c) entails that psychologically complex structures cannot be innate. What these problems imply is that the primitivist account is uninformative [(a)] or mistaken [(b) and (c)] about what psychological structures are innate. I spell out each problem in turn.

Consider (a). Whether a structure is innate, on Samuels's official view, turns on there not being any correct psychological theory that explains its acquisition. Given that even by the most optimistic of lights there remain significant and fundamental disagreements about the acquisition of almost every postulated cognitive structure, one might wonder just what structures a correct psychological theory will postulate as acquired. Even the current best candidates for innate psychological structures, such as those governing intuitive mechanics and intuitive mathematics, have previously been viewed within developmental psychology as having been acquired. Without a more complete psychological theory it is difficult to see what their status will be when psychologists call it a day.[21]

Suppose that the sorts of inferences that we can make from the current state of psychological theorizing to a complete and correct psychological theory are limited, as I am implying. Then the primitivist account is about as useful as an analysis that claims that innate cognitive structures are those that God planted in our minds. Either (even both) of these analyses might be true, but it seems desirable for a correct analysis to be more informative than is either. To draw a contrast, it may be that the fundamental physical entities in the universe are those posited in a correct physical theory. But this constitutes a part of an insightful characterization of (say) physicalism only given that current physical theory provides us with some solid epistemic guidance as to what a correct physical theory looks like, and what it postulates as fundamental. I suggest that this is precisely what we lack at the moment in the cognitive sciences.

Let us put the official talk of "correct psychological theories" to one side and turn to the heart of the primitivist view, which holds that "innate cognitive structures are ones that are not acquired by any *psychological* process or mechanism."[22] Since there is an obvious sense in which all of an organism's structures are acquired – all organisms develop (say, from a fertilized egg) from not having to having them – this account implies that innate structures are those acquired by nonpsychological mechanisms. This brings me to (b): that the account presupposes a neat alignment between the nativist status of psychological structures and the kinds of mechanisms that mediate their acquisition. Those acquired by nonpsychological mechanisms are innate, while those acquired by psychological mechanisms are not innate. This appears to make good sense of cases where the mechanisms responsible for some psychological structure are paradigmatically psychological (for example, learning) or paradigmatically nonpsychological (for example, genetic). But I think the appearances are misleading, and very few, if any, structures are acquired exclusively either by psychological or by nonpsychological mechanisms.

Consider first the sorts of case where the primitivist account might seem adequate. For example, the psychological structures needed to read or to consciously solve algebra problems are not innate because they are acquired via the psychological mechanisms or processes of teaching, learning, study, and instruction. And the psychological structures needed for identifying something as a physical object are innate because they are acquired simply by "triggering," that is, by brute-causal exposure. Reflection on both of these cases, however, brings out why (b) is a problem.

Reading and consciously solving algebra problems are abilities acquired through psychological processes, no doubt. But their acquisition usually also involves mechanisms that are "social," such as instruction, and others that are "biological," such as those for neural storage. Indeed, any psychological ability that is acquired once a child already has acquired a lot of psychological structures and mechanisms, however they are acquired, is likely itself to be acquired through the agency of both psychological and nonpsychological mechanisms. One reason for the importance of psychological mechanisms is the way in which they are integrated with both "higher" and "lower" level mechanisms, and thus the correspondence proposed by Samuels between "noninnate" and "psychological" acquisition processes is problematic.

There is much the same problem for the proposed link between "innate" and "nonpsychological" mechanisms of acquisition. Brute-causal exposure to physical objects and how they behave mechanically may be sufficient to activate the intuitive physics of normal newborn or very young infants. But there are surely psychological mechanisms that such infants are equipped with that play a role in how their intuitive physics operates. For example, they have mechanisms that govern how they behave, such as how long they look at a stimulus, or other mechanisms that govern preferential looking or other forms of early behavior. That such behavior is subject to psychological generalizations, such as "Find unexpected stimuli interesting" and "Look longer at objects that are interesting," is presupposed in the very experimental paradigms for the investigation of the proposed innate structures.

The basic problem that the entwinement of psychological and nonpsychological mechanisms poses for the primitivist account is as follows. If innate psychological structures are those acquired without any psychological mechanisms or only by nonpsychological mechanisms, then no or almost no psychological structures are innate. Importantly, the psychological structures that are our current best bet for being innate turn out not to be innate, on such a view. And that suggests that the primitivist account is mistaken.

Finally, consider (c). By a "psychologically complex" structure I mean a structure that has parts whose coordination and integration physically constitute that structure, and which themselves are psychological. For such a structure, there may be a correct psychological theory for how that structure is acquired that is given in terms of how those parts develop and are integrated. But for all that, the structure itself might be innate. Indeed, those who view such structures as innate often take the prime psychological task to be to functionally decompose the structure and understand the developmental trajectory of those parts. For example, Leslie and Baron-Cohen posit an innate theory of mind module (ToMM), and propose a particular compositional structure that is psychological and whose development is explained in psychological and neuropsychological terms. This is true more generally of domain-specific theories (for example, of physical objects, of biology), which are complex in something like the sense above but whose innateness is not thereby ruled out. One could insist that the psychologists who talk in this way are mistaken. But such insistence would seem high-handed in much the way that I claimed was Khalidi's account.

6 SATISFYING SOME DESIDERATA

Samuels also posits five constraints on the adequacy of any account of innateness. These can be summarized group-wise as follows:

(i) X is innate $\rightarrow$ (X is not learned) and $\not\rightarrow$ (the environment plays no role in the acquisition of X)
(ii) the account must make sense of the chief arguments for/against nativism, of why nativism matters to cognitive science, and of the logical geography to the debate.

Samuels calls (i) "conceptual constraints" and views them as more foundational than (ii). All five desiderata that Samuels offers are plausible constraints, and despite its other problems, the primitivist account prima facie does a reasonable job of satisfying all of them.

But the two-dimensional view does a better job here. Since the external minimalism thesis is one dimension in the view, that view readily meets both of Samuels' "conceptual constraints." I have already directed much of my discussion at showing how the two-dimensional view fares regarding the "logical geography" of the debate over nativism. The only point I shall underscore here is that, in contrast to both one-dimensional views we have considered in the previous section (as well as others we have not), the two-dimensional view makes it natural to see why "being innate" is not an all-or-nothing matter, and how we might distinguish between stronger and more moderate positions in the debate over nativism.

The two remaining constraints that Samuels posits are that the account make sense of the arguments for nativism and that it show why the debate over nativism is of significance within cognitive science. Samuels himself considers two chief arguments for the innateness of cognitive structures – those which appeal to the poverty of the stimulus and to early development – and to those we might add arguments that cite the universal manifestation of a psychological structure, or that invoke neurally specific deficits. Endorsement of the external minimalism thesis itself would make each of these arguments plausible ones to offer, but it is worth noting that the internal richness thesis also provides grounding for at least these arguments. For if that thesis is true and cognition is governed by internally rich units, then we might expect normal environments to be sufficient for the abilities that those units imbue an organism with, and so develop early and universally, and for those abilities to be subject to impairment through specific neural damage. As importantly, the two-dimensional view also helps to explain why rejecting one of these arguments – say, the poverty of the stimulus argument – or viewing it as having limited significance, need not involve a wholesale abandonment of nativism or the arguments for it.

Finally on the desiderata front, the two-dimensional view suggests that part of the significance of the debate over nativism lies in arguments over two distinct issues: the causal role of the environment in the acquisition of cognitive structures and abilities and the nature of the internal resources that individuals antecedently bring to bear on cognitive tasks. Positions on each of these issues carry with them methodological implications for how to study cognition and particular cognitive abilities. One's view of the external minimalism thesis affects (or reflects) whether one thinks that cognitive scientists should invest their time in studying the structure of an organism's environment. And one's view of the internal richness thesis does the same with respect to the sort of internal structures posited or sought inside the cognizer. The two-dimensional view also makes it clear that a position on one of these issues leaves room for disagreement about the other.

7 BUT COULD TWO DIMENSIONS BE ENOUGH?

Suppose that we need two dimensions to understand the debate over nativism and that these are the two that I have suggested. Might we need more? For example, one might think, particularly if one views evolutionary psychology as a strong nativist position, that one should also view the claim that rich mental structures are evolved or encoded in the genome as part of the commitment of a strong nativist view. The developmental psychologist Frank Keil has suggested that it is an important part of nativist views of the mind that what is "built in" to an organism is not simply perceptual abilities but something intuitively further "upstream" from perception, something "more central," something genuinely cognitive. The philosopher Fiona Cowie has also identified five theses as constituting the "Chomskyan nativist" view of language acquisition. On Cowie's view, a firm commitment to one of these, the thesis that the constraints and principles that define the domain of language and its subdomains be identified with universal grammar is what distinguishes the Chomskyan view from all nonnativist alternatives. Perhaps we should view this dimension as essential to characterizing nativism more generally. Since we double the grain of our analysis for each additional dimension we add, why not enrich our view of the nativism debate further? Why think two is enough, especially if one leads to impoverishment? But I think we should be cautious here. More is not always better. Consider briefly the proposals of Keil and Cowie in turn.[23]

Keil's suggestion is motivated by the recognition that all parties to the debate over nativism acknowledge that some psychological structures are innate and that what distinguishes nativists from empiricists, apart from their views of what I am calling internal richness, is how far upstream this structure is to be found. I think that there is some truth in this and that this dimension of "centrality" is important in understanding why nativist views come in degrees. In some specific debates over nativism, such as that over concepts, this dimension of centrality may be significant enough to be added as a distinct dimension. So it may be that it is useful to add this as a third dimension to the account I have proposed, at least for some purposes.

Yet this dimension doesn't help much in understanding other particular nativist debates, such as that over the theory of mind, which occur, so to speak, at one place in the stream, and whose alternatives to strong nativism include positions that identify intuitively more central abilities, such as metarepresentation (Perner) or theory construction (Gopnik), as underlying the theory of mind. Moreover, I am inclined to think that this dimension is subsumable under or derivable from the internal richness thesis insofar as stronger nativist views tend to view the thesis as having a greater range of applicability to cognition than do weaker nativist views. Although it is conceptual open space to hold the internal richness thesis about just "upstream" structures, and not those together with sensory-perceptual structures, as a matter of fact no one holds such a view. Thus, all views that apply the thesis more widely also apply them to central cognitive processes.

My view of Cowie's proposal is similar. Recognizing the distinctness of a "universal grammar" dimension to the nativism debate over language may be useful – for example, in distinguishing Chomskyan nativism from lexical functional grammar. Yet it is unclear how such a distinct dimension helps elucidate such debates in all areas of language, such as that over the acquisition of the semantics of words, or more importantly, beyond language and in other cognitive domains where the combinatorial, generative role that universal grammar plays in language is less significant. For example, in cognitive development, there is a focus on the principles that govern infants' recognitional and behavioral interactions with physical objects – naive physics – and on those that govern their interactions with persons – naive psychology and naive sociology. Yet such principles constitute a grammar or a syntax in only a metaphorical or extended sense. The problem of how to apply this dimension to other nativist and empiricist views is exacerbated once we turn from cognitive to noncognitive domains.

In summary, while further dimensions may shed light on nativism in specific contexts, they do not appear to have the generality that make them useful in understanding nativist views more generally. We can see this more clearly by turning from the cognitive to the biological domain.

8 NATIVISM ABOUT COGNITION AND BIOLOGY

The renewed attention that the concept of innateness has received of late within the cognitive sciences, typified by the authors I have thus far discussed or mentioned, is marked by a near-exclusive preoccupation with what it means for a cognitive or linguistic structure or process to be innate, and arguments for thinking that many such structures and processes are innate. This focus neglects the other primary domain in which the concept of innateness has been deployed, that of the biological sciences. As I suggested in Chapter 1, an additional virtue of the two-dimensional account is that it sheds some light on nativism debates within this domain, and so reveals an affinity between the largely disjoint cognitive and biological literatures. Here I shall focus on the two areas of biology to which nativism – under the heading of innateness – has been perhaps most central. I shall expand on my compressed remarks about inheritance and development from Chapter 1 and will also briefly consider innateness in behavioral ecology. Rather than talk of strong antinativist views here as empiricist, which would sound idiosyncratic, I shall refer to such views as *strong externalism* about organismic development.[24]

There has been some skepticism about whether the notion of innateness any longer plays a useful role within either of these areas of biological science. For example, the biologist Patrick Bateson has distinguished seven different things that "innate" means when applied to behavioral phenotypes in the study of animal behavior, and suggested that only confusion arises from continuing to use "innate" to mark all of these correspondingly distinct contrasts. The philosopher of biology Paul Griffiths has said that in "molecular developmental biology innateness seems as antiquated a theoretical construct as instinct and equally peripheral to any account of gene regulation or morphogenesis," and has claimed that since there is no univocal sense in which biologists speak of a trait as being innate, we should retire talk of innateness.[25]

These views of Bateson and Griffiths do, I think, provide reason to be skeptical about traditional one-dimensional analyses of innateness in biology, including recent analyses in terms of canalization and generative entrenchment. But I also think that the two-dimensional approach shows why Bateson and Griffiths are perhaps overly pessimistic about the prospects for making sense of innateness in the biological sciences.[26]

While both the internal richness thesis and the external minimalism thesis are formulated with respect to cognitive phenomena, processes, and abilities, it is a trivial matter to apply them both to biological structures and processes. In Chapter 1, I pointed out that within genetics and developmental biology questions of the innateness of any given trait are often cast in terms of the role and conception of genes in organismic development. The dominant view here accepts genes as being at the core of the rich, internal structure that guides organismic development, with the environment subsequently playing some kind of secondary causal role. Both classical and molecular genetics, both cast since the 1940s largely in terms of the complex metaphor of a genetic program – blueprints, codes, instructions, reading frames, executive controls – adopt this view of organismic development in general, and as such represent strongly nativist views of this process. Genes code for phenotypes – the traits are "in the genes" – and organismic development is the unfolding of this preformed structure in the genetic code. But as a global view of development, such a view has been challenged in whole and in part, and the two-dimensional view of nativism captures much of the character of such challenges.[27]

While no one denies the existence of intracellular units, genes, both their role in organismic development and the relationship between "genetic" and "environmental" causes of development have been contested. For example, as I said in Chapter 1, Richard Lewontin has rejected the idea that genes are the primary agents of development, as well as the claim that genetic and environmental "factors" can be partitioned in ways so as to meaningfully inform us about the causes of particular developmental changes. In effect he rejects both the internal richness and the external minimalism theses, and his views represent a strong externalist position within genetics and developmental biology. Thus, the two-dimensional view captures both strong nativist and strong externalist positions in genetics and development. But it is important to see that, as is the case with cognition, the two-dimensional analysis also provides for the representation of less extreme views of genetics and development.

For example, some developmental biologists, such as Brian Goodwin, share Lewontin's skepticism about the roles ascribed to genes as agents of development, and have argued for the primacy of other internal factors, such as morphogenetic fields, as such agents. In seeing development as endogenously driven, however, Goodwin accepts the external minimalism thesis. What he rejects is the particulate view of the engines of development entailed by the internal richness thesis. Rather than a gene-centered developmental biology, Goodwin advocates a revival of what he calls a *rational morphology*, with a primary role for developmental constraints and morphogenetic fields in explaining development. This is one way to depart from both strong nativist and strong externalist views of organismic development.[28]

Proponents of developmental systems theory, by contrast, maintain that the agents of development are *developmental systems*. Developmental systems are series of processes that use a range of developmental resources, including genes, chromatin markers, and organelles within the organism. Those developmental systems theorists, such as Eva Jablonka and Marion Lamb, who emphasize cellular, epigenetic inheritance systems have some affinity with Goodwin's rational morphology insofar as they identify nongenetic, internal causes of development. But others, such as Susan Oyama and Russell Gray, take there to be an important symmetry

Nativism on My Mind

External Minimalism Thesis

	YES	NO	
	Classical and Molecular Genetics	Developmental Systems Theory	**Internal Richness Thesis** (YES)
	Rational Morphology	Radical Interactionism	(NO)

FIGURE 3.3. Nativism, Genetics, and Development

between developmental resources that lie within and beyond the boundary of the organism, and so view developmental systems as often extending beyond that boundary. Such views reject the external minimalism thesis, and often share much with Lewontin's criticisms of gene-centered views that predominate within molecular genetics. Figure 3.3 shows how the two-dimensional analysis applies to this range of positions about nativism and development.[29]

I have so far been suggesting that the two-dimensional view of nativism sheds some light on the "logical geography" of recent views within genetics and developmental biology. The basic idea of this view – to treat internal structural richness and external causal role as independent dimensions – is particularly apt for genetics and development within the biological domain. This is because much of the debate over innateness within biology has turned on questioning traditional dichotomies – innate versus acquired, genetic versus environmental – that deny their independence.

These two dimensions also subsume many of the characterizations that have been given of innate traits in behavioral ecology. For example, consider the seven senses that Patrick Bateson detects in uses of "innate" in describing animal behaviors. Innate behaviors have been claimed to be those that are (a) present at birth, (b) not learned, (c) adapted over the course of evolution, (d) unchanging through development, (e) shared by all members of the species, (f) part of a distinctly organized system of behavior driven from within, (g) caused by a genetic difference when two organisms differ with respect to them. Bateson points out that there are many cases in which these criteria come apart and argues that this limits the usefulness of the concept of innateness within behavioral ecology.

If we view these, however, not as criteria for innateness but as seven evidential bases for ascriptions of innateness in a given case, then I think the two-dimensional approach is well placed to explain their appeal. Of these, (a) and (b) are readily explained as a result of the external minimalism thesis, and (f) and (g) by acceptance of the internal richness thesis. The remaining three, (c), (d), and (e), could be explained by either or both theses. Likewise, the idea of innate traits as those that are developmentally fixed and so relatively insensitive to environmental variation (canalized), or those that are "generatively entrenched," can be understood in terms of the external minimalism thesis. In short, the two-dimensional approach to nativism within biology makes sense not only of the major positions that have been held within genetics and developmental biology, but also explains the sorts of characterization of, and evidence for, claims of innateness in particular cases within behavioral ecology.

9 CONCEPTUAL ANALYSIS AND NATIVISM

Identifiable nativist views have been with us for a long time, and they are likely right about at least some of our cognitive structures and abilities, just as they are right about some of our bodily phenotypes and behaviors. The two-dimensional approach to the debate over nativism developed in this chapter allows us to pinpoint what nativists are claiming about cognition, and how one might go part of the way toward accepting the strongest nativist views and so accommodate their insights without simply assimilating cognitive development to bodily growth.

I also argued in the previous chapter, however, that the "short history, long past" view of the debate over nativism should not blind us to the significant differences between what has been central to that debate during distinct historical periods. In particular, there are several important discontinuities between traditional rationalism and empiricism and contemporary nativist and nonnativist views. The two-dimensional approach aims to capture something important about the contemporary debate, and I think we do less justice to that goal by either insisting on or implicitly assuming a continuity thesis that is insensitive to the distinctive preoccupations of seventeenth- and eighteenth-century thinkers.

It has been part of my plaint over an extended period of time that philosophers (as well as zealous cognitive psychologists, linguists, and computer scientists) should learn to say "some." In philosophy we too often yearn for something close to a standard conceptual analysis of a concept or term in attempting to understand what is a corresponding complicated and messy reality. If we view one-dimensional accounts of nativism about the mind as suffering from the inability to "say 'some,'" and from focusing largely or exclusively on nativism about the mind, then it is a fair question to ask how well the two-dimensional analysis itself scores on both of these fronts.

The two-dimensional view is an analysis of sorts, but one aimed primarily at understanding the debate over nativism, rather than our "ordinary" concept of innateness (if there be such), and that is primarily descriptive rather than normative or revisionary. Although I have responded briefly to a few "higher dimensional" accounts of the debate, and attempted to show how further or other dimensions that have been proposed are, in some sense, derivative from the two-dimensional approach, it seems to me less significant whether one is right here than it is to be right about the need to move beyond one-dimensional accounts. This is in part because nativist views are enmeshed with a variety of other views about cognition – about its evolution, about its relationship to culture, about its demarcation from both the noncognitive and the nonpsychological. And where one's nativism ends and such other views begin seems difficult to pronounce on, and unwise to build on.

One final point. One implication of the two-dimensional view is that strong nativist views will likely be defensible for a more limited range of traits than have often been thought of as innate. But this is neither because there are more traits for which strong antinativism or externalism hold, nor because "innate" is a term that has outlived its usefulness. Rather, the two-dimensional view requires that we consider both internal and external dimensions to the acquisition and development of any given trait. Often enough, we will have to be satisfied with a view of acquisition and development that departs from strong forms of both nativism and empiricism.

PART TWO

INDIVIDUALISM AND EXTERNALISM IN THE PHILOSOPHY OF MIND AND THE COGNITIVE SCIENCES

4

Individualism

Philosophical Foundations

1 MAKING SENSE OF THE INDIVIDUALISM-EXTERNALISM DEBATE

Individualism about the mind was introduced as a form of methodological solipsism in considering the nature of an individual's mental life and how we ought to theorize systematically about it. I drew on this conception of individualism in discussing the disciplining of psychology as a field of inquiry and nativism about the mind. In Chapter 1, I also provided a more precise characterization of individualism in psychology in terms of the notion of supervenience.

It is sometimes unclear to those outside of the philosophy of mind just how either the methodological solipsism or the supervenience formulation of individualism could give rise to a substantive debate about the mind and its study. Consider construals of each of these formulations that make either individualism or externalism seem trivially true.

Methodological solipsism in psychology is the view that psychological states should be construed without reference to anything beyond the boundary of the individual who has those states. It is in light of this view that the debate between individualists and externalists has sometimes been glossed in terms of whether psychological or mental states are "in the head." But to the initiated and uninitiated alike, that is likely to sound like a puzzling issue to debate: Of course mental states are in the head! ("Where else could they be?," as Robert Stalnaker once asked.) So this construal of individualism makes externalism a nonstarter, and so individualism seem trivially true.

The supervenience formulation of individualism says that mental states supervene on the intrinsic, physical properties of the individual who bears them. The most important feature of the central relation of supervenience, at least for our purposes, is that it is *determinative*: According to the individualist, an individual's mental states are determined by her intrinsic, physical states. But then it seems that it is individualism that is a nonstarter, for surely, one might think, there are many environmental factors that determine what mental states an individual ends up having. What visual perceptions you have are determined by what's in front of you, and what sorts of things you think about is determined, in part, by what concepts you have been taught – by parents, siblings, peers, and teachers. So construed, externalism seems trivially true.

Both of these construals of the issue that separates individualists from externalists about the mind are, of course, caricatures. But I hope they will be useful caricatures, in several ways. First, the ease with which they can be derived from each characterization of individualism poses a challenge: to articulate a version of the issue separating individualists and externalists that makes more perspicuous why individualism is a substantive and thus potentially controversial claim about the nature of the mind and how we should study it. Second, I think that each caricatured objection contains more than a grain of truth about, respectively, externalism and individualism. And so part of our task in responding to the above challenge in the next section will be to sift the true from the false in each of these caricatures.

In section 3, I review the original arguments for externalism provided by Hilary Putnam and Tyler Burge. Both of these arguments appeal to claims about intentionality or content, but the issues they raise concern psychological states in general. Burge concluded his original critique by saying that the "sense in which man is a social animal runs deeper than much mainstream philosophy of mind has acknowledged." In section 4, I show just how externalism suggests that the social aspects to having a mind run deeper than a range of influential philosophical views have allowed.[1]

If individualism is as entrenched a view in the history of philosophy and in contemporary analytic philosophy as sections 2–4 imply, then it should be no surprise that the conclusions from the Putnam-Burge arguments have been resisted. The remainder of the chapter aims to give the flavor of this resistance. In section 5, I discuss perhaps the most prevalent response, one that draws on a distinction between narrow and wide content. The idea that there is some notion of content immune to the Putnam-Burge arguments is popular even if problematic. Section 6 explores the prima facie relationship between individualism, functionalism, and physicalism, and section 7 reviews and critiques an influential argument for individualism that purports to show that individualism in psychology is mandated by reflection on the nature of scientific taxonomy more generally.

The chapter as a whole aims to be a self-contained philosophical primer on individualism about the mind. At its end, we will be in a better position to lay the foundations for an alternative, externalist conception of the mind.

2 INDIVIDUALISM, TAXONOMY, AND METAPHYSICAL DETERMINATION

Externalists don't claim that mental states are somewhere other than in the head, and individualists don't think that what is outside the head has nothing to do with what ends up in the head. The key to understanding the debate between individualists and externalists about the mind lies in grasping the notion of determination at its heart.

Individualists and externalists agree that an individual's environment is a causal determinant of that individual's thoughts and thus mind. Agents causally interact with their world, gathering information about it through their senses, and through their communicative interactions with others. Thus, the nature of their minds, in particular what their thoughts are about, is in part causally determined by the character of their world. That is, the world is a contributing or efficient cause to the content of one's mind, to what one perceives, desires, and thinks about. This is just to say that the content of one's mind is not causally isolated from one's environment. Separating individualists and externalists is the question of whether there is some deeper sense in which the nature of the mind is determined by the character of the individual's world. It is this "deeper sense" of world-mind determination that we need to articulate further.

We can approach this issue by extending the brief discussion of the idea that the content of the mind is in part causally determined by the agent's environment to explore the conditions under which a difference in the world implies a difference in the mind. Individualists hold that this is so just in case that difference in the world makes some corresponding change to what occurs inside the boundary of the individual. Externalists deny this, thus allowing for the possibility that individuals who are identical with respect to all of their intrinsic features could nonetheless have psychological or mental states with different contents. And, assuming that mental states with different contents are ipso facto different types or kinds of states, this implies that an individual's intrinsic properties do not determine or fix that individual's mental states.

Thus, to state individualism adequately, we need to draw a distinction between *causal* and *metaphysical* determination. Individualists claim, and externalists deny, that what occurs inside the boundary of an individual *metaphysically* determines the nature of that individual's mental states. The individualistic determination thesis, unlike the causal determination thesis, expresses a view about the nature or essence of mental states, and identifies a way in which, despite their causal determination by states of the world, mental states are, according to individualists, autonomous or independent of the character of the world beyond the individual.

I said in the previous section that the relation of supervenience was determinative, and we can now see that the appropriate notion of determination is that of metaphysical, rather than causal, determination. The determining properties – for individualists, the intrinsic, physical properties of the individual – are called the *subvenient* or *base* properties. Once subvenient or base properties are fixed or held constant across two or more situations, the supervening properties – in this case, the individual's mental properties – are also fixed across those situations.

Many of an individual's properties, particularly those that are of interest to scientists, supervene on that individual's intrinsic, physical properties. To take a simplistic example, it is plausible to suppose that your mass supervenes on the mass of all of the particles in your body. Given this supposition, it doesn't matter what else is true or false about you: If the mass of all the particles in your body is 100 kg, then your mass is also 100 kg. Provided that the mass of all the particles in your body is 100 kg, it doesn't matter whether you are on Earth or on Mars, whether you are stationary or in motion, or whether you are rapidly losing or gaining particles, your mass is also 100 kg. There is no way to change your mass from 100 kg that does not also change the mass of all of the particles in your body. We might say that mass is compositionally or mereologically determinative: The masses of any physical thing's components or parts metaphysically determine its mass.

If you are injected with a liquid that increases your mass by 1 kg, that action also increases the mass of the particles in your body by a total of 1 kg – say, either because of the mass of the particles in the liquid itself, or because of the way in those particles interact chemically with the particles already in your body, thus adjusting their number and mass. Alternatively, if someone removes a part of your body – an arm, a kidney, a piece of skin – and your mass is subsequently reduced, the mass of all of the particles in your body is reduced by precisely the same amount. At least this is true provided that a person's mass supervenes on just the mass of all of the particles in her body, that is, that mass is mereologically determinative. (Note that the theory of special relativity provides no grounds for denying this view, since although an object's mass varies with its velocity, so too does the mass of its physical parts.)

The idea that mental properties should be assimilated to properties explored in the physical sciences, such as mass, has provided one powerful reason for thinking that psychology must or should be individualistic. If the distinction between causal and metaphysical determination shows what is mistaken about the caricature of individualism that makes externalism appear trivially true, then this assimilation between the mental and the physical helps to explain the grain of truth in the caricaturized picture of individualism with which we began. For what is right about the caricature is that individualists do place an emphasis on the central role that an individual's intrinsic, physical properties play in the causal economy of that individual. This has sometimes been summarized in the slogan "No mental difference without an intrinsic physical difference." That many see such a slogan as derivative from the physicalist slogan "No difference without a physical difference" points to one perceived link between individualism and physicalism. (More of which in sections 6 and 7.)

Let us return to the methodological solipsism formulation of individualism to see both why externalists are not committed to denying that mental states are "in the head," and why this caricature nonetheless captures something true about externalism. Even though individualism is a thesis of metaphysical determination, it is also a claim about how psychological states ought to be individuated or taxonomized. This is implicit, I think, in the adjectives "mental," "psychological," or "cognitive," predicated of states, properties, processes, or events, since to talk of (say) mental properties is already to talk of properties as being of a certain kind or type. Thus, although individualists and externalists agree that mental states are "in the head" – just as they agree that they are causally determined, in part, by what lies beyond the head – they disagree about how mental states should be individuated or taxonomized. What is right about the caricature of externalism, then, is that externalists do think that what lies beyond the head, what is not "in the head," is relevant to psychological taxonomy. What psychological kinds an individual instantiates is not metaphysically determined by what is in the head.

We can now put together the point about determination with this claim about taxonomy. Individualism is the view that mental states, qua mental states, are metaphysically determined by an individual's intrinsic, physical properties. The expression "mental states" is used in a general sense, and encompasses determinate forms of states, properties, processes, and events that are termed "mental," including kinds that are motivational, cognitive, and perceptual.

Individualists are right to think that the cognitive and physical sciences are subject to similar constraints, but wrong, I shall suggest, about what those constraints are. Talk of mental states being "in the head" may be a useful shorthand for the individualistic claim that such states, insofar as they feature in psychological taxonomies, are metaphysically determined by an individual's intrinsic, physical properties. But we should also be wary of how readily this locational metaphor can mislead us about the nature of the debate between individualists and externalists.

As a thesis of metaphysical determination, individualism implies that two individuals identical in their intrinsic respects must have the same psychological states. The modal aspect to this implication makes supervenience an appropriate concept to use in stating individualism more precisely. This implication, and indeed the debate over individualism, is often made more vivid through the fantasy of *doppelgängers*, molecule-for-molecule identical individuals, and the corresponding fantasy of Twin Earth. I turn to these dual fantasies next.

3 GETTING TO TWIN EARTH: WHAT'S IN THE HEAD?

Hilary Putnam's "The Meaning of 'Meaning'" introduced both fantasies in the context of a discussion of the meaning of natural language terms. Putnam was concerned to show that "meaning" does not and cannot jointly satisfy two theses that it was often taken to satisfy by then prevalent views of natural language reference: the claim that the meaning of a term is what determines its reference, and the claim that knowing the meaning of a term is simply a matter of being in a particular psychological state. This latter claim is sometimes glossed by saying that meanings are "in the head," and it is an individualistic claim in that it implies that knowledge of meaning is metaphysically determined by what's inside the head.

These theses typified descriptive theories of reference, prominent since Gottlob Frege and Bertrand Russell explicitly formulated them, according to which the reference of a term is fixed or metaphysically determined by the descriptions that a speaker attaches to that term.

Such views were central to both ordinary language philosophy and logical empiricism, two encompassing frameworks that had much influence in English-speaking philosophy throughout the 1950s and 1960s. The basic idea of descriptive theories of reference is perhaps best conveyed through an example.[2]

Suppose that I think of Aristotle as a great, dead philosopher who wrote a number of important philosophical works, such as the *Nicomachean Ethics*, and who was a student of Plato and teacher of Alexander the Great. These are the descriptions that I associate with the name "Aristotle." Then, on a descriptivist view of reference, the reference of my term "Aristotle" is just the thing in the world that satisfies the various descriptions that I attach to that term: It is the thing in the world that is a great philosopher, is dead, wrote a number of important philosophical works, was a student of Plato, and was a teacher of Alexander the Great. Such descriptivist views of the reference of proper names were the critical focus of Saul Kripke's influential *Naming and Necessity*, while in his attack on this cluster of views and their presuppositions, Putnam focused on natural kind terms, such as "water" and "tiger." Both Kripke and Putnam intended their critiques and the subsequent alternative theory of natural language reference, the causal theory of reference, to provide another, general way to think about the relationship between language and the world. As Gary Ebbs has pointed out, in at least Putnam's case, this theory was part and parcel of a more wide-ranging critique of the notions of analyticity, the *a priori*, and reductionism associated with empiricist views in metaphysics and the philosophy of science. But let us stay close to Putnam's argument in "The Meaning of 'Meaning'" and draw out its connection to individualism.[3]

Consider an ordinary individual, Oscar, who lives on Earth and interacts with water in the ways that most of us do: He drinks it, washes with it, and sees it falling from the sky as rain. Oscar, who has no special chemical knowledge about the nature of water, will associate a range of descriptions with his term "water": It is a liquid that one can drink, that is used to wash, and that falls from the sky as rain. On a descriptive view of reference, these descriptions, what the logician Gottlob Frege called the *sense* and Rudolph Carnap the *intension* of the term, determine the reference of Oscar's term "water." That is, the reference or extension of Oscar's term "water" is fixed by the set of descriptions he attaches to the term as part of his grasp of its sense. And since those descriptions, so grasped, are "in the head," natural language reference on this view is individualistic.[4]

To continue Putnam's argument, now imagine a molecule-for-molecule *doppelgänger* of Oscar, Oscar*, who lives on a planet just like Earth in all respects but one: The substance that people drink, wash with, and see falling from the sky is not water (that is, H_2O), but a substance with a different chemical structure, XYZ. Call this planet "Twin Earth." This substance, XYZ, is called "water" on Twin Earth, and Oscar*, as a *doppelgänger* or twin of Oscar, has the same beliefs about it as Oscar has about water on Earth. (Recall that Oscar, and thus Oscar* as his twin, have no special knowledge of the chemical structure of water.) Oscar* believes that it falls from the sky as rain, is drinkable, can be used for bathing, is found in rivers and streams – all the things that Oscar believes of water. Oscar and Oscar* associate just the same descriptions with the term "water": their term has the same meaning, sense, or intension, where these are conceived individualistically. On a descriptive theory of reference, since meaning determines reference, their terms "water" should have the same reference.

But there are several reasons to resist the claim that Oscar and Oscar* have a term, "water" with a common reference. First, recall that Twin Earth is introduced as being just like Earth, except that it has another substance, XYZ, in place of water, that is, H_2O. If this is right, then it is hard to see how Oscar* could come to refer to water, since there is no water on his planet. Twin Earth has what we might call "*twin-water*" or "*twater*" on it, not water, and it is twater that Oscar* interacts with, not water. Second, Oscar and Oscar* stand in the same relation to their respective environments, which suggests a certain parity in their cases. Given that Oscar's term "water" refers to or is about water, then Oscar*'s term "water" refers to or is about twater. Putting these two points together: If Oscar uses "water" to refer to water because that is the stuff that is in his local environment, then Oscar should use "water" to refer to twater, for just the same reason. That implies that Oscar and Oscar* have natural language terms that differ in their reference. And this is so despite the fact that their terms agree in their in-the-head meaning. By hypothesis, Oscar and Oscar* are *doppelgängers*, and so are identical in all their intrinsic properties, and so are identical with respect to what's "in the head." Thus, Putnam argues, the reference of the natural language terms that Oscar uses is not metaphysically determined by what is in Oscar's head.

Putnam's target was a tradition of thinking about language that treated the meanings of natural language terms and language more generally in ways that supposed that the world beyond the individual language user did not exist. With a focus on natural kind terms, and the broader, naturalistic alternative that Putnam saw himself as offering, Putnam's views here became associated with scientific and metaphysical realism, whereby the referents of those terms, natural kinds, had underlying essences that were discovered *a posteriori* through scientific methodology. Thus, part of the interest that Putnam's views have generated, and some of the controversies they have engendered, turn on these broader features of his views.[5]

Since Putnam's chief point is one about natural language terms and the relationship of their semantics to what's inside the head, one needs at least to extend his reasoning from language to thought to arrive at a position that denies individualism about the mind itself. Indeed, there are various points at which Putnam himself seems to presuppose individualism about the mind in making his case against descriptive theories of reference. For example, Putnam says, in reference to a *doppelgänger* of his whose word "elm" refers not to elms but to beech trees that "[i]t is absurd to think *his* psychological state is one bit different from mine: yet he 'means' *beech* when he says 'elm' and I 'mean' *elm* when I say elm. Cut the pie any way you like, 'meanings' just ain't in the *head*." It is precisely the view that Putnam labels as absurd here, however, that is expressed by anti-individualists about the mind.[6]

The term "individualism" itself, and the development of a series of thought experiments that made a case against individualism and which in many ways paralleled Putnam's Twin Earth thought experiment, were introduced by Tyler Burge in "Individualism and the Mental." Burge identified individualism as an overall conception of the mind prevalent in modern philosophical thinking at least since Descartes in the mid-seventeenth century, and argued that our common sense psychological framework for explaining behavior, our folk psychology, was not individualistic. Importantly, Burge was explicit in making a case against individualism that did not turn on perhaps controversial claims about the semantics of natural kind terms. He developed his case against individualism using agents with thoughts about arthritis, sofas, and contracts, and so his argument did not presuppose any type of scientific essentialism about natural kinds. Like Putnam's argument, however, Burge's argument does presuppose some views about natural language understanding.[7]

The most central of these is that we can and do have incomplete understanding of many of the things that we have thoughts about and for which we have natural language terms. Given that, it is possible for an individual to have thoughts that turn on this incomplete understanding, such as the thought that one has arthritis in one's thigh muscle. Arthritis is a disease only of the joints, or as we might put it, "arthritis" in our speech community applies only to a disease of the joints. Consider an individual, Bert, with the thought that he would express by saying "I have arthritis in my thigh." In the actual world, this is a thought about arthritis; it is just that Bert has an incomplete or partially mistaken view of the nature of arthritis, and so expresses a false belief with this sentence. But now imagine Bert as living in a different speech community, one in which the term "arthritis" does apply to a disease both of the joints and of other parts of the body, including the thigh. In that speech community, Bert's thought would not involve the sort of incomplete understanding that it involves in the actual world; in fact, his thought in such a world would be true. Given the differences in the two speech communities, it seems that an individual with thoughts about what he calls "arthritis" will have different thoughts in the two communities. In the actual world, Bert has thoughts about arthritis. In the counterfactual world he has thoughts about some other disease, what we might refer to as "tharthritis" to distinguish it from the disease that we have in the actual world.

In principle, we could suppose that Bert himself is identical across the two contexts, that is, that he is identical in all intrinsic respects. Yet we attribute thoughts with different contents to Bert, and seem to do so solely because of features of the language community in which he is located. Thus, the content of one's thoughts, and so how we taxonomize those thoughts as intentional states, is not metaphysically determined by the intrinsic properties of the individual. And again taking a difference in the content of two thoughts to imply a difference between the thoughts themselves, this implies in turn that thoughts themselves are not individuated individualistically.

One contrast sometimes drawn between the externalist views of Putnam and Burge is to characterize Putnam's view as a form of *physical* externalism and Burge's view as a form of *social* externalism. According to Putnam, it is the character of the physical world – the nature of water itself – that, in part, metaphysically determines the content of one's mind, while according to Burge it is the character of the social world – the nature of one's linguistic community – that does so. While this difference may serve as a useful reminder of one way in which these two views differ, we should also keep in mind a social aspect to Putnam's view of natural language that I have not yet mentioned: his division of linguistic labor. What allows individuals to use natural kind terms to refer to objects in the world despite those individuals not necessarily having identifying descriptions, according to Putnam, is their ability to borrow reference from others, experts, who are able to reliably pick out those referents through their knowledge. Important to both Burge's and Putnam's views is the idea that language users and psychological beings depend and rely on one another in ways that are reflected in our everyday, common sense ways of thinking about language and thought. Thus, there is a social aspect to the nature of meaning and thought on both views, and this is in part what justifies the appropriateness of the label anti-*individualism* for each of them.[8]

I close this section with a parenthetical observation that raises one issue for further thought. The contrast between the individual and the social is built into the debate between individualists and externalists. This contrast takes different forms in the cognitive, biological, and social sciences, but it remains central to influential views in all three. We might wonder just why, whether there is a theoretically illuminating account of the distinction itself, and why it has such ubiquitous appeal in the fragile sciences.

4 THE SOCIAL ASPECT TO HAVING A MIND

In retrospect, Putnam's conclusions about natural language meaning should have been no real surprise since there is obviously a social dimension to language. While the fact that language is used in a social context, and that one of its chief functions is to communicate between individuals or groups of individuals, have rarely been completely ignored in the philosophy of language and cognitive science, a range of dominant views about language have, however, downplayed these aspects of language and treated them derivatively.

For example, Paul Grice had proposed that we understand what a speaker means by an utterance in terms of a complex set of intentions that that speaker has, and that we then understand what a sentence means in terms of this notion of speaker meaning. Part of the complexity to a speaker's intention was that it was an intention to effect a change in the mental states of one or more hearers. Thus, there clearly is a communicative aspect to Grice's proposal. But like descriptivism about reference, this is an account of meaning given primarily in terms of what happens in the head of a given speaker, with shared, interpersonal meaning – what Grice called *sentence meaning* – analyzed in terms of individual speaker meaning. Thus, as Burge pointed out in "Individualism and the Mental," the Gricean program in semantics is individualistic. The same general point holds of a range of other influential views of natural language: for example, David Lewis's account of convention, Noam Chomsky's conception of linguistics, and early attempts to develop natural language understanding programs in artificial intelligence.[9]

Although I have emphasized the similarities between the externalist arguments of Putnam and Burge, the corresponding moral that can be drawn from Burge's argument – that there is a social dimension to having a mind – remains striking, even in retrospect. This is so not only because one might think that mental states do not primarily serve a communicative function, but also because the very enterprise of understanding the complexities of the mind directs us inside the head. I want to point to two distinct features of minds and how they have been thought about to illustrate this.

First, the cognitive sciences have developed an elaborate conception of cognition as a form of computation. In what has become known as "classical" cognitive science, the focus here has been on the specific algorithms governing state-to-state transitions between internal, mental symbols. On this conception, any social aspect to cognition would need to be secondary or derivative in some way, since computation itself is fundamentally asocial. This asociality assumption would also seem to be shared by connectionist variations on the computational theme, whereby cognition is at bottom the adjustment of connection weights between idealized, neurally inspired nodes. Likewise, the development of computational techniques within cognitive neuroscience, from single-cell computation to computation in relatively large-scale units, such as columns and modules, has not provided any reason to give up this assumption. As a kind of computation, cognition is not social at all.

Second, we are conscious of many of our mental states. There is something it is like to have them, a phenomenology to our mental lives that is "had," that is experienced, from a particular perspective, that of the first person. If we focus not on the intentionality of mental states but on their phenomenology, there seems less room for an externalist – let alone a social – dimension to mentality. There is simply an asymmetry between how I know about my own mental life and how I find out about those of others, one that makes it difficult to see how externalism could be true of that part of mentality of which we are conscious. We are intimate with some of our own mental states, and have a knowledge of them whose directness and noninferential nature seem hard to reconcile with the idea that such states are metaphysically determined by factors beyond the head, including social factors. Indeed, our first-person knowledge of the mind has been thought to be incompatible with externalism in general.[10]

Other aspects of our conception of minds, however, are more conducive to the claim that the mind is itself social in nature. The first of these is the idea that cognition is situated or embedded in a particular social environment. The social embeddedness of thought is apparent both in Putnam's appeal to the division of linguistic labor and Burge's reliance on incomplete understanding, and some work in the cognitive sciences has adopted this sort of view of cognition from a more general and developed interest in embedded cognition. Work on embedded or situation cognition has often focused on how individuals become tightly coupled to their physical (rather than their social) environments, concentrating on the role of the physical environment in enhancing or even constituting individual performance. But other people and the artifacts, institutions, practices, and interactions they both create and inherit are the most significant feature of any individual's environment for her mental life, a point being slowly taken up within some areas of the cognitive sciences.[11]

A second way in which sociality has been thought to permeate the psychological is via the idea that the mental is normative. The source of this normativity is the social world: from interpersonal relationships, to institutional roles, conventions, and institutions themselves. Ordinary folk psychological ascriptions carry with them normative and not just descriptive implications, those concerning justification, rationality, and appropriateness of what those states are about. This normative dimension to the mental has been recognized in the idea that mental states provide reasons for acting, and it forms the backbone of interpretationist views of folk psychology, such as those of Donald Davidson and Daniel Dennett. These discussions predate, and have been largely orthogonal to, the individualism-externalism debate.[12]

Others have made a more direct connection between these two issues, normativity and externalism. Burge himself has argued for externalism via an appeal to the normativity of perception and the general idea of intellectual norms. And normative requirements have been taken by some to provide the basis for distinguishing between simple intentionality and truly thoughtful intentionality of the type that human beings possess. Philip Pettit, for example, has argued that rule following is one thing that distinguishes true thinkers from merely intentional creatures, and that rule following requires what Pettit calls an *ethocentric* conception of thought, one which views thought as involving activity in an interpersonal, social world.[13]

It is in what we might call the "Pittsburgh school" of thought, rooted in Wilfrid Sellars' classic attack on the Myth of the Given, that links between the normativity of the mental and externalism about the mind have been forged most thoroughly, particularly in the work of John McDowell, Robert Brandom, and John Haugeland. What these views share, and what allows them to carve a path from normativity to externalism, is the idea that in order to have at least certain kinds of meaningful, content-laden, intentional states one must be subject to rules, standards, and conventions, where these are not metaphysically determined by intrinsic facts about the bearer of those states. Recognizing the normative dimension to the mental takes one beyond the individual and into the social.[14]

5 NARROW AND WIDE CONTENT

Thus far, I have barely paused to register the various twists and turns to the debate over individualism in psychology. But there is one twist that can't be missed: the distinction between narrow and wide content. One intuitive response to the initial Putnam-Burge arguments against individualism has been to concede that while there is a sense in which even *doppelgängers* can have mental states with different content, there is an equally important sense in which they must have mental states with the same content. That content, content shared by *doppelgängers* no matter how different their environments, is *narrow content*.

We can use Putnam's own example to illustrate what narrow content is, and why we might insist on its existence and importance. While we might distinguish between the meaning of Oscar's term "water" and Oscar*'s term "water" (which we designate with "twater" in part to highlight its distinct extension), we could equally ascribe a common, shared meaning to their terms "water," one neutral between H_2O and XYZ. Precisely because there is so little difference between Oscar's term "water" and Oscar*'s term "water," it is plausible to view those terms, even if not strictly identical, as sharing so extensive a common core of meaning that we would be overlooking something important were we simply to treat them as semantically independent terms. Putting this in terms of the psychological states that Oscar and Oscar* are in, we can say that although there is some difference between the psychological ascriptions we would make for each – those that employ a notion of wide content – there is also much intentional psychology that is shared between Oscar and Oscar*. In fact, there is much shared in their intentional psychology even when we are considering their thoughts about "water." In order to express these common, intentional psychological states, we need some notion of content that Oscar and Oscar* share. That is, we need some notion of narrow content. Whatever else it is, narrow content is the type of content that physical twins must share, however different their environments.

In introducing narrow content in this way, I have presupposed that the Twin Earth thought experiments have shown that Oscar and Oscar*'s psychological states differ in their propositional content. More generally, I am assuming that our ordinary, pretheoretical notion of propositional content, of intentionality, is not narrow but wide. This is certainly the received view both of what the Twin Earth arguments show and of intentionality, but it will pay to spell out just why.

The intentional content of so-called folk psychological states, such as belief and desire, is what is specified by the that-clause of an ascription of propositional attitudes. For example, when we say that

Peter thinks that wolves are placental

the content of Peter's thought is that *wolves are placental*. This is what his thought is about, what it represents as being the case. Peter's thought is about wolves, that is, he has some internal representation that refers to wolves. At least when we are considering folk psychology, there is no intrinsic feature of the representation itself that makes the corresponding thought one about wolves. In fact, note that we can understand what the content of Peter's thought is in this example although I have said nothing at all about the nature of the representation in and of itself. Even supposing that Peter's thought is instantiated in Peter in virtue of there being a token of the sentence "wolves are placental" inscribed in Mentalese in a place in Peter's brain that we can call his "thought box," this fact about Peter's internal organization is not sufficient for Peter's thought to be about wolves being placental. Someone else with just that inscription in her thought box could have a distinct thought were facts outside of Peter other than they are. In particular, this could be so were "mere" social facts, facts about our communal conventions for the use of natural language terms, different. The claim that this is true not just of particular in-the-head facts but of all of them considered together – they do not suffice to fix the contents of one's thoughts – is another way to express the conclusions drawn from the Putnam and Burge thought experiments.

The intentionality of folk psychological states seems intuitively to lie in some sort of relation between what is inside the head and what is outside of it. It is plausible to think that this relationship is, broadly construed, causal in nature. It is *wolves* that Peter thinks about because wolves are what causally impinge on Peter either through his own direct sense experience of wolves or through his causal location in a linguistic community. This is not to imply that being in such a causal relationship is sufficient for one to have thoughts about wolves, but that these causal relationships are partial metaphysical determiners of the intentionality of one's thoughts. And this shows that folk psychological content is wide, since individuals who differ in this beyond-the-head relation may have thoughts with different contents, no matter how similar they are in in-the-head respects.

Consider now the claim that there is, in addition, a notion of narrow content to be found lurking in folk psychology. The intuition that the above appeal to the twins Oscar and Oscar* drew on was that their internal identity does generate some sort of intentional identity or similarity, even if there is also a respect in which they have thoughts with different contents. One way to express this is to say that their "water" thoughts share a narrow content, even though they differ in their wide content. There are three motivations for the notion of narrow content, none of which presupposes individualism about the mind, and each of which has served as the basis for a distinct proposal about what narrow content is.

The first concedes that individual representations do not suffice to fix content but challenges the generalization of this point to cover all individual representations considered together. That is, once one considers not just the intrinsic features of a given mental token but its relations to other mental tokens, the claim that what's inside the head doesn't fix any notion of content loses its plausibility. This is because there can be, and typically is, a rich causal structure to mental representations that begins and ends in the head, a point that can be drawn from the functionalist view of mental states. For example, Oscar and Oscar* both have a mental token, "water," that is causally related to the same perceptual and mental inputs, and the same mental and behavioral outputs, and this is sufficient for those tokens to share some type of content. The proposal that the narrow content of a mental state is its narrow conceptual role is a development of this idea, an idea also developed independently as procedural semantics in early work in artificial intelligence.

The second motivation reaches back to the origins of the anti-individualistic perspective on mental content in the philosophy of language to recognize two aspects to the meaning of natural language terms and the close relationship between them. Consider the term "I." The referent of a spoken token of the word "I" is the speaker, and that is who is meant by that utterance; the referent of "you" is the person being spoken to. But "I" also has a common meaning when different people utter it, as does "you": "I" refers to the speaker, and "you" refers to the hearer. Thus, when two people say to each other "I love you," there is clearly some sense in which they have said exactly the same thing to one another, and we need some notion of content that captures this. What these utterances and their constituents share is their narrow content, this being a sort of rule – such as that "I" refers to the speaker/writer – that is then contextualized to derive the referent of a particular token utterance or inscription. This has generated the proposal that narrow content is a function from contexts to truth conditions: Take the narrow content of a mental state, add a beyond-the-head context, and one arrives at its (wide) propositional content.

The third motivation is the idea that the beyond-the-head differences between Oscar and Oscar* make absolutely no difference to how the two see the world: Their worlds are phenomenologically identical. But how the world appears to one is mediated not just by any old internal machinery but by content-laden internal machinery, such as concepts and ideas, thoughts and beliefs. If physical twins are phenomenologically identical, then they must be intentionally identical in some sense, and the corresponding notion of content must be narrow, not wide. This has given rise to the idea that narrow content is phenomenological content.[15]

These various proposals provide a strategy for limiting the significance of the Putnam-Burge arguments against individualism since they suggest that there is no strict incompatibility between an individualistic and an intentional psychology. Thus, one can concede the conclusions that Putnam and Burge draw but rely on the presence of some notion of narrow content either to continue with content-laden psychology as it has developed to date, or to look for alternative ways to connect computational and neuroscientific approaches to the mind with our existing folk psychology.

6 FUNCTIONALISM, PHYSICALISM, AND INDIVIDUALISM

For many philosophers of mind, individualism has been attractive because of a perceived connection between that view and physicalism and functionalism in the philosophy of mind, both of which have been widely accepted over the last thirty years. Physicalism or materialism has been expressed in various ways, perhaps most commonly in terms of the notion of supervenience that we have already met: All facts, properties, processes, events, and things supervene on the physical facts, properties, processes, events, and things, as they are posited in elementary physics.

This ontological formulation of physicalism is often accompanied by an explanatory thesis, which states that physical explanations are, in some sense, the ultimate explanations for any phenomenon whatsoever.

Individualism has been thought to be linked to physicalism since it implies, via the supervenience formulation, that there is no psychological difference without a corresponding difference in the intrinsic, physical states of the individual. Those rejecting individualism have sometimes been charged with endorsing a form of dualism about the mind, or making a mystery of mental causation by ignoring the causal powers that mental states have. Causation between mental states and events is ultimately causation between an individual's physical states and events, and scientific taxonomies of mental states must respect this feature of mental causation. Connecting this up with the methodological formulations that have had influence in cognitive science itself, individualism has been claimed to be a minimal constraint on arriving at psychological explanations that locate the mind suitably in the physical world, a psychology that taxonomizes its entities by their causal powers.

Individualists themselves disagree about what this implies about the substantive nature of psychology. For example, Jerry Fodor thinks that this minimal constraint provides a way of seeing how folk psychology in particular and the notion of mental content more generally, have a respectable place in the cognitive sciences. At one point Fodor held that this in turn required cognitive science to use a notion of narrow content, although in subsequent work he has suggested that ordinary propositional content was adequate for the tasks of theory construction in the cognitive sciences. Stephen Stich, by contrast, thinks that individualism implies that cognitive science should jettison both folk psychology and the notion of mental content altogether, arguing that content should be eliminated from psychology, in part because of the conflict between the minimal constraint of individualism and the wide nature of intentionality. Both of Fodor's positions suggest that individualistic approaches to psychology will be recognizable descendants of current work in the cognitive sciences that incorporates and builds on folk psychology. Stich's view, by contrast, implies that an individualistic cognitive science will be divorced from folk-contaminated research traditions, being fashioned instead from the cloth of computational intelligence or cognitive neuroscience.[16]

Functionalism is the view that psychological states and processes should be individuated by their causal or functional roles, that is, by their place within the overall causal economy of the organism. It has been common to suppose that these functional or causal roles are individualistic.

Functionalism was introduced in the 1960s as a way of understanding the relationship between mental and physical states that meshed with two perspectives on minds then nascent but that have had a lasting effect on how philosophers think about the mind.

The first was the rise of the computer metaphor and particularly the analogy between the distinctions between software or program and hardware, on the one hand, and mind and brain, on the other. Mental states were not strictly identical to brain states but involved some sort of abstraction from those states, in much the way that computer programs were not strictly identical to electronic states of computers but abstractions from them. Functionalism cohered with the computer metaphor because it offered a causal understanding of the mind that also seemed to characterize the less mysterious relationship between programs and computers. Functionalism also provided the conceptual underpinnings for the very idea of artificial intelligence.

The second was the related desire for an account of the mind that made it perspicuous how creatures very much unlike us in many respects could nonetheless share a mental life very much like our own. This desire was typically expressed as the requirement that an account of the mind must allow mental states to be multiply realized, that is, instantiated even when the underlying physical realizers vary. The most attention-grabbing, putative cases of multiple realization were those where the physical realizers for mental states were extremely different – humans versus possible silicon-based Martians. But there was also a concern for biological differences across species, as well as physical differences in a given individual over time.

Functionalism seemed perfect for an account of the mind that meshed with these two perspectives, since what mattered for functional identity was not the nature of the physical stuff but, rather, the way in which that stuff was structured or organized. Again, the computer metaphor was apt here: The very same program could be instantiated on physically quite different machines. Crucial to program identity is functional organization; likewise for minds.

Functionalism has been understood in various ways, but the two ways most prevalent in cognitive science – in terms of the notion of computation, and in terms of the idea of analytical decomposition – both lend themselves to an individualistic reading. Computational processes, conceived as operating solely on the syntactic properties of mental states, have been plausibly thought to be individualistic. And it is natural to view analytical decomposition as beginning with a psychological capacity, such as memory or depth perception, and seeking the intrinsic properties of the organism that create and constitute that capacity.

7 THE APPEAL TO CAUSAL POWERS

Appeals to an entity's causal powers are prominent in discussions of the relationship between individualism and physicalism, and they are at the core of an influential and persistent argument for individualism first offered by Jerry Fodor. Although criticisms of the argument seem to me decisive in showing the argument to be fatally flawed, the argument itself taps into a cluster of intuitions that run through deep philosophical waters. Perhaps for this reason the argument has continued to inspire individualistic appeals to causal powers, despite acknowledged problems with Fodor's original statement of it.[17]

The basic version of the argument itself is easy to state. Taxonomy or individuation in the sciences in general satisfies a generalized version of individualism about psychology: Sciences taxonomize the entities they posit or discover by the causal powers that those entities have. Psychology and the cognitive sciences should be no exception here. But the causal powers of anything supervene on that thing's intrinsic, physical properties. Thus, science taxonomizes entities by properties that supervene on the intrinsic, physical properties of those entities. Science, and so psychology, is individualistic.

One way to identify the problem with this argument is to ask what it is that makes the first premise about scientific taxonomy in general true. Given the naturalistic turn supposedly embraced by those working in contemporary philosophical psychology, one would think that support here would come from an examination of actual taxonomic practice across the sciences. However, once one does turn to look at these practices, it is easy to find a variety of sciences that do not taxonomize "by causal powers": They instead individuate their kinds relationally, where often historical relations determine kind membership. Examples often cited here include species in evolutionary biology, which are individuated phylogenetically (and so historically), and continents in geology, whose causal powers are pretty much irrelevant to their individuation as continents.

The problem is particularly acute in the context of this argument for individualism, since a further premise in the argument states that a thing's causal powers supervene on that thing's intrinsic properties. Thus, one cannot simply save the first premise in the argument by stipulating that individuation in these sciences is "by causal powers," using some extended or nonstandard sense of that notion. For example, Fodor has acknowledged that scientific taxonomy is often relational, and his attempt to show how his argument accommodates this fact involves precisely this sort of broadening of what it means to individuate "by causal powers." If one operates with an extended notion of individuation by causal powers, however, then "causal powers" no longer supervene on an individual's intrinsic, physical properties. In adjusting the sense of "causal powers" to accommodate relational taxonomies in science, we make the other chief premise in the argument, that causal powers supervene on intrinsic, physical properties, false. We can take causal powers to be intrinsic properties, in which case they do supervene on what's inside an individual, but then the claim about scientific taxonomy being "by causal powers" is false. It is for this reason that the argument from causal powers equivocates on the crucial term "causal powers." I have argued elsewhere that this equivocation permeates all versions of the argument. I also think that we have reason to be skeptical about views of explanation and taxonomy that place more weight on the distinction between intrinsic and relational properties than it can bear.[18]

The cluster of intuitions that persists despite an acknowledgment that the argument itself is flawed in something like this way revolve around the idea that an entity's causal powers are central to both the place of that entity in the causal nexus and in how that entity is or should be taxonomized. To get from this somewhat vague idea to the individualistic claim that scientific taxonomy is "by causal powers," one has to establish some sort of asymmetry between properties that are metaphysically determined by what lies within the boundary of the individual, and those, like relational properties, that are not. For example, one could claim that only an entity's intrinsic properties feature in causal laws governing that thing's behavior, or that the causal efficacy that any relational property has depends only on the intrinsic properties of the entities it relates.

Such views are part of the smallist legacy of corpuscularianism that I identified at the end of Chapter 1. The basic problem with them is very much that with the claim they are invoked to defend: That once one turns to taxonomic and explanatory practice in a range of sciences, one finds many examples in which the putative asymmetries – between causal powers and other properties, or between intrinsic and relational properties – do not exist. This is a developed form of the prima facie general problem that, I claimed in Chapter 1, smallist views in metaphysics and the philosophy of science face, and we will encounter it again in the next two chapters in critiquing the standard view of realization.

Nonetheless, the intuition that individualism does articulate a constraint for the explanation of cognition that sciences more generally satisfy, one that would make for a physicalistically respectable psychology, persists. My view is that this intuition itself seriously underestimates the diversity in taxonomic and explanatory practice across the sciences, and that it simply needs to be given up. Attempts to revitalize this sort of argument for individualism proceed by making the sorts of *a priori* assumptions about the nature of scientific taxonomies and explanations that are reminiscent of the generalized, rational reconstructions of scientific practice that governed logical positivist views of science. This should sound alarm bells for any self-professed naturalist.

There is one difference between individualists and externalists about psychology emerging from reflection on the argument from causal powers worth keeping in mind as we think about the individual in the fragile sciences more generally. Individualism, especially as it has been articulated by those proposing or defending this particular argument, is touted as a global thesis about individuation in psychology that follows from an even more general thesis about individuation in science. Externalism, especially as defended by those attacking the argument from causal powers, is accompanied by a more pluralistic view of psychological taxonomy. This view allows some place for the causal powers of individuals but also sees scientific (and so psychological) taxonomy in many cases as being determined by an entity's relational and even historical properties. These individuative theses carry with them normative visions about what good and bad scientific taxonomy, and thus explanation, is like in particular sciences. In psychology, individualism implies that folk psychology, together with the vast tracts of psychology proper that incorporate or develop folk psychology – including much of social psychology, cognitive developmental psychology, and work on decision making – involves a problematic taxonomy of mental states. It also implies that the way to repair such problematic taxonomies is to modify them to reconcile them with individualism. Hence, the narrow content program. Externalists are likely to view scientific taxonomies and scientific explanation as being sensitive to a range of factors, and to be skeptical about the prospects for any recipelike prescription regarding proper scientific taxonomy of the sort that individualists propose.

8 METAPHYSICS AND THE FRAGILE SCIENCES

What of the more general, putative connection between physicalism and individualism? If the denial of individualism could be shown to entail the denial of a plausibly general version of physicalism, then externalism would itself be in real trouble. But like the individualist's appeal to causal powers and scientific taxonomy, the move from the general intuitions that motivate such an argument to the argument itself will likely always prove problematic. For example, externalists can respect the physicalistic slogan "no psychological difference without a physical difference" because the relevant physical differences lie beyond the boundary of the individual; attempts to refine this slogan (for example, no psychological difference without an *intrinsic* physical difference) are likely either to beg the question against the externalist, or to invoke a construal of physicalism that is at least as controversial as individualism itself.

Externalists have not been as attentive to the metaphysical notions at the core of contemporary materialism as they could have been, however. When they have so attended they have often opposed prevalent physicalist or materialist views without offering a substantive, alternative metaphysics of the mind. Tyler Burge is the most prominent externalist of whom this could be said. In his original discussion of the implications of individualism for related views about the mind, Burge claimed that the rejection of individualism implied the rejection of widely accepted token-token identity theories of the mind. In much of his later work, he has also pointed to inadequacies both in arguments from physicalist assumptions to individualism and to materialist conceptions of the metaphysics of mind. At the same time, Burge has emphasized the need to focus on explanatory practice in order to properly understand metaphysical issues, such as the nature of mental causation, saying that such reflection "motivates less confidence in materialist metaphysics than is common in North American philosophy."[19]

The most underdiscussed, relevant metaphysical notion is that of realization. The next two chapters offer an extended treatment of the concept of realization. They will not only prove useful in our current focus on individualism in psychology, but also in thinking about the role of the individual in the fragile sciences more generally.

5

Metaphysics, Mind, and Science

Two Views of Realization

1 THE METAPHYSICS OF MIND AND THE FRAGILE SCIENCES

It is commonplace for materialist philosophers of mind to talk of mental states as being realized in states of the brain. So much so, that realization has become part of the very framework in terms of which many conceptualize the metaphysics of mind. However, while the concept of realization has been invoked in the philosophy of mind and psychology for over forty years, it has only recently become the subject of direct philosophical theorizing. Implicit in the literature on the metaphysics of mind is the idea that realization is a general relation, rather than one invoked solely to answer the mind-body problem. In this chapter, I shall argue that making this assumption explicit provides reason to rethink the concept of realization. By the end of this chapter, I hope to have shown how the metaphysical foundations of the cognitive sciences are intertwined with broader metaphysical and methodological issues in other parts of the fragile sciences.

Psychologists and cognitive neuroscientists do not, for the most part, talk of realization, but of the *neural correlates*, or of the *neural mechanisms* for psychological functions and capacities. Cognitive capacities are localized in states of the brain. It is part of philosophical lore that such talk is loose-speak for the more metaphysically loaded discussions within the philosophy of mind cast in terms of supervenience and realization. This lore is what justifies the sense that philosophical discussions of the metaphysics of mind are continuous with and contribute to the cognitive sciences, even though one does not hear "realization" in the mouths of cognitive scientists themselves. It is part of the self-image of naturalistic philosophy of mind.

Challenging this self-image is no part of my aim in this chapter, though developing how I think realization should be conceptualized requires some discussion of the network of concepts to which realization is related, including those of mechanism and localization within the cognitive sciences. Having spent the core of the chapter discussing realization in general, I shall conclude it by returning to neural realization. I begin with a brief history of how realization came to be so central to the metaphysics of mind.

2 REALIZATION WITHIN THE PHILOSOPHY OF MIND

Contemporary views of realization in the philosophy of mind can be traced to Hilary Putnam's use of the notion in the context of his appeal to Turing machines in discussing the mind-body problem. Putnam argued that the relationship between mental and physical states should be no more puzzling (and no more interesting) than the relationship between the abstract states of a given Turing machine and the structural states of the device realizing that Turing machine. Putnam drew a distinction between the logical description of a Turing machine and the physical states that realize the states to which that description refers, the idea being that we see mental states as realized by physical states of the brain in just this sense.

Accompanying this idea were two claims that have had far-reaching consequences for how philosophers have thought about the mind over the last forty years: first, that systems adequately characterized by Turing machine descriptions can be multiply realized by physical states; and second, that there are no significant barriers to identifying mental states with brain states. Within a few years, the first of these ideas, that of the multiple realizability of mental states, had become a central reason for rejecting the second of them, the mind-brain identity thesis, largely through Putnam's own influence. Thus arose the functionalist view of the mind that, despite its critics (including a later timeslice of Putnam himself), has survived as the dominant "ism" in contemporary philosophy of mind.[1]

With the rise of functionalism, the claim that mental states are realized in physical states of the brain became part of the received wisdom on the mind-body relationship. Indeed, the concept of realization, particularly that of multiple realization, is well-entrenched in the articulation, explanation, and defense of nonreductionist forms of physicalism. Yet perhaps the most sustained discussion of realization itself, that of Jaegwon Kim, advocates reductionism about the mind on the basis of what Kim thinks is a proper understanding of the metaphysics of realization. The dissonance here derives in part from the neglect of the concept of realization that I mentioned in the opening paragraph, a point that Terence Horgan observed in a "state of the art" review of the concept of supervenience for the journal *Mind* over a decade ago.[2]

In this respect, at least, the state of the art has changed only very recently. My read on the current state of the art is that the standard view of realization, shared by reductionists and nonreductionist alike, is deeply flawed, and that there exists a general alternative to that view, which accords a central place to the idea that realization is essentially and irreducibly *context sensitive*.[3]

3 A SKETCH OF TWO VIEWS OF REALIZATION

As a way of outlining the chief contrast between these two views of realization, I begin with a first approximation of what I take to be the standard view of realization as used in the philosophy of mind. While there is a recognition both of realization as the (two-place) relation that holds between mental and physical states, and of realizations as the physical states that occupy the realizer place in this relation, it is the latter of these that has been the focus of discussion. Intrinsic, physical states of individuals – more particularly, of the central nervous systems of individuals – are the physical realizations of an individual's mental states, and these realizers are metaphysically sufficient for the presence of the states they realize. This is what makes realization a metaphysically robust relation simultaneously suitable and problematic for underwriting an account of mental causation: Suitable because metaphysical sufficiency would seem to have the strength to underwrite an account of mental *causation*; and problematic because, so-construed, physical realizer states, themselves being physical, seem to leave no room for distinctly *mental* causation. Thus, while the intuitions about psychological explanations generated by the Putnam-Burge thought experiments that we recounted in Chapter 4 may indicate ways in which our concept of the mental is sensitive to beyond-the-head factors, such as the nature of the physical environment or facts about one's social location, a proper understanding of the metaphysics of realization points one to an individualistic or internalist view of mental states.

By contrast, the view of realization that I shall propose in this chapter, and articulate and defend at greater length in the next, takes the context-sensitive character of mental states to be inherent to their nature, since realization itself is a context-sensitive notion. More poignantly, the claim at the core of the standard view of realization – that realizers are metaphysically sufficient for the properties or states that they realize – drives one to this view. This presents those adopting the standard notion of realization with a dilemma: Either give up or soften this claim of sufficiency (but at the expense of a range of further physicalist claims), or admit that realization, and so the metaphysics of the mental, is ineliminably context sensitive. Either way, some widely held physicalist views need to be revised or rejected.

In the next section, I offer a more rounded characterization of the standard view of realization that brings out more explicitly two theses at the heart of that view. This will make my chief objection to the standard view easy to state and set the scene for an exploration of some context-sensitive alternatives to it.

4 THE STANDARD VIEW (I): REALIZERS AS METAPHYSICALLY SUFFICIENT

A widespread view amongst physicalists in the philosophy of mind, whatever their other differences, is that realizers satisfy what I shall call the metaphysical sufficiency thesis:

Metaphysical Sufficiency Thesis: Realizers are metaphysically sufficient for the properties or states they realize.

I want to say something about why this thesis is implicit in standard conceptions of realization, particularly those used in the philosophy of mind.[4]

One reason is historical. As materialists came to be influenced by the way in which the computer metaphor suggested that mental states were multiply realized in physical states, rather than strictly identical to those states, the claim that physical states were metaphysically necessary and sufficient for particular mental states, appropriate when considering an identity theory, was weakened to one of sufficiency only.

A second reason is that many statements of what it means for mental states to be realized by physical states presuppose or imply this claim. For example, it is common to think of realization as a relation of determination (of mental states by physical states), and the sufficiency thesis is at least a necessary condition for such determination. Also, in explaining the one-many relationship between mental and physical states allowed by the notion of multiple realization, it is common to point out not only that this is not to be confused with the claim that there is a many-one relationship between mental and physical states, but that such a possibility would call physicalism itself into question. This possibility, that of emergent realization, that is, of a physical realizer for a given mental property that could realize some other mental property were the world different in various ways, is precisely what is ruled out by the sufficiency thesis, because such realizations would not in and of themselves determine the properties they realize.

A third reason is that the sufficiency thesis is needed to make sense of many of the positions that physicalists have adopted and the arguments they have offered in support of them. An intuition at the core of physicalism is that all the relevant physical facts fix all the nonphysical facts, and the notions of supervenience and realization have both been used to articulate this intuition further. Supervenience, in all its varieties, is itself a relation of determination, and if one thinks of realization as a correlative notion, then it too must be determinative. (Alternatively, if one holds that the physical realization of a given property is typically a subset of the subvenient base properties, realizations are, at most, partial determinants of the properties they realize, a view I return to discuss in Chapter 6.) And as already bruited above, the sufficiency thesis not only seems necessary for reductively identifying mental and physical states in views such as Kim's, but it also generates the recent wave of what Jerry Fodor calls epiphobia among nonreductionists – epiphobia being the fear that one is becoming an epiphenomenalist.

5 THE STANDARD VIEW (II): REALIZERS AS PHYSICALLY CONSTITUTIVE

Kim's own reductionism about the mind is also guided by a second thesis, one at least implicitly shared by many others, including Richard Boyd, David Lewis, and Sydney Shoemaker. I shall call this thesis the physical constitutivity thesis:

Physical Constitutivity Thesis: Realizers of states and properties are exhaustively physically constituted by the intrinsic, physical states of the individual whose states or properties they are.

I understand this thesis broadly such that stronger and weaker versions of it could be articulated in terms of the notions of supervenience, type-identity, or token-identity. In the philosophy of psychology, this thesis might be thought to have its methodological counterpart in the popular endorsement of the idea that homuncular functionalism and functional analysis involve the decomposition of psychological capacities into their constituent capacities, a claim we will have reason to consider more carefully later.[5]

Since physical realizations have been claimed to provide a metaphysical and explanatory basis for the higher-level properties they realize, it is not surprising that these links between functionalism, realization, and constitution structure (or perhaps derive from) a broader physicalist metaphysics, one that accords microstructure a central role. As Kim says, speaking in the first instance of our common sense conception of chemical kinds, but clearly with a more general view in mind:

> ...many important properties of minerals, we think, are supervenient on, and explainable in terms of, their microstructure, and chemical kinds constitute a microstructural taxonomy that is explanatorily rich and powerful. Microstructure is important, in short, because macrophysical properties of substances are determined by microstructure. These ideas make up our 'metaphysics' of microdetermination for properties of minerals and other substances, a background of partly empirical and partly metaphysical assumptions that regulate our inductive and explanatory practices.

As Kim says a little later, "[t]o have a physical realization is to be physically grounded and explainable in terms of the processes at an underlying level." Such a view is also manifest in Kim's one-time enthusiasm for the prospects of understanding "mind-body supervenience as an instance of mereological supervenience," that is, the supervenience of wholes on their parts.[6]

6 SMALLISM, THE STANDARD VIEW, AND THE FRAGILE SCIENCES

In Chapter 1, I claimed that smallism, discrimination in favor of the small, lurked in the background of contemporary individualism and nativism. The constitutivity thesis is certainly smallist, but it might well be thought that its influence is quite limited. After all, recall that "realization" is a term of art with its contemporary origin in a specific literature – that of Turing machine functionalism in the philosophy of mind. However, I want to suggest that an explicit appeal to or an implicit reliance on that concept and thus on the sufficiency and constitutivity theses can also be found in a range of positions in the fragile sciences, and thus the reach of the standard view of realization is significantly broader than one might think. Consider two such positions that are related to issues first discussed in Chapter 1 in exploring the relationship between various forms of individualism and nativism.

One such position concerns the status of collectives in the social sciences. It is common to hold that social-level entities, such as electorates, institutions, practices and rituals, and their properties are, in some sense, nothing over and above the individuals who are involved in them and their properties. Although this relationship has seldom been expressed in terms of the standard notion of realization, that concept would seem ideal to capture the "nothing over and above" aspect to that relationship: Individuals physically constitute social-level entities, and having the individuals and their particular properties is metaphysically sufficient to have social-level entities and their corresponding properties. Consider an electorate that is in the process of voting in a new government. That electorate is physically realized at that time by a given number of individuals, and by a certain majority of those individuals voting for the opposition party the electorate is thereby in the process of voting in a new government. Nothing more than these mundane individual-level facts is needed for this to be true. Nonreductionists and reductionists about collectives disagree about what this realization relation implies about social ontology. As we saw in Chapter 1, methodological individualists in the social sciences hold that the relevant properties of individuals are psychological and that this has methodological implications for how to do social science.

A quite distinct arena in which there is likewise an implicit reliance on the conjunction of the constitutivity and sufficiency thesis is developmental biology in which genes are held to be both physically constituted and metaphysically determined by particular DNA sequences. The claim about the physical constitution of genes in general – they are strings of DNA – is one of the triumphs of twentieth-century biology. That particular sequences of DNA are held to be metaphysically sufficient for the presence of a given gene underlies not only comparative molecular phylogenetic inferences that identify the same (or, as it is typically put, a homologous) gene across organisms belonging to two different species or other cladistic groups, but also the robustness of the appeal to DNA sequences in talking of the "gene for" a given phenotypic trait. Thus, it would make sense of key aspects of how genes are conceptualized in genetics and developmental biology to say that they are *realized in* sequences of DNA, even if that is not how biologists have in fact put the matter. In fact, I shall argue later in this chapter that talk of realization has quite general application within the biological sciences.

While I think that it is the constitutivity thesis that is problematic in talk of the realization of mental states, my general challenge is to the conjunction of the sufficiency and constitutivity theses for at least a variety of properties and states, including not only mental properties and states but those from across the fragile sciences. Context can feature in an account of realization in a number of ways, but feature it must, and I see no way of adequately representing the role of context in such an account that does not undermine either the sufficiency thesis or the constitutivity thesis. A bald statement of my chief objection to the standard view of realization is that the sufficiency and constitutivity theses are typically not true of the same putative realizers. Often the realizations that are metaphysically sufficient for the properties they realize are not exclusively physical constituents of individuals with those properties. Conversely, sometimes the physical constitution of an individual with a given property is not metaphysically sufficient for that property to be present. Mental properties are no exception here.

Physicalists who understand realization as a relation of metaphysical determination, as most do, should embrace the idea that at least some states and properties, including mental states and properties, have realizers that extend beyond the individual instantiating them. States and properties that have what I shall call a wide realization are prevalent in both common-sense thinking and in the biological and social sciences. Perhaps because there has been no general framework for such a view of realization, this view has not been explicitly endorsed in the literature on mental properties, although it is the view of realization that makes most direct metaphysical sense of the widespread recognition that a range of mental properties are not individualistic, and a view that externalists should readily agree with. This advocacy of wide realizations represents one way of developing a context-sensitive notion of realization.

There are initially less striking ways in which realization is context sensitive, however, and I shall discuss two of them next.

7 CONTEXT-SENSITIVE REALIZATION AND METAPHYSICAL SUFFICIENCY

As a way of introducing the idea that realization is context sensitive, consider the mental state of pain and the Ur-example of its realizer, C-fiber stimulation. As Sydney Shoemaker has pointed out, C-fiber stimulation is, at best, a partial realization of pain. What Shoemaker calls a *core* realization of that mental state is the specific part of the central nervous system most readily identified as playing a crucial, causal role in producing or sustaining the experience of pain. But when an individual is in pain other parts of her central nervous system are also activated, and the activity of these parts is crucial for C-fiber stimulation to play the causal role that is, according to functionalists, definitive of pain.[7]

In general, the physical states that are partial realizations of a property or state will be *metaphysically* context sensitive in that they will realize that property or state only given their location in some broader physical system. Considered just in themselves, they do not satisfy the sufficiency thesis. Additionally, in the special case of the core realization of a property, conceived of as the most salient part of some larger system in which that property is instantiated, we have an *epistemic* dimension to the context sensitivity of the realization. What we find of greatest causal salience depends on our conceptual and perceptual abilities. It also depends on the questions we ask, the background information we have, and, more generally, our epistemic orientation.

The context sensitivity of partial and core realizations should be uncontroversial, but might be thought of as having little relevance here because such realizations do not and have never been claimed to satisfy the sufficiency thesis. Even if core realizations of a property are what we most readily call to mind in thinking of the realization of that property, there are more complete physical states of which core realizations are a part that do satisfy the sufficiency thesis. Any interesting context-sensitivity thesis about realization should apply to them, not simply to core or other partial realizations. Following Shoemaker, we might define a total realization of a property as just such a state of a system.

By talking of a given higher-level property, P, and the system, S, in which P is realized, we can characterize the general distinction between core and total realizations as follows:

(a) core realization of P: a state of the specific part of S that is most readily identifiable as playing a crucial causal role in producing or sustaining P
(b) total realization of P: a state of S, containing any given core realization as a proper part, that is metaphysically sufficient for P

In particular cases, "S" is to be replaced by the appropriate system, whether it be psychological, biological, economic, computational, chemical, and so on, and their more determinate forms. In the case of pain, the appropriate system is the nociceptive system, containing mechanical and polymodal nociceptors in the skin (muscles and viscera), myelinated and unmyelinated axons (the latter being the famed C-fibers), spinal neurons, parts of the brainstem and thalamus, and the somatosensory area of the cerebral cortex. More generally, while P is a property of some individual entity, such as an organism, S need not be identical to that entity but, as in the example of pain, may form a part of it. Paradigms of such systems are those in which bodily functions and their associated properties are realized – for example, the respiratory system, the digestive system, the circulatory system – that are a part of each creature with the respective properties. Total realizations of P are exhaustively constituted by a core realization of P plus what I will refer to as the *noncore part* of the total realization.

While total realizations are in some sense complete states of S, they are incomplete in two important respects. First, the distinctness of S and the subject or bearer of P entails that total realizations do not include all states of those subjects or bearers, for not all states a subject or bearer is in form part of the system specified. For example, a person's having a toenail of two centimeters, while a property of that person, is not a property of that person's digestive or respiratory systems; "x has a toenail of two centimeters" expresses a property of persons, not of digestive or respiratory systems. Second, the total realization of P excludes the background conditions that are necessary for there to be the appropriate, functioning system. While these may themselves be necessary for a given entity to have P, since they are not states of S, they are no part of the total realization of P. Thus, total realizations should be distinguished from the broader circumstances in which they occur.

To illustrate these points, consider the mammalian circulatory system, which is made up of various parts – such as the heart, arteries, capillaries, arterioles, venules, and blood. Various states of these parts, considered together, determine what circulatory properties one has at any given time. Related common sense and medical theories about circulation specify what the circulatory system includes and excludes, but it is clearly a (proper) part of an organism. For a given circulatory property – say blood pressure – not all parts are of equal causal importance. From an intuitive point of view, one's blood pressure is most saliently determined by the condition of one's heart and arteries. Thus, the core realization of, say, having blood pressure of 120/80 would be identified with a state of these parts of the circulatory system – for example, having clogged arteries and a strong heart. Yet such states do not by themselves and independent of the state of the rest of the circulatory system guarantee blood pressure of 120/80 in a person. Rather, they need to be located in a certain way within the rest of the person's circulatory system. A total realization of having blood pressure of 120/80 is a state of the circulatory system, including the states of having clogged arteries and a strong heart, that determines the presence of that property. Excluded from total realizations are both properties instantiated by the individual that are not properties of the circulatory system at all (such as her having brown hair, or being six feet tall), as well as broader features of the individual's environment that are necessary for her to have a functioning circulatory system (such as there being oxygen in the environment and the world's persistence through time). Such background conditions are no part of the total realization of the corresponding property since they are not properties of the circulatory system at all.

Since the distinctness between a total realization and background conditions is important for the general alternative to the standard view of realization that I want to present, consider one other way of coming to this distinction. Consider our common sense view of circulation and how we would expect it to be modified by the findings of circulatory physiologists. While we would expect physiologists to offer a more precise specification of both the core and total realizations of the properties of this system, we wouldn't expect them to contribute much to our understanding of the nature of the background conditions of these realizations. (For this, they defer to other scientists, or to common sense.) In investigating a given biological system, scientists examine both the core and noncore parts of a total realization, but the boundaries of the system to a large extent delineate the boundaries of their inquiry.

Strictly speaking then, it is only the physical states constituting a total realization *together with the appropriate background conditions* that metaphysically suffice for P. Our paradigms for the relevant systems are functioning, integrated physical systems, and without the appropriate background conditions in each case there would be no such systems. This might be taken to suggest that even total realizations, considered simply as complex configurations of physical matter and energy, are metaphysically context sensitive in much the way that partial realizations are.

Let us catch our breath. In this section I have identified two types of realization that violate the sufficiency thesis and pointed to ways in which realizations are context sensitive. First, core realizations in and of themselves are not metaphysically sufficient for the properties they realize, but must be part of some larger functional system. This point is of some significance because even if no one really believes the sufficiency thesis to be true of core realizations, it is physical structures that in fact are core realizations – such as C-fiber firings – that are typically invoked in discussions of reductionism, realization, and functionalism. This is especially true in discussions of mental states. By the end of this chapter I hope to have made a prima facie case that this is also true of the sorts of example, those of genes in developmental biology and collectives in the social sciences, that I mentioned in the previous section. Second, since total realizations are physical states of larger functional systems, and there are background conditions necessary for their functioning, strictly speaking even total realizations themselves do not satisfy the sufficiency thesis. The significance of these points will unfold over the next few sections.

8 PHYSICAL CONSTITUTIVITY AND WIDE REALIZATIONS

Thus far, I have assumed the constitutivity thesis and thus individualistic realizations, having used that assumption to challenge the sufficiency thesis. That thesis was challenged as a global view of realizations in two ways by realizations that are context sensitive: Core realizations are both metaphysically and epistemically context sensitive, and in presupposing background conditions necessary for the existence and functioning of the corresponding system, total realizations are metaphysically context sensitive. I turn now to the constitutivity thesis and how it is undermined by a more far-reaching type of context sensitivity. Here the sufficiency thesis will be my ally, and I shall return to focus initially on mental properties in particular.

I begin by elaborating on my claim, made in section 5, that homuncular functionalism is often construed as a methodological counterpart to the constitutivity thesis. The idea of the prevalent strategy of homuncular decomposition in cognitive science is to explain complex, intelligent, representational capacities by functionally analyzing them into simpler (but typically more numerous) capacities, and then reapplying this first step recursively until we have simple abilities that require neither representation nor intelligence. If each homuncular level of analysis provides a realization of the level above it, and realizations satisfy the constitutivity thesis, then any view of homuncular functionalism that purports to be a physicalist view should proceed via physical decomposition.[8]

The constitutivity thesis itself implies that realizations of mental properties are individualistic, in that two molecularly identical individuals must also share the same realizations of mental properties. And if realizations are determinative of the properties they realize, mental properties must be individualistic, too. Indeed, reflection on the relationship between the above bodily systems and the individuals to whom they belong supports this as a general view of realizations: Because bodily systems are parts of individuals, there is no way for molecularly identical individuals to differ in the bodily systems that each has.

This general view overlooks, however, that there are two species of total realizations, only one of which can be understood in terms of the notion of constitution above. While it is often the case that S is a part of the individual that has P, there are a variety of examples in which the converse is true, examples in which *the individual that has P is a part of S*. These are cases in which S extends beyond the boundary of the individual, and I shall call the type of total realization that exists in such cases a *wide realization*.

Let IB be the subject or *i*ndividual *b*earer of P. In constitutive decomposition, of which homuncular functionalism is often construed as a paradigm, S is a part of IB. By contrast, in cases of integrative synthesis IB is a part of S; in these cases, P has total realizations that are wide. We can summarize the distinction between wide and entity-bounded realizations in terms of the location of the noncore part of a total realization as follows:

(c) entity-bounded realization: a total realization of P whose noncore part is located entirely within IB, the individual who has P
(d) wide realization: a total realization of P whose noncore part is not located entirely within IB, the individual who has P

Figure 5.1 provides a simplified depiction of the metaphysical parallels between these two forms of realization, as well as of the differences between the corresponding strategies of constitutive decomposition and

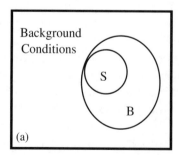 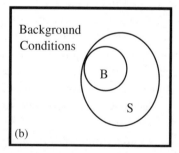

FIGURE 5.1. (a) Constitutive Decomposition. Involving Entity-Bounded Realization (b) Integrative Synthesis Involving Wide Realization. S = system B = bearer

integrative synthesis. As a species of total realization, wide realizations satisfy the sufficiency thesis. Yet since they extend beyond the physical boundary of the individual, they are not exhaustively constituted by the intrinsic, physical properties of the individual subject, and so do not satisfy the constitutivity thesis.

The concept of a wide realization allows us to make metaphysical sense of Putnam-Burge externalism introduced in the last chapter. On this view the propositional attitudes are not individualistic, or they at least have a nonindividualistic aspect. The propositional attitudes have a physical total realization, albeit one that is not entity bounded. The realization of particular folk psychological states is wide, and given the framework I am proposing that entails that those states should be understood by using integrative synthesis to locate their bearers in some broader system, presumably one that involves social relations between individuals. I shall call this our *folk psychological system.*

The width of our folk psychological system is not anomalous in psychology; in fact, the strategy of integrative synthesis also applies readily to computational psychology. Many computational systems that govern cognition are themselves wide, where the computational system S extends beyond the boundary of IB, the individual who bears the psychological properties, and the appropriate type of total realization is a wide realization. I have previously argued that we should expect wide computational systems of cognitive states just when there has been sustained mind-world constancy over evolutionary time of the type that one finds in the case of many perceptual and behavioral systems. Such systems include our mechanisms for form perception and the navigational systems that ants and bees deploy.[9]

This general view – that world-mind constancy creates the opportunity for cognitive loads to be shifted from inside the head to beyond it – has been advocated by a number of others. The philosopher Andy Clark's 007 Principle – "know only as much as you need to know to get the job done" – was postulated to explain informational off loading and exploitation by organisms in an evolutionary context, and his more recent endorsement of "the extended mind" has taken this perspective as implying that the mind itself may include parts of the world beyond the individual. Likewise, the cognitive scientist Edwin Hutchins views his study of navigation as a socially distributed, embedded computational process to imply that the computations performed during navigational tasks extend beyond the boundary of the individual. I shall discuss wide computationalism and these sorts of views of computation and cognition more generally and in more detail in Chapter 7. My chief point here is that the invocation of wide realizations is not an ad hoc ploy introduced simply in order to add a supporting metaphysics to our externalist intuitions about folk psychology.[10]

9 WIDE REALIZATIONS IN THE BIOLOGICAL AND SOCIAL SCIENCES

The further point that I do want to develop in some detail is that there is nothing strange about mental states having wide realizations. In fact, I think that wide realizations are typical across the fragile sciences, at least if we assume a notion of realization that satisfies the sufficiency thesis. The claim that wide realizations are ubiquitous in such sciences should sound, at least prima facie, counterintuitive. But once we make the connection between the metaphysics of realization and the methodologies of constitutive decomposition and integrative synthesis, this point becomes easier to understand. Let me begin with the biological sciences.

A variety of evolutionary and ecological properties themselves have wide realizations and are profitably understood through the strategy of integrative synthesis, that is, by locating their bearers in the corresponding wide systems. Such properties include fitness, being highly specialized, and being a predator, properties of individual organisms or even species; and properties of phenotypic traits or behaviors, such as being an adaptation, a homology, or a spandrel.

Consider first the example perhaps closest to common sense, the property of *being a predator*. Predators play a certain role in an ecological system, occupy a particular ecological niche, that of preying on other living things, typically (but not solely) for nourishment. This property is relational in at least two ways: An organism is a predator for certain other living things (their prey), and in certain types of environments (their ecosystem). Here the relevant system is the predator-prey system, whose dynamics are captured, in part, by the Lotka-Volterra equations and is explored in population ecology. While the part of this system most readily identifiable as playing a crucial causal role in producing or sustaining the property of being a predator, that is, that property's core realization, might be thought to be contained within the organism that is a predator, the total realization for this property is clearly wide. Since being a predator involves a relation between an organism and something beyond its organismic envelope, what is metaphysically sufficient for particular bodily or behavioral phenotypes to be those of a predator extends beyond the boundary of the individual.

The same is true of *being highly specialized*, the ability to occupy a relatively small number of the ecological niches available. Although we can and do speak of certain organisms as realizing this property by virtue of having specific intrinsic properties, at best these states or properties of individual organisms are core realizations for that property, and the corresponding total realizations include physical configurations in a system that extends beyond the individuals who are highly specialized, that is, the total realizations are wide. In the case of being highly specialized, the relevant system is the ecosystem, this being the system in terms of which ecological niches are defined.

Finally, an organism's *fitness* is its propensity to survive and reproduce in its environment. We can represent the former as a probability between 0 and 1 (the organism's viability), and the latter as a number greater than or equal to 0 (the organism's fertility) where this number represents the organism's expected number of offspring. In either case, although fitness is a dispositional property of individual organisms (or even whole species), this disposition is not individualistic, since physically identical organisms may differ in fitness because they have been or are located in different environments. That is, the numbers that represent viability and fertility may vary solely because of an organism's environmental location. This relational aspect to the property of fitness is often masked by the fact that an organism's environment usually plays (in effect) the role of a constant in many of the contexts in which the concept of fitness is put to work. Yet the properties "has a probability of surviving of 0.7" and "has an expected number of offspring of 2.2" are incompletely specified in a way that makes them meaningless without an implicit reference to an environment. What metaphysically suffices for a given organism to have a specific level of fitness is not instantiated entirely in that organism: The total realization of fitness (and its determinate forms) is wide, not entity bounded. Here the relevant wide system is the organism plus something like its niche, broadly construed to include its location within a particular population structure and other relational aspects of its existence.

The same general point is true of the total realizations of properties of phenotypic traits or behaviors. What makes a trait or behavior an adaptation, for example, is something about its evolutionary history, not just something about the individual who has the trait or exhibits the behavior. Whatever we want to say about their core realizations, their total realizations are not individualistic. There are ontological and not just epistemic reasons why you can't tell whether, say, the presence of wings in a given species is an adaptation or a spandrel simply by inspecting existing members of that species. These examples show that total realizations do not always satisfy the constitutivity thesis; they also exemplify the pattern of integrative synthesis depicted in Figure 5.1b.

To this point I have been discussing wide realizations whose core part is located within an individual. One might well wonder whether a realization could be wide in that not only does its noncore part extend beyond the individual *but so too does its core part*. I shall call this type of wide realization a radically wide realization:

(e) radically wide realization: a wide realization whose core part is not located entirely within IB, the individual who has P

The clearest examples of radically wide realizations are those of social actions that involve engaging with the world and have further social and institutional background conditions. And so we move from the biological to the social sciences in thinking about the full range of the applicability of the notion of wide realization in the fragile sciences.

Consider actions such as making a withdrawal from a bank, committing a felony, or voting, each of which we might do by signing a piece of paper in certain circumstances. Here, not only the noncore part of the total realization extends beyond the individual agent, but so too does the most natural candidate for the core realization of these actions, signing a piece of paper. This is an action of an individual, as are the actions that it in turn realizes, even though the action extends beyond the boundary of that individual. The relevant system explored via integrative synthesis – whether it be the banking system, criminal justice system, or electoral system – likewise extends beyond the boundary of the individual agent and has its own background conditions.

The same is true of many of our ways of classifying agents in the social sciences. In economics, for example, individuals are homeowners, consumers, and wage earners. Someone falls into one of these categories by virtue of the relationship between what she does and has done, and the legal, social, and economic institutions and practices to which she is subject. An agent enters into verbal and written agreements, hands over cash, or performs certain tasks. Since these actions or behaviors themselves, however, literally extend into the world beyond the individual, their realizations do not stop at the skin. Behavior that stops at the skin – the bodily movement that an agent makes – is not even a core realization of the corresponding action.

10 TWO VIEWS RECONSIDERED

This chapter has been more analytical in its style than those that have preceded it, and it may pay to step back from the thick of it all to see where we are. In sections 7–9, I have proposed four ways in which physical realizations are context sensitive, the first and second of which challenge the sufficiency thesis (section 7), and the third and fourth of which challenge the constitutivity thesis (sections 8 and 9). In section 8, I introduced (ordinary) wide realizations, exemplified by intentional mental properties, whose noncore part extends beyond the boundary of the individual who has those properties. What I called radically wide realizations in section 9 are wide realizations whose core part also extends beyond that boundary, and I suggested that they were exemplified by social actions. My more general point in these sections was that there are a range of properties in the cognitive, biological, and social sciences that have wide rather than entity-bounded realizations.

Those who would like to salvage the standard view of realization can shuffle where they locate the particular examples I have introduced in this four-fold schema. But since the sufficiency and constitutivity theses are jointly satisfied in none of the four forms of context-sensitive realization, there will remain a problem for the standard view, no matter how much shuffling is done.

On the standard view, realizations are metaphysically determinative of the properties they realize and physically constitutive of the individuals who possess those properties. On the alternative, context-sensitive view, realizations are located within systems, and those systems in turn are located within broader environments. The context-sensitive view allows that the standard view is correct about some properties and their realizations – what I called entity-bounded realizations in section 8 – but insists that there is a wide variety of properties that involve another form of context sensitivity, that of wide realization. In effect, the context-sensitive view implies that there is an individualistic bias within the standard view, one that not only ignores the perhaps innocuous context sensitivity of both core and total realizations, but that overlooks or denies the possibility of wide realizations altogether.

I have been arguing that there is a sort of tension within the standard view. For what metaphysically determines the presence of a property often extends beyond the boundary of the individual who has that property. And, conversely, the physical constituents of the individual bearing a given property are often contributory or partial realizers rather than metaphysically determinative realizers of that property. In advocating the endorsement of wide realizations, I have, in effect, embraced the sufficiency thesis over the constitutivity thesis.

With this in mind, let us return to reconsider the relationship between realization and appeals to neural correlates, mechanisms, and localizations within the cognitive sciences. Just as realizations are assumed to be entity bounded on the standard view, psychologists and cognitive neuroscientists typically think of neural mechanisms as entity bounded, and so have an individualistic view of these. Given the tension within the standard view of realization, it seems clear that cognitive scientists themselves would likely opt for the constitutivity over the sufficiency thesis. If that is so, however, then this represents a significant break from the physicalist orthodoxy that "the physical" determines "the mental."

Whatever psychologists and cognitive neuroscientists think about rejecting the metaphysical sufficiency thesis, it is a view deeply embedded within the philosophy of mind. First, the view of realization as a determinative relation is one that reductionists such as Jaegwon Kim have readily drawn on in arguing that nonreductionists occupy an unstable position; this view is not likely to be seen by such reductionists as merely optional. Second, while this view of realization can be seen as just one of a cluster of intuitions, it is central to understanding the cluster as a whole. For example, it connects the notion of realization with that of supervenience in an intuitive way by treating them as correlative notions; it explains the sense in which realizers provide a metaphysical basis for what they realize; and it allows us to make sense of the idea that mental properties are "nothing over and above" their realizations. This should make us wary of simply dropping the sufficiency thesis. To settle, in effect, for partial realizations, such as core realizations, as the metaphysical substrate to the mind that best corresponds to what cognitive scientists mean in talking of the neural correlates, mechanisms, and localizations for cognitive capacities would require a softening in how physicalism is to be understood.

The context-sensitive view suggests another option, one that both preserves this determination thesis in some cases and brackets consideration of it in others. While some cognitive capacities and the mechanisms that realize them have an entity-bounded realization, others merely have neural correlates, or are partially realized in specific regions of the brain. The context-sensitive view of realization goes hand in hand with a context-sensitive view of mechanisms. Neural mechanisms, like neural realizations, are always located within cognitive systems, and I have argued that cognitive systems sometimes extend beyond the boundary of the individual property bearer. When they do, the mechanisms are wide in that they are individuated by reference to this system.

By identifying a range of properties across the cognitive, biological, and social sciences to which the context-sensitive view seems to naturally apply, I have made a start on articulating that view and perhaps shown its promise. I have thus far, however, done little by way of really defending the context-sensitive view of realization. That is the task of Chapter 6.

6

Context-Sensitive Realizations

1 ADJUSTING ONE'S METAPHYSICS

The argument of Chapter 5 provides a prima facie motivation for further exploring the context-sensitive view of realization. In this chapter, these explorations are of three kinds. Together they constitute a philosophical defense of the context-sensitive view.

The first concerns context-sensitive realizations and physicalism. Physicalism has been formulated both as a view of the mind-body relation and as a more encompassing metaphysical position, and one of the aims of the next three sections is to show what the context-sensitive view of realization implies about both of these forms of physicalism. Although there are forms of physicalism in the philosophy of mind that allow for the possibility that minds have nonphysical realizations, it remains true that the expression "physical realization" would be regarded by many as containing a redundancy. There are, so far as I know, no developed proposals for just how the mind would be "realized in" nonphysical stuff. Given this, it is important to show that the context-sensitive view of realization preserves robust forms of physicalism, even if, as I shall argue, it requires either giving up or revising several strands to physicalist thought. In this regard, in sections 2–4, I shall discuss, respectively, the thesis of microphysical determinism, the nature of dispositions, and nonreductive materialism.

The second kind of exploration considers modifications to the standard view of realization that address its putative shortcomings as a general view of realization. One general suspicion often directed at so-called novel views is that their putative insights can be captured within the framework of existing views. Being clear about the overall metaphysical package that the context-sensitive view brings with it is one way to dissipate such suspicions, or at least to shift the burden of proof to those who harbor them. In section 5, I shall examine a version of the standard view that ascribes a key role to background conditions without buying into the full range of context-sensitive realizations advocated within the context-sensitive view. Here it will be useful to introduce an analogy to the place of background conditions in discussions of causation, both in articulating the modified standard view and in revealing its shortcomings relative to the context-sensitive view. In section 6, I examine several other attempts to encompass some form of context sensitivity within the framework provided by the standard view. If the argument of these sections is correct, then the context-sensitive view offers a richer metaphysical package than do these variations on the standard view.

Finally, I shall conclude this exploration of the context-sensitive view by tacking to the other side, responding to claims that taking the context-sensitive view seriously requires adjustments in our conception of minds, individuals, and the relationship between them that are considerably more radical than those I have entertained thus far. Those who both are sympathetic to the context-sensitive view and think that it marks an important departure from business as usual in a materialist metaphysics of mind might well think that my own remarks in this chapter and the last are unduly conservative or restrictive. If mental states have wide realizations, for example, why think that we can (or would want to) hold onto realism about the mind, or restrict ourselves to viewing subjectivity as attaching to individuals at all? I shall take up these sorts of question in sections 7–9.

2 MICROPHYSICAL DETERMINISM, RELATIONS, AND SMALLISM

In the previous chapter, I said that the fragile sciences often traffic in properties that have a wide realization, and that when they do the corresponding methodology of integrative synthesis is appropriate. These sciences often investigate the relational properties of individuals, and what (totally) realizes such properties is not contained within the boundary of those individuals. Thus, an adequate account of the metaphysics of realization must move beyond the standard view with its exclusive focus on entity-bounded realizations and the constitutivity thesis.

We could express this criticism of the standard view by saying that it manifests a form of smallism, discrimination in favor of the small, in directing us always to what constitutes individuals, rather than what they in turn constitute. Indeed, the problem that relational properties and kinds pose for the standard view of realization is a version of what in Chapter 1 I identified as a general problem for smallist views in metaphysics. Since the standard view has been developed as part of a physicalist view of the mind, and the problem that I claim exists is quite general, we should consider the resources that physicalists have to address it.

A common expression of physicalism says that all facts, properties, events, processes, states, and entities supervene on the totality of physical facts, properties, and so on. This is to say that there is some or other physical base set of facts, properties, and so on for any fact, property, and so on. It is a short step from here to gloss physicalism as the view that there are physical realizations for all of the facts, properties, and so on that there are. Given that both supervenience and realization are relations of metaphysical determination that, so to speak, run in opposite directions – from realizations and to supervening properties – we might be tempted to think of these two metaphysical notions as getting at much the same idea. If that is so, then our short step is very short indeed, perhaps just that of logical closure.

Physicalism, thus, is a kind of thesis of metaphysical determination: The physical determines the rest. *Micro*physical determinism is a more specific form of physical determinism, one more directly relevant to assessing smallist metaphysical views. It is the general view that the properties of microphysical entities metaphysically determine all the properties there are. Consider two forms of microphysical determinism that, in different ways, attempt to accommodate relational properties.

The first of these Paul Teller has called *local physicalism* and *particularism*. It holds that the microphysical properties of any given entity, plus the microphysical properties of all those entities to which it is related, determine all of the properties it has, including its relational properties. Suppose that Tom has a mass of 100 kg and Susan has a mass of 80 kg, and so Tom has the relational property of being heavier than Susan, and stands in the relation "is heavier than" to Susan. On the particularist version of microphysical determinism, this relational property of Tom's, and the corresponding relation between Tom and Susan is determined by Tom's microphysical properties and Susan's microphysical properties. What particularism preserves from corpuscularianism is the idea that an individual thing's intrinsic properties determine its place in the causal nexus. The relational properties of that individual thing are determined not simply by its intrinsic properties, but by those together with the intrinsic properties of the object(s) to which it is related. For the particularist, it is still the intrinsic properties of individual things that realize all the properties that there are. There are several problems that particularism faces.[1]

First, it suggests a general strategy for understanding relational properties that presupposes that they can always be viewed in terms of relations to particular objects. Certainly, many relational properties in science can be so understood. But many of the examples I provided in the last chapter invoke relational properties that lack this feature, and this points to the difficulty of assimilating all relational properties and kinds to particularism. For example, thinking about water, having a specific level of evolutionary fitness, and being a homeowner are relational properties, but not ones that simply involve relations between the bearer of the property and one or more particular objects. Rather, they are relational in presupposing the location of the bearer of the property in some larger system or environment: in a certain linguistic community, in an (usually, the actual) environment, in a society with certain kinds of institutions and practices. An individual's intrinsic properties do not determine whether it has any of these relational properties. Simply adding to these the intrinsic properties of other individuals (and which individuals?) does not lead to a realization base for these relational properties.

Second, since particularism is a form of microphysical determinism, then not any old intrinsic property serves as the ultimate realizer for all the properties that there are, but just those intrinsic properties that very small things have. Thus, as well as having to show how all relational properties are determined as are quantitative comparative properties, particularism must also show how all of the intrinsic properties of the not so small are realized by just the intrinsic properties of the smaller things that constitute them. But this seems false even at the chemical level, let alone at "higher" levels. For example, salt is physically composed of sodium and chlorine. Yet its having the chemical composition that it does, and the intrinsic properties it has, is not realized simply by the intrinsic properties of sodium and the intrinsic properties of chlorine, but those together with the realizing sodium and chlorine standing in a certain relation to one another (in particular, having a certain spatial relation to one another). Likewise, organism-level properties that we might think of as intrinsic, such as running speed or bodily musculature, are not simply realized by the intrinsic properties of organism parts, but by those plus how those parts are put together.

This second problem suggests a natural revision to particularism. Rather than try, as does particularism, to maintain that there is a sense in which the relational properties of individual things can be reduced to intrinsic properties, instead consider relations themselves as ontologically basic. Call such a view *relational microphysicalism*. This form of microphysicalism plays up the fact that when we turn to physics itself we find a range of relations: from states of entanglement in quantum mechanics, to space-time separation in relativity theory, to velocity in good old Newtonian mechanics. Since such relations occur within physics – indeed, within microphysics – the idea is to take them, together with the intrinsic properties of the physically very small, as what realizes all other properties and relations. Relational microphysicalism thus abandons the putative asymmetry between intrinsic and relational properties, but as a form of smallism maintains the asymmetry between the small and the not so small.

David Lewis's endorsement of the thesis of Humean supervenience expresses a form of relational microphysicalism. In the preface to the second volume of his collected papers, he expresses this doctrine as holding that

> all there is to the world is a vast mosaic of local matters of particular fact, just one little thing and then another . . . we have local qualities: perfectly natural intrinsic properties which need nothing bigger than a point at which to be instantiated. For short: we have an arrangement of qualities. And that is all.[2]

Lewis's references to mosaics and arrangements suggest relational microphysicalism rather than particularism.

Relational microphysicalism avoids the second problem facing particularism, but still faces the first problem, that of accounting for relational properties that do not involve relations between objectlike relata. In addition, while it countenances relations, it is only those relations that, in the first instance, hold between the smallest objects there are. Relational microphysicalism is aggregative about what there is in the world in that any composite entity, C, and its properties, are realized by its physical constituents, A and B, their properties, together with the relations between A and B. Yet C's relational properties will not be realized by A, B, and the relations between them. So quite apart from relational properties that do not involve objectlike relata, and even granting that relational microphysicalism is true of an object's intrinsic properties, provided that aggregated individuals have some relational properties, relational microphysicalism remains unable to account for all there is.

3 DISPOSITIONS AND SCIENCE

Many properties in the physical sciences are dispositional: They are tendencies that the objects that have them manifest in certain circumstances or under certain conditions. While the manifestation of dispositional properties may require those circumstances or conditions to obtain, what is often called the categorical base of the disposition surely does not: The base is intrinsic to the bearer of the disposition. Given this popular view of dispositions, one might develop several lines of argument that indicate that there is something mistaken about the context-sensitive view of realization. For example: Because dispositions are intrinsic and important in the physical sciences, in trafficking in unanalyzed relational properties the fragile sciences lose their mooring from physicalism. Or: Because we can understand the role of "context" in making sense of dispositions without abandoning the idea that intrinsic properties are ontologically basic in some sense, we can preserve the asymmetry between intrinsic and relational properties, or between the small and the not so small, that I have used the context-sensitive view of realization to challenge. Or: Even if there is a sense in which some properties have a wide realization (manifestation), there is a deeper sense in which they must have an entity-bounded realization (base).[3]

These lines of argument could, no doubt, be filled in. But my aim here, in part, is to suggest that to do so will be in vain, for the view of dispositions on which they rest is problematic for much the reason that the standard view of realization is problematic. Many of the dispositional properties in the physical sciences are themselves relational.

Consider dispositional properties in chemistry, such as acidity and miscibility. Whether a given liquid is acidic or miscible, whether it has those properties, is metaphysically determined by more than facts about its constitution. This is because these dispositional properties are dispositions to have certain effects on other chemical substances and kinds, and so whether they have those dispositions – not just whether they manifest these dispositions – is determined by facts about those other substances. The very presence of the disposition, not just its manifestation, involves the physical configuration of the world beyond the bearer of the disposition. These dispositional properties have a wide physical realization.

This point is reflected in standard definitions of an acid – for example, as a proton donor, or as an electron-pair acceptor. For whether a substance with a given physical structure has the disposition to donate protons or to accept electron pairs depends on facts about the broader chemical system in which that substance exists. If just these facts were different, a liquid that is actually an acid could lose this disposition, and could do so even were its chemical composition to remain unchanged. While successive concepts of an acid – from that of Arrhenius, to those of Bronsted and Lowry and Lewis – characterize an acid in what we might think of as increasingly "purely" dispositional terms, that is, in ways that increasingly abstract away from the environment, all of these concepts make either explicit or implicit reference to the character of solvents and bases which lie beyond the boundary of the thing that is an acid. Whether, say, hydrogen chloride has the disposition to donate a proton when placed in water depends not only on the intrinsic chemical character of HCl, but also on that of H_2O and the relative strengths of the forces governing their interaction. A physical duplicate of a substance that is acidic in the actual world may not be acidic in a world where the facts about solvents and chemical forces are different, even once we hold fixed the facts about other physical forces as "background conditions." In this respect, acidity is like fitness or any of the other examples I have cited from the fragile sciences. The salience of the fact that the core realization of acidity is individualistic should not obscure our view of what metaphysically determines the presence of the disposition itself.[4]

This requires endorsing that at least some dispositions, including some dispositions in the physical sciences, are extrinsic or wide: Whether something has the disposition is not determined by that thing's physical constitution. Jennifer McKitrick has recently provided a range of commonsense examples of extrinsic dispositions – the power to open a door, weight, the disposition to dissolve the contents of my pocket, vulnerability, visibility, and recognizability – and defended this idea against the predominant view of dispositions as essentially intrinsic. Some of the examples that I provided in Chapter 5, such as fitness and folk psychological states, are often understood as dispositional properties within the fragile sciences, and if they are relational, as I have argued, then they must be wide dispositional properties. In fact, such wide dispositions are widespread across the sciences, including acidity, miscibility, and solubility in the chemical sciences; conductivity, heat sensitivity, and rigidity in the electronic and engineering sciences; and trustworthiness, fertility, and stability in the cognitive, biological, and social sciences.[5]

Part of my point here is to preempt a series of connected ideas concerning dispositional properties and the nature of science: that an appeal to dispositional properties provides refuge for the standard view of realization and the asymmetry that it posits between intrinsic and

TABLE 6.1 *Realization in Psychology, Biology, and Chemistry*

	Psychological	Biological	Chemical
P	being in pain	being a predator	being acidic
S	nociceptive system	predator-prey system	system of acids and bases
IB	person	animal	liquid solution
core R	certain level of activity in C-fibers	having sharp claws	having an empty electron shell
total R	CR plus states of rest of nociceptive system (thalamus, somatosensory cortex, etc.)	CR plus properties of rest of IB and prey that, together with IB, constitutes the predator-prey system (e.g., speed of IB, body of prey)	CR plus properties of rest of IB plus those of solvents and bases that, together with IB, constitute the acid-base system
BCs (holding of world beyond S)	conditions of rest of the nervous system of IB and IB's other bodily systems	conditions of the ecological environment occupied by S, e.g., terrain, seasonal conditions	conditions of the physical world more generally, e.g., particular value of gravity on Earth, laws of nature

relational properties; that "real dispositions" in the fundamental sciences are intrinsic; that the ubiquity of relational properties in the fragile sciences casts them under some sort of metaphysical shadow dissipated by the light of the physical sciences. But part of my point is to extend the reach of the context-sensitive view of realization and the framework that I introduced in articulating it. To this end, Table 6.1 shows the parallels between psychological, biological, and chemical examples that I have discussed within this framework. As this table shows, both entity-bounded and wide realizations can be readily represented within this framework. The symmetry between the two, I want to suggest, implies the irrelevance of the boundary of the individual as a constraint on what counts as a realization of a given property.

Relational properties permeate the fragile sciences, and we should thus expect wide realization and the strategy of integrative synthesis to be the norm across those sciences. While entity-bounded realization and constitutive decomposition are certainly predominant in the physical sciences, there is also a range of properties in these sciences that are, like those in the fragile sciences, implicitly relational, including many dispositional properties. Such properties and kinds have wide realizations. This is one more reason to be skeptical of the idea that there is anything metaphysically distinctive or suspicious about the fragile sciences, or about the nature of relational properties and kinds.

Having examined aspects of physicalism in general in light of the context-sensitive view of realization, let us return to focus on physicalism about the mind in particular. Rejecting the constitutivity thesis in the standard view most obviously requires giving up reductive physicalist views of the mind. But this also has perhaps less obvious implications for nonreductive materialism.

4 NONREDUCTIVE MATERIALISM

The conflict between the sufficiency and constitutivity theses provides a novel way of expressing a long-acknowledged tension between externalism and reductive forms of physicalism in the philosophy of mind. Yet it also points to largely unrecognized inadequacies in a number of ways of expressing nonreductive materialism. Those expressions have also relied on the standard view of realization.

Nonreductive materialism has sometimes been formulated in terms of the acceptance of a "token-token" identity thesis or via a compositional view of realization. Consider each view in turn and what our discussion thus far implies about it.[6]

The token identity theory claims that tokens of mental and physical states may be identical even if types of mental and physical states are not identical, where the relevant physical states are intrinsic states of the brain. Since such states are intrinsic, physical states of the individual, the token identity theory, at least as usually articulated, presupposes the constitutivity thesis. But our exploration of the varieties of realization suggests that this thesis and the token identity theory are false for at least many mental states. This is because the total realizations of a range of mental states are wide and thus are not intrinsic states of the brain. At most, it is the core realizations of mental and physical states that are identical, but this does not help us identify mental and physical states.

Compositional views of realization, and thus physicalism, likewise take the relevant composed entity to be the individual or her central nervous system, and in so doing, rely on the constitutivity thesis. They thus face the same problem that the token identity theory faces. In short, both of these common expressions of nonreductive materialism have relied on the standard view of realization. If either view allows the relevant tokens or composed entity to be larger than the individual who instantiates the corresponding mental properties, and so in effect gives up the constitutivity thesis, it must be revised in fairly significant ways: We are no longer talking of token physical states of the brain, or compositional states of individuals.

Nonreductionist forms of physicalism are also often expressed in terms of there being "higher" and "lower" levels of explanation, the latter of which provide a metaphysical but not a reductive basis for the former. Whether we can adequately conceptualize mental states in general as being (totally) realized by "lower level" states seems to me doubtful. Those articulating this idea further using a constitution-based conception of realization either will be hard pressed to maintain the view that realization is determinative, or will, in effect, concede that lower levels do provide a reductive base for higher levels. Neither option should be attractive to a nonreductionist. To tackle the first horn of this dilemma would require a fairly radical rethinking of the concept of realization, one that gives up on the sufficiency thesis altogether. Tackling the second horn threatens to locate the site of one's nonreductionism solely within the realm of explanation, a threat exploited by Jaegwon Kim in his attacks on nonreductive physicalism.[7]

We can make this point in another way and more positively. I said in Chapter 5 that a constitution-based conception of realization appears to provide the metaphysical grounding for the explanatory strategy of homuncular functionalism. In the language of higher and lower levels, this is the idea that things and properties specified by lower-level homuncular descriptions physically constitute those specified by higher-level homuncular descriptions. If we grant that at least these latter things and properties are often relationally individuated, this relation of constitution can be determinative only if the former things and properties are likewise relationally individuated. This is to say that the things and properties specified by lower-level homuncular descriptions may be relationally individuated. And if the relevant relations for the higher-level properties extend beyond the boundary of the individual, so too must those for the lower-level properties. So while there may be some sense in which lower levels "constitute" higher levels, neither need be exhausted by the subject or bearer's intrinsic, physical properties, that is, by those properties usually taken to physically constitute an individual.

In effect, a homuncularly decompositional view that takes relational individuation seriously entails rejecting a premise crucial to a smallist, individualist view of the mind, a variation on the sufficiency thesis, viz., that the physical constitution of an individual metaphysically determines what mental properties that individual has. There are metaphysical and not merely pragmatic grounds for construing homuncular functionalism as an externalist view, and proponents of homuncular functionalism need to transcend their implicit reliance on the standard view of realization.

5 THE MODIFIED STANDARD VIEW: CAUSATION AND REALIZATION

One might reasonably doubt whether the context-sensitive view is needed to accommodate the ubiquity of relational properties and kinds in the fragile sciences. Perhaps a view of realization embracing both the sufficiency and constitutivity theses can be modified in ways that "take relational individuation" seriously. In this section I shall consider the idea that a modified standard view can do so simply by acknowledging a greater and more explicit role for background conditions than the standard view itself has.

Central to the standard view is the intuition that total realizations themselves always satisfy the constitutivity thesis. Even if ascribing any property to an entity, including mental properties, presupposes that certain beyond-the-individual background conditions hold, realizations themselves do not extend into the world beyond the individual. That is, all total realizations are in fact entity bounded, and we should not mistake some of these background conditions for (parts of) the realization itself. Putative examples of wide and radically wide realizations should be reinterpreted within the parameters of the standard view, modified to acknowledge just the first two forms of context-sensitive realization that I identified in Chapter 5: That of core realizations, which are realizations only insofar as they are part of a total realization, and the context sensitivity of total realizations to background conditions. On this modified standard view, realizations are entity bounded and the background conditions necessary for the realization of a given property may be more extensive than initially envisaged.

We can support the modified standard view by drawing a parallel between how we should think about realization and how we often think about causation. We talk of Sarah's carelessly discarding an unfinished cigarette as the cause of the bushfire even though we recognize that this event is not itself sufficient to bring about that effect. It does so only given that a variety of conditions are in place: that it not rain, that the cigarette make contact with combustible material, that the cigarette is not extinguished before it ignites that material, and so on. This distinction between causes and conditions is commonplace in theories of causation, but it provides little reason, one might argue, for viewing causes as either "wide" or "radically wide." It is Sarah's action itself that is the cause, this event here and now. Likewise, one might think, for realizers.

I shall return to the analogy to causation in a moment, but let us consider the modified standard view itself. Of critical importance is how well that view allows us to make sense of the full range of properties and kinds posited across the various sciences, as well as those found in our commonsense discourse. Both social actions and mental states pose problems here, and suggest that utilizing only an entity-bounded notion of realization, together with background conditions, requires abandoning some central intuitions about realization.

Most problematic are social actions, which not only have political, economic, and legal background conditions but also literally extend into the world beyond the individual who enacts them. For these actions, such as signing a check, even what I have been calling their core realizations do not stop at the skin. On the modified standard view, realizations must be entity-bounded, and so the realization of the action of (say) signing a check can appeal to holding a pen and writing on paper only as background conditions for the realization of the action. Yet these are things that the agent herself does, part of the action itself, not (like other background conditions) merely general features of the social and institutional environment in which she acts, and this distinction is ignored within the modified standard view. The real problem here is that viewing something like the bodily movements of the agent as the total realization of her signing a check is to collapse the distinction between total and core realizations. In effect, it is to accept that the realization of the action is not determinative of the action, and so give up the sufficiency thesis. Thus, this is not a view that modifies the standard view, but abandons it.

There is a fundamental, general problem for the modified standard view here. Treating total realizations as core realizations makes them not simply metaphysically context sensitive but also *epistemically* context sensitive. That is, what counts as the realization for any given property or kind depends on us. On the modified standard view, any part of the world that is a physical constituent of the bearer of the property or kind can be the realization of that property. All that needs to be done is simply to assign the state of the remainder of that individual to the set of background conditions that, together with that physical constituent, are metaphysically sufficient for the property or kind. In the case of signing the check, we could push further back into the body (in fact, all the way to the brain) in identifying the realization of that action, provided that we make a corresponding addition to our set of background conditions. Thus, on the modified standard view there is no mind-independent way to specify the realization for any given property or kind.

Consider another example that illustrates the general problem here. Any psychological state, such as being in pain, involves the firing of thousands if not millions of neurons. In a given case, which of these is the realization of the pain? On the modified standard view, a single neuron firing in my head at a given time could realize this state, provided that we specify the remaining neural firings as part of the background conditions for that single neuron's firing to metaphysically suffice for the individual to be in pain. To put it the other way around, on the modified standard view there is no answer to the question of what the realization of a given psychological state is that is independent of our decision of how extensive to make the corresponding background conditions. This constitutes a *reductio* of the modified standard view.

We can return to the analogy to causation to diagnose what has gone wrong here. Causes have the effects they do only given the presence of certain conditions. If one has only the distinction between causes and the conditions that necessitate their effects, then the arbitrariness of the distinction between cause and condition lead one to consider the individuation of some events as causes and others as conditions as dependent on us. Essentially, this is the situation that the modified standard view of realization is faced with.[8]

On the context-sensitive view that I have introduced, background conditions are necessary for there to be a functioning system that (totally) realizes an individual's properties. To be a realist about properties, and so about mental properties, is to be a realist about at least their total realizations, and thus about the systems with respect to which total realizations are defined. Thus, the distinction between background conditions and noncore parts of total realizations is required by the realism implicit in the view I have defended (see also section 7). We simply don't get to decide where to draw the line between realization and background conditions, and in particular it is not simply up to us to decide that the realizations must be entity bounded, or where within an individual they begin and end. Given that the systems in terms of which realizations are characterized are robust entities either that form parts of individuals or that individuals form a part of, background conditions have a more restricted role to play than this modification of the standard view suggests.

In terms of the analogy to causation, this is something like introducing the notion of a causal chain between that of cause and condition. Causes form causal chains, and these causal chains have their effects only given the existence of certain other conditions. When we single out some event as the cause of another, we imply that it is the most salient part of that causal chain, and that without the remainder of that causal chain linking cause and effect, "the cause" would be no cause at all. But it is also true that without further supporting conditions, even the whole causal chain would not produce that effect. What distinguishes causal chains from the conditions against which they bring about their effects is a further issue. But the parallel between two trichotomies – between cause, causal chain, and conditions; and core realization, system (total realization), and background conditions – is useful in understanding why one needs to go beyond the resources of the modified standard view of realization.

6 CONTEXT SENSITIVITY WITHIN THE STANDARD VIEW

Several physicalist proposals that can be viewed as sympathetic to the standard view have acknowledged the role of context in the metaphysics of mind, and in this section I shall consider whether they adequately accommodate the sort of context sensitivity introduced in Chapter 5. These proposals are Terry Horgan's idea of regional supervenience, and Denis Walsh's more recent defense of a view he calls wide content individualism or alternative individualism.

While neither of these proposals focus on the notion of realization per se, both views can be construed as attempts to provide a role for context that maintain some version of the constitutivity thesis about realizations. Their shortfalls qua modifications of the standard view are my concern here.

Terry Horgan introduced a thesis of supervenience that he later christened *regional supervenience*:

There are no two P-regions [spatio-temporal regions of a physically possible world] that are exactly alike in all qualitative intrinsic physical features but different in some other qualitative intrinsic features.[9]

Regional supervenience was introduced to account for what Horgan calls an individual's context-dependent properties. Horgan's examples include being president of the United States, being a bank, and knowing that Oscar Peterson is a jazz pianist. It does so by extending the subvenient base to the spatio-temporal region that contains that individual so as to include the relevant contextual factors in that base.

If we were to consider this as extending the realization base beyond the individual, we would have something like a wide realization (though note that I have defined these in terms of entitylike systems, rather than spatio-temporal regions). But this, of course, would be to give up the constitutivity thesis. Horgan himself thinks that the realizers for such properties are typically narrower than the corresponding subvenient base, suggesting that he views realizations as satisfying the constitutivity thesis. Such a view, however, can claim only that realizations, together with the larger spatio-temporal region of which they are a part, determine the properties they realize, giving us realizers that by themselves do not satisfy the sufficiency thesis. What satisfies the sufficiency thesis is the regional supervenience base, but clearly that does not satisfy the constitutivity thesis. In short, I think that we see here something like the tension between the two halves of the standard view of realization, with what metaphysically determines the presence of a property in an individual being something more than what is bounded by that individual.

This argument can be reexpressed as follows. Consider the constitutivity thesis reformulated with respect to the entire region, rather than the individual in that region who has the properties:

regional physical constitutivity thesis: Realizers of states and properties are exhaustively physically constituted by the intrinsic, physical states of the region containing the individual whose states or properties they are.

The metaphysical sufficiency thesis and the regional physical constitutivity thesis are jointly satisfiable. But since the latter thesis explicitly requires going beyond the boundary of the individual property bearer in order to specify what determines that property, realization on this view is not entity bounded.

Denis Walsh's chief aim is to articulate a position that resolves what he calls the antinomy of individuation: that combining the plausible claim that thoughts of the same psychological kind have the same (wide) contents with individualism entails the implausible conclusion that "thoughts which are instances of the same physiological kind have the same wide contents." Walsh's solution to the antinomy is to reformulate each of these three claims so as to make explicit the way in which psychological states are context sensitive. These three principles, which together he calls wide content individualism, are:

(1) Necessarily, if individuals have thoughts of the same psychological kinds with respect to a context, then their thoughts have the same (context-sensitive) content with respect to that context.
(2) Necessarily, states of the same physiological kind which share a context realize states of the same psychological kind with respect to that shared context.
(3) Necessarily, states of the same physiological kind which share a context realize thoughts with the same content with respect to that context.

As in the original antinomy of individuation, (1) and (2) entail (3), but (3), Walsh suggests, is true.[10]

(2) implies that identical intrinsic, physical states of individuals in the same context realize the same psychological states with respect to that context, while (3) spells out this implication for the special case of intentional psychological states. While Walsh suggests that this is a way of reconciling wide content with individualism, consider (2) and especially (3) in light of the constitutivity and sufficiency theses and my argument thus far. In effect, (3) says that if you take realizers that satisfy the constitutivity thesis and fix their context you have realizations of intentional states with the same content. Yet, as the conjunction here makes clear, the realizers that satisfy the constitutivity thesis themselves satisfy the sufficiency thesis with respect to intentional states only in conjunction with their context. Thus, we lack any one realizer that satisfies both theses. This is precisely the problem that we identified with Horgan's regional supervenience thesis considered as an attempt to develop the standard view of realization in a way that accounts for the context sensitivity of mental states.

Furthermore, since Walsh's view invokes a relatively unconstrained notion of context, it would seem subject to a variation on the single neuron objection that I introduced in section 5. Walsh says that he thinks of a context "as corresponding to a set of properties of an individual's environment." If we make the context rich enough – including, for example, properties that might normally be determined by an individual's internal, functional organization – then two identical single neurons that realize the same physiological state (and so satisfy the constitutivity thesis) could also realize the same psychological state relative to that context. But this does not so much provide us with a realization that also satisfies the sufficiency thesis as indicate a problem with drawing on such a notion of context as a way of saving the standard, entity-bounded view of realization.[11]

Walsh's defense of the view that he calls *alternative individualism* is also an attempt to build context sensitivity into individualistic views in the metaphysics of mind and science. At the heart of alternative individualism is a context-sensitive version of the local supervenience thesis – that causal powers supervene on the intrinsic, physical properties of individuals – that serves as a premise in the argument from causal powers that we discussed in Chapter 4. Walsh takes alternative individualism to preserve what is right in that argument. The supervenience thesis that Walsh defends claims that for all contexts, necessarily an object's having a given context-sensitive, intrinsic property in a context fixes the psychological properties it has in that context. We can express this succinctly in symbols, where "γ" denotes a context, "Φ" an intrinsic, qualitative property, and "Ψ" the psychological kinds that an individual, x, instantiates:

$$\forall \gamma. \ \forall x (\Phi x, \gamma \supset \Psi x, \gamma).$$

Walsh's claim here generalizes beyond psychology to scientific taxonomy in general.[12]

The problem with this supervenience thesis, as with that of the orthodox individualist (which simply drops the contextual relativization) is that there is a wide range of examples for which it simply does not hold. Two organisms that were intrinsically identical in the very same context could belong to different species because of ancestral, phylogenetic differences between them; two viral infections in a person – and thus, I assume, in one context – that give rise to identical symptoms and are treated by the same regimen can be of different types because they are caused by different viral agents. In both cases, the identity of consequences in a given context (or even of intrinsic, qualitative properties) does not determine sameness of scientific kind because the corresponding kinds are individuated, in part, by causal antecedents, not causal consequences.

Recall that my basic criticism of the original argument from causal powers was that it equivocates on "causal powers," invoking a wider notion in one premise (individuation in science is by causal powers) and a narrower notion in another (causal powers supervene on an individual's intrinsic properties). Alternative individualism is an attempt to salvage both of these premises, but it too involves much the same equivocation, this time on context-sensitive causal powers. The fundamental problem in all of these cases is brought out once one asks whether relational properties are, or are fixed by, such powers: They must be if the premise about scientific individuation is to be true, but they can't be if the premise about supervenience is to be true. Alternative individualism, like Walsh's wide content individualism, is committed to causal powers (within a context) subsuming relational properties – that is what allows these modifications of individualism to be seemingly compatible with externalism about the mind – but this is precisely what calls into question whether such powers are a subset of (or supervene on) an individual's intrinsic properties.

Considered in tandem, Horgan's and Walsh's views illustrate that the basic tension in the standard view of realization between the constitutivity and sufficiency theses is not easily relieved. Placing emphasis on the need to move beyond the boundary of the individual subject in order to have a determinative base for mental states, as Horgan's regional supervenience does, highlights the point that realizations satisfying the sufficiency thesis do not themselves satisfy the constitutivity thesis. And emphasizing that realizations that satisfy the constitutivity thesis determine mental states only given a shared context, as Walsh's (2) does, suggests, conversely, that in at least some cases realizations that satisfy the constitutivity thesis do not satisfy the sufficiency thesis.

Those willing to hum along to the tune of the context-sensitive view of realization and entertain the idea that properties at least sometimes have wide realizations might reasonably wonder whether the context-sensitive view has more radical implications for how we think about the mind. I want to take up one such putative implication in each of the next three sections, beginning in each case with a question.[13]

7 KEEPING REALISM AFLOAT

Why doesn't the context-sensitive view lead to an irrealist position on mental states? Consider the very idea of a physical state's being "metaphysically sufficient" for a given mental state. We have seen that, strictly speaking, metaphysical sufficiency requires both that some physical system be in a certain state (a total realization) and that certain background conditions hold, thus making even total realizations *metaphysically* context sensitive. This view of total realizations underwrites the realism about mental properties that I invoked at the end of section 5, a realism that would be called into question if it could be shown that total realizations are also *epistemically* context sensitive, as core realizations are.

In fact, we might well think that total realizations must be epistemically context sensitive if the core realizations they contain as proper parts are.

A sufficient condition for an individual having a given mental state is that there be a total realization of that state whose core part lies wholly or in large part in that individual. But if that core realization is epistemically context sensitive, then so too is that state itself, and mental states start to sound more like merely ascribed states of individuals.

Moreover, it seems doubtful that there is any particular physical state that is "the" total realization for a given property, for what we pick out as a total realization will depend on what we count as part of the corresponding system and what counts as background conditions to it. Once one "goes wide" about realization, as perhaps one should, we have a plurality of ways of thinking about the underlying metaphysics, a pluralism that suggests that we, rather than the world, are the source for how the distinctions between individual, system, and background conditions are drawn. Thus, the context-sensitive view is faced with much the problem that we posed for the modified standard view in section 5, and an externalist metaphysics pushes us toward irrealism about mental states.

Reply: This irrealist challenge can be arrested. In the first place, epistemic context sensitivity is not simply inherited by total realizations from their constituent core realizations. A total realization of P could have been defined simply as a state of S that is metaphysically sufficient for P, that is, by dropping the relative clause that refers to core realizations, without significantly changing the view that I have defended. There can be multiple total realizations not because there are multiple core realizations that are epistemically context sensitive, but because, given the complexity of the sort of systems there are, there will at least typically be many ways in which those systems can be arranged or instantiated, each of which will metaphysically suffice for P.

Consider pain. The nociceptive system that realizes pain is complicated, and there are many states it can be in that would metaphysically suffice for an individual organism to be in pain. Even given a particular core realization of pain – say, as a particular instance of C-fiber stimulation – there remain multiple total realizations for that core realization because there are various noncore parts of the realization that could suffice for the mental state of pain, even though just one of these will, in any given instance, form part of the total realization in that case.

As with the boundary of individuals, the line between what falls within any given system (and so can be part of the total realization for any properties it realizes) and the background conditions for its existence and operation can be fuzzy. The individuation conditions for systems are likely to be more coarse grained than for individuals, but in both cases these are highly constrained by the physical facts. It is not simply up to us to determine what constitutes a system or the system of relevance. Like the individuals that they either constitute (in cases of entity-bounded realization) or that constitute them (in cases of wide realization), systems have individuation conditions, and these depend only in minor ways on our epistemic proclivities and fancies. Whether teeth or saliva form part of the digestive system, or operate as background conditions for its operation, may be "up to us." But the facts about how food is broken down into energy and distributed within the body, and how waste products are formed and removed from the body determine what, plus or minus a bit, constitutes the digestive system. We discover, rather than invent, what physically constitutes the digestive system; the same is true of cognitive systems, whether they be entity bounded or wide.

Cases in which there are genuinely alternative systems which we could, plausibly, identify as the locus of a given total realization are likely to be rare. Again, focus on the physiological systems that are a paradigm here. While it is logically and metaphysically possible that there be two or more candidate systems for the realization of any biological function – respiration, circulation, digestion, reproduction – the requisite complexity to each of these systems in practice makes it relatively unproblematic to single out what "the" relevant system is for any given property. The same is true of cognitive systems. As with any systematic theorizing, in science or elsewhere, this theorizing about both entity-bounded and wide cognitive systems is subject to error, modification, and revision, but this is not the sort of epistemic context sensitivity that would undermine a realist view of the ascription of psychological states.

8 PLURALISM ABOUT REALIZATION

Why don't all mental states have wide realizations? Suppose that we accept the view that at least some mental states have wide realizations. Might we replace the existential by the universal quantifier here, and suggest that the moral of the story so far is that mental states have individualistic core realizations but wide total realizations?

Reply: Given that social actions appear to have radically wide realizations, and the ways in which at least our common-sense conception of the mind is linked to such actions via the idea of a reason for acting, we might have pause about the latter of these two views. Here I want to make some brief comments about the former claim, the idea that all mental states might have wide realizations.

One natural thought in response to the claim that all mental states have wide realizations is that mental properties typically denoted by monadic predicates – such as pain – surely have entity-bounded realizations, since their presence at a time or over a time interval is determined solely by what is going on within the boundary of the individual who has the property. In fact, given that the nociceptive system is a proper part of an individual organism, as I suggested earlier, this conclusion about pain seems inescapable. More generally, there would seem to be a range of mental states and processes that form part of entity-bounded systems: Good candidates include fear, motor imagery, and haptic perception.[14] It has been the working assumption of much traditional cognitive science, committed as it has been to individualism about the mind, that all mental states and processes can be viewed in such a way, with the task of cognitive science being to uncover what these entity-bounded systems are. While I think that there is little reason to think that such a general view of cognition can be sustained, my point here is that the individualistic view of at least some cognitive processes does seem correct. This suggests a general conclusion – that whereas some of our mental states have an entity-bounded realization, others have a wide realization.

It may be worth briefly pondering the more radical alternative that I am somewhat casually rejecting here, that no psychological states actually have entity-bounded realizations. This would imply that the cognitive sciences are radically mistaken, and supposing that many noncognitive, biological systems are individualistic, suggests that there is a radical juncture between the cognitive and the biological. The scale and nature of the error within the cognitive sciences would be akin to that which behaviorists charged introspectionists with in the first part of the twentieth century: There simply aren't the kinds of internal states (systems) that such a psychology seeks. While I don't say that it is impossible that the cognitive sciences have gone awry so radically, it does seem that one would want very strong reasons for thinking that the smallist metaphysics on which they have relied is completely mistaken. It seems to me clear where the burden of proof lies on this issue.

If some of our mental states do have entity-bounded realizations, while others have wide realizations, then there is a respect in which the standard way of characterizing (total) realizations via the Ramsey-Lewis method for defining theoretical terms – a method commonly used to characterize functionalist views in the philosophy of mind – is both restricted and misleading. Ramsey-Lewis sentences purport to represent complete theories for a given domain and are constructed by conjoining all of the truths specified by such theories; one derives the total realization for a particular property or state by conjoining its core realization to the realization of complete theories for the domain. Here let us simply grant that such a conception of folk and scientific theories is coherent and a close enough approximation of the theories we have actually developed to model those theories usefully. Now, if some part of psychology is wide, then since the total realization for a complete psychology will be a wide realization, that for any particular psychological state will also be wide. Given the wide nature of the propositional attitudes and at least some subpersonal psychological states, the goal of characterizing a complete psychology implies that the total realizations of any psychological state must be wide.[15]

If we follow Brute Intuition and our brief reflection on cognitive science, and insist that surely some psychological properties have entity-bounded realizations, and thus accept my claim that not all psychological properties are wide, the Ramsey-Lewis method appears to provide us with no way to represent a significant distinction. The most obvious modification to the standard Ramsey-Lewis view – to attempt to define properties like being in pain by reference to a theory of pain, and properties like believing that p by a theory of belief – fails, because each of these theories will almost certainly mention terms from the other, and so will not allow one to define properties with entity-bounded realizations. I leave further exploration here to those more enamored with the Ramsey-Lewis method than am I.

9 ABANDONING THE SUBJECT?

Why aren't subjects or bearers of mental states themselves wide? The characterization of wide realizations preserves the idea that properties with such realizations are still properties of individual subjects. Thus, fitness remains a property of individual organisms even though its realization is wide. And my belief that Paris is the capital of France remains *my* belief even though it has a wide realization. I think that this is also true even in cases of radically wide realization, paradigmatically in those involving social actions: Jane's signing a check is her action because the core realization of that action is realized in large part (even if not wholly) by something that she does, such as moving her pen-grasping hand over a piece of paper. To my mind this is a desirable feature of my view of mental properties in particular because it preserves what is right about an individualistic view of subjectivity and in so doing readily allows for both third- and first-person perspectives on the mind.

One might well challenge this aspect of my view as unnecessarily and unjustifiably conservative. Unnecessarily, for once realization goes wide surely we are on our way to undermining subjectivity and the misplaced position of privilege that the individual subject has in our thinking about the mind. And unjustifiably, since in at least some cases of wide realization, particularly those of radically wide realization, there is no nonarbitrary way to single out individuals as the subjects or "owners" of the corresponding mental properties. If we have wide realizations of mental states, and thus wide mental states, so too we should have "wide subjects" of those states. Andy Clark and David Chalmers suggest something like this view of the self as a consequence of their endorsement of what they call "the extended mind," a view that Clark has more recently applied in his thinking about rationality and human intelligence.[16]

Reply: There may, of course, be interesting science fiction or other fanciful examples that pull our intuitions toward such radical conclusions, but it is important not to lose sight of the fact that, at least in the world that we actually inhabit, and being the creatures that we actually are, there is a basis for marking out individuals as the subjects of properties, even those properties with wide realizations. Individuals – and here, as always, our paradigms are individual people and individual organisms – are spatio-temporally bounded, relatively cohesive, unified entities that are continuous across space and time. Recall that the possibility of wide systems was modeled on the actuality of systems that formed part of such individuals as exemplified by the variety of physiological systems theorized about in biology and medicine. While these narrow systems (for example, the circulatory system) share some of the features that make individuals metaphysically distinctive and certainly have their own properties, they are not themselves individuals, and it seems strained or at best derivative to view them as the subjects of the sorts of properties that we would intuitively ascribe to the individuals they constitute. For example, the visual system and its parts can be lesioned, can have imbalances in levels of neurotransmitters, and have certain of its pathways blocked (either experimentally or "in the wild"). But it is the individual who perceives, who suffers from a visual agnosia, who experiences a hallucination. The same is true of wide systems, and this provides a principled basis for ascribing mental properties in particular to individual subjects rather than the wide systems of which those subjects are a part. In the actual world, it is individuals who form and maintain beliefs, experience emotions, and wonder about what will happen next, even if those individuals form part of what I have called folk psychological systems.

10 PUTTING OUR METAPHYSICS TO WORK

The chief goal in this chapter and the last has been to develop a metaphysics suitable for an externalist view of psychology. Importantly, the argument here is not driven by a consideration of the special nature of mental properties but, rather, by reflection on a range of properties from the cognitive, biological, and social sciences.

Although the considerations here have been metaphysical, the externalist metaphysics developed is one with explicit ties to a strategy of explanation, integrative synthesis. The explanatory strategy of constitutive decomposition is well entrenched in a range of physical sciences, and philosophers of the cognitive sciences have been quick to point to ways in which it can be and has been used with respect to cognitive capacities. Integrative synthesis, while not quite novel, has a patchier history, particularly in the physical sciences. But again, if I am correct about the significance of the notion of wide realization, then we should expect integrative synthesis to find a central place in the sciences, including the fragile sciences.

In the following chapters we will see where all of this philosophical footwork leads us. In the next chapter I consider externalist views of the mind in action in the cognitive sciences, extending this to noncomputational approaches to cognition in Part Three.

7

Representation, Computation, and Cognitive Science

1 THE COGNITIVE SCIENCE GESTURE

We saw in Chapter 4 that as individualism was coming under attack in the late 1970s, it was also being defended by Jerry Fodor and Stephen Stich as a view of the mind particularly apt for a genuinely scientific approach to understanding cognition. In contrast to the original externalist papers of Putnam and Burge, those in which methodological solipsism and the principle of autonomy were introduced focused on the relevance of individualism for explanatory practice in cognitive science. They appealed to the computational nature of cognition in arguing that cognitive science should be individualistic, and for substantive conclusions about its scope and methodology.

These early individualist arguments of Stich and Fodor invoked what I shall call the *cognitive science gesture*. They primarily pointed to general features of cognition and theory in cognitive science that, they claimed, revealed their individualistic nature. Cognition was computational (and computation was individualistic), or cognitive processing was mechanistic (and such mechanisms were individualistic). Neither used a sustained, detailed examination of particular theories and explanations in cognitive science to argue that they were or must be individualistic. Instead, Fodor and Stich were, at least initially, content with a more abstract, general gesture toward features they took to be central to cognitive science.

The gesture toward the developing cognitive sciences in defending individualism served both to motivate and to buttress more purely philosophical considerations in favor of individualism, such as appeals to functionalism, physicalism, and causal powers, and the idea that it was narrow content that was truly explanatory of cognition. We have also seen how the standard view of realization reinforced the idea that the metaphysics of the mental pushed one to an individualistic view of the mind. Thus, a happy confluence: A range of broadly accepted views of the place of the mind in the world and empirical practice in our nascent sciences of the mind both imply individualism about the mind. The rejection of individualism was to come both at the expense of an acceptable metaphysics and at the cost of forgoing the prospects for a thoroughly naturalistic treatment of the mind.

My chief aim in this chapter will be to undermine this view of the relationship between individualism and cognitive science. After outlining the accepted view of individualism and the cognitive sciences in sections 2–3, in sections 4–7 I turn to reshape the received wisdom here, using David Marr's celebrated theory of vision as a focus. At the core of my discussion are the notions of *exploitative representation* and *wide computationalism*. With an alternative vision of representation and computation in hand, I shall conclude by considering some recent examples within the cognitive sciences that exemplify externalism in practice.

2 INDIVIDUALISM IN COGNITIVE SCIENCE

Although the cognitive science gesture is a gesture – rather than a solid argument that appeals to empirical practice – it is not an empty gesture. Even if the original arguments of Fodor and Stich did not win widespread acceptance amongst philosophers, they struck a chord with psychologists, linguists, and computer scientists. Indeed, the dominant research traditions in cognitive science have been at least implicitly individualistic. Consider three expressions of individualism that bring out its attractions for many cognitive scientists.

One attraction of individualism is its perceived connection to the representational theory of mind, which holds that we interact with the world perceptually and behaviorally through internal mental representations of how the world is (as the effects of perceiving) or how the world should be (as instructions to act). Ray Jackendoff expresses the connection between such a view and individualism when he says:

Whatever the nature of real reality, the way reality can look to us is determined and constrained by the nature of our internal mental representations.... Physical stimuli (photons, sound waves, pressure on the skin, chemicals in the air, etc.) act mechanically on sensory neurons. The sensory neurons, acting as transducers in Pylyshyn's (1984) sense, set up peripheral levels of representation such as retinal arrays and whatever acoustic analysis the ear derives. In turn, the peripheral representations stimulate the construction of more central levels of representation, leading eventually to the construction of representations in central formats such as the 3D level model....

Jackendoff calls this view the "psychological" (versus philosophical) vision of cognition and its relation to the world. His central claim in this paper is that only the psychological vision directs us to viable research programs in cognitive science. Provided that the appropriate, internal, representational states of the organism remain fixed, the organism's more peripheral causal involvement with its environment is irrelevant to cognition, since the only way in which such causal involvement can matter to cognition is by altering the internal mental states that represent that environment.[1]

Jackendoff's skepticism about the "philosophical" vision parallels the disdain for "philosophical" approaches to language that Noam Chomsky has expressed in drawing the distinction between the "I-language" and the "E-language," and Chomsky's insistence that only the former is suitable as an object of scientific study. Chomsky takes a conception of language as "internal," "intentional," or "individual," as opposed to "external," "extensional," or "social," to be a condition of a serious, empirical investigation of language, to be what makes the question of language use and acquisition a problem rather than a mystery. The I-language is the generative procedure inside the individual that is causally responsible for that individual's linguistic output, and it is what linguists attempt to reconstruct when they postulate features of a universal grammar. Given Chomsky's nativism about language in particular and cognition more generally, and in light of the relationship between strong nativism and individualism that I argued for in Chapter 1, we could expect Chomsky to be critical of externalist views of the mind. Indeed, in "Language and Nature" Chomsky turns directly to the philosophical tradition of externalism, beginning with the Putnam-Burge thought experiments, to argue for the poverty of the resulting conception of language and how it should be studied, a critique extended in his *New Horizons in the Study of Language and Mind.*[2]

To take a third example of an individualistic perspective on cognition, consider this extract from the Foreword to Simon Baron-Cohen's *Mindblindness*, written by Leda Cosmides and John Tooby:

Although it is a modern truism to say that we live in culturally constructed worlds, the thin surface of cultural construction is dwarfed by (and made possible by) the deep underlying strata of evolved species-typical cognitive construction. We inhabit mental worlds populated by the computational outputs of battalions of evolved, specialized neural automata. They segment words out of a continual auditory flow, they construct a world of local objects from edges and gradients in our two-dimensional retinal arrays, they infer the purpose of a hook from its shape, they recognize and make us feel the negative response of a conversational partner from the roll of her eyes, they identify cooperative intentions among individuals from their joint attention and common emotional responses, and so on.[3]

While Cosmides and Tooby clearly do assign the environment of the organism a role in the evolutionary history of species-typical capacities, the cognitive capacities themselves are individualistic.

These individualistic conceptions of the mind ultimately shape the direction of research within the cognitive sciences through their effect on how central notions in the field are construed. The most central of these are representation and computation.

3 MENTAL REPRESENTATION AS ENCODING

Underlying the visions of cognitive science exemplified by Jackendoff, Chomsky, and Cosmides and Tooby are what I will call *encoding views* of mental representation. Simply put, encoding views hold that to have a mental representation, M, is for M to encode information about some object, property, event, or state of affairs m. A well-known, protean version of the encoding view is the picture or copy theory of mind, where to have a mental representation of m is to have a mental picture or image of m in your head, where the picture is "of m" just because it looks like m. A version of the encoding view more prevalent in cognitive science and that builds more elaborately on the idea that cognitive processing takes place in a sort of code is the language of thought hypothesis: To have a mental representation of m is to have a token in your language of thought, M, that stands for or refers to m. Unlike the copy theory of mental representation, on this view there need be no resemblance between representation and represented, with the role that resemblance plays in the picture theory being played by some putatively natural relation, such as causation or teleology. At the time that the debate between individualists and externalists was being articulated, the language of thought hypothesis laid some claim to be "the only game in town" for understanding mental representation and computation.

On either view, discrete mental representations encode information about particular aspects of the world. Existing in a code, mental representations are governed by a kind of syntax that determines how basic units in the code can be manipulated, combined, and decomposed. Thus, cognitive scientists can and should explore the properties of representations and the rules that govern their internal dynamics, rather than the relationships that exist between organisms and environments.

There are two prima facie problems with encoding views of mental representation. These problems constitute not so much knockdown arguments against such views as open the way to reconceptualize representation in ways amenable to externalist visions of cognition.

The first is that although there may have been a time at which encoding views were the only game in town, over the last twenty years a number of views of mental representation have been developed that break from the encoding tradition. One of these is the notion of distributed representation. On this view of representation, what is in the head are not discrete symbols, each encoding their own piece of information, but less content-laden nodes which, in combination with the connection strengths linking them, collectively represent information about the world. A related alternative conception of representation is that of subsymbolic computation, whereby the units over which the computations are defined, the representations, are not themselves symbols (that is, codes). Both of these conceptions of representation were developed within a connectionist framework, but they have a basis in dynamic approaches to cognition more generally. What these views share is the idea of thinking about representation as fleeting, situated, dynamic, and interactive, and as such they mark a departure from encoding views of representation.[4]

These alternatives to encoding views constitute an important break in the grip that those views have had on how cognitive scientists have conceptualized mental representation. (So much so, in fact, that they were sometimes overenthusiastically characterized as dispensing with the notion of representation altogether.) But as they have been developed primarily within individualistic frameworks, they do not themselves constitute externalist views of representation. They reconceptualize the internal form and dynamics of mental representation, but do little by way of viewing mental representation as, in some sense, essentially embodied or embedded.[5]

The second problem for encoding views is that the medium for encoding is, by definition, some type of code, and codes themselves need to be interpreted. By virtue of what is such interpretation performed? Either by virtue of some other type of code – in which case we face the same question again – or by virtue of some brute noninterpretative and so noncoding process – in which case it is difficult to see what role the initial appeal to codes (and thus interpretation) is doing. Thus, the appeal to mental encoding either leads to a regress or it was not necessary in the first place. This dilemma constitutes an objection to the use of the notion of encoding in understanding mental representation in particular, rather than representation in general, since in other cases one or the other of these two horns can be grasped.

For example, consider public codes, such as communicated natural language and Morse code. These are interpreted by people, with interpretation mediated by knowledge of the conventions governing those codes. Since public codes are not self-interpreting, grasping the first horn of the dilemma above to explain how public codes are interpreted is unproblematic.

Alternatively, consider computational languages. These can be layered on top of one another through compilation and translation, with the most basic language, the machine language, engineered directly in the circuitry of the machine. Although this combination of compilation and engineering is sometimes taken as an analogue for how our language of thought is instantiated in the head, note that while compilation does involve encoding a higher-level language in a lower-level language, engineering does not. Electronic circuitry is not a code for machine languages but an implementation of them. Thus, in this case we can grasp the second horn of the dilemma, but we do so by giving up the metaphor of encoding.

One might object that this dilemma argument takes too literally what is only an analogy between mental representations and codes. All that cognitive scientists mean by mental representations are discrete, internal structures that correspond to things in the world. It is these structures and their properties, not objects in the world and their properties that cognitive processes are sensitive to. Things in the world drop out as irrelevant for cognitive processing once the structures to which they correspond are formed or activated, and it is for this reason that cognition is methodologically solipsistic.

This view can be expressed by saying that the process of mental representation is simply that of symbol formation, and that cognition is symbol crunching. Mental representations are natural symbols in that they are generated by brute causal relations between cognizers and their worlds, and so unlike conventional symbols, such as those on road signs and in written languages, they do not require interpretation to be symbols.

I suspect that this softening of the parallel between mental representations and codes accurately captures what many cognitive scientists (particularly psychologists) think is right about the language of thought hypothesis. Such softening, however, raises or leaves open questions that strict encoding views close off.

First, by placing more weight on mental representations as natural symbols, it weakens the connection between mental representation and computation, since computational symbols are conventional. If mental representations are not, strictly speaking, codes but natural symbols, then we need some account of the basis for not only how the symbols get their original meaning, but for the syntax that governs their processing. Fodor's own view – that the symbols in the language of thought, concepts, are innate, as is the syntax of that language – is one answer to this question, but one that very few cognitive scientists have been prepared to swallow.[6]

Second, this softening also highlights the question of what is special about *mental* representation. There are myriad causal dependencies between an organism's internal structures and states of the world. Why are those involving my perceptual apparatus and my mind symbolic, while those that concern my digestive system or the state of tension in the muscles in my leg merely causal? Philosophical projects in "psychosemantics," such as informational semantics and teleosemantics, have inevitably appealed to other forms of representation in articulating their vision of mental representation. The paradigm of these have been the internal states of measuring instruments, such as thermostats and fuel gauges, in the former case and the functioning and products of biological organs and behaviors, such as the heart and the dance of bees, in the latter. This deflationary naturalism about mental representation is no doubt a good thing. But whether such views can be happily married to something like encoding views of representation, or to an individualistic view of the mind, seems far from clear.[7]

4 THE DEBATE OVER MARR'S THEORY OF VISION

Thus, individualism receives some support from the computational and representational theories of mind, and so from the cognitive science community in which those theories have been influential. But I have also indicated that the claim that a truly explanatory cognitive science will be individualistic has an epistemic basis more like a gesture than a proof. One way to substantiate this second view in light of the first is to turn to examine the continuing philosophical debate over whether David Marr's celebrated theory of early vision is individualistic.

Marr's theory occupies a special place in cognitive science as well as in the individualism-externalism debate. Marr was trained in mathematics and theoretical neuroscience at Cambridge in the 1960s, and spent most of the 1970s at both the AI Lab and the Department of Brain and Cognitive Sciences at MIT before dying, tragically, of leukemia at the age of thirty-five. His ability to draw on and contribute to neuroscience, artificial intelligence, psychology, and philosophy exemplified cognitive science at its best. Although many of the specific algorithms that Marr and his colleagues proposed have been superceded by subsequent work in the computational theory of vision, the sweep and systematicity of Marr's views, especially as laid out in *Vision: A Computational Investigation into the Human Representation and Processing of Visual Information*, have given their views continuing influence in the field. The importance of Marr's theory for the individualism-externalism debate can perhaps best be understood historically and in light of the cognitive science gesture made by individualists in the late 1970s.[8]

In the final section of "Individualism and the Mental," Burge had suggested that his thought experiments and the conclusion derived from them – that mental content and thus mental states with content were not individualistic – had implications for computational explanations of cognition. These implications were twofold. First, purely computational accounts of the mind, construed individualistically, were inadequate. Second, insofar as such explanations did appeal to a notion of mental content, they would fail to be individualistic. It is the latter of these ideas that Burge pursued in "Individualism and Psychology," in which he argued, strikingly, that Marr's theory of vision was not individualistic. This was the first attempt to explore in detail a widely respected view within cognitive science vis-à-vis the individualism issue, and it was a crucial turning point in moving beyond the cognitive science gesture toward a style of argument that really does utilize empirical practice in cognitive science itself.[9]

What is called "Marr's theory of vision" is an account of a range of processes in early or "low-level" vision that was developed by Marr and colleagues, such as Ellen Hildreth and Tomas Poggio, at the Massachusetts Institute of Technology. These processes include stereopsis, the perception of motion, and shape and surface perception, and the approach is explicitly computational. Marr's *Vision* became the paradigm expression of the approach, particularly for philosophers, something facilitated by

Marr's comfortable blend of computational detail with broad-brushed, programmatic statements of the perspective and implications of his approach to understanding vision. For example, in his first chapter, entitled "The Philosophy and the Approach," Marr recounts the realization that represented a critical breakthrough in the methodology of the study of vision, as follows:

> The message was plain. There must exist an additional level of understanding at which the character of the information-processing tasks carried out during perception are analyzed and understood in a way that is independent of the particular mechanisms and structures that implement them in our heads. This was what was missing – the analysis of the problem as an information-processing task.... if the notion of different types of understanding is taken very seriously, it allows the study of the information-processing basis of perception to be made *rigorous*. It becomes possible, by separating explanations into different levels, to make explicit statements about what is being computed and why and to construct theories stating that what is being computed is optimal in some sense or is guaranteed to function correctly.[10]

Over the last twenty years, work on Marr's theory of vision has continued, extending to cover the processes constituting low-level vision more extensively. By and large, the philosophical literature on individualism that appeals to Marr's theory has been content to rely almost exclusively on Marr's *Vision* in interpreting the theory.

As the passage from Marr quoted above suggests, critical to the computational theory that Marr advocates is a recognition of the different levels at which one can – indeed, for Marr, must – study vision. According to Marr, there are three levels of analysis to pursue in studying any information-processing device. First, there is the level of the computational theory (hereafter, the *computational level*), which specifies the goal of the computation, and at which the device itself is characterized in abstract, formal terms as "mapping from one kind of information to another." Second is the level of representation and algorithm (hereafter, the *algorithmic level*), which selects a "representation for the input and output and the algorithm to be used to transform one into the other." And third is the level of hardware implementation (hereafter, the *implementational level*), which tells us how the representation and algorithm are realized physically in an actual device.[11]

Philosophical discussions, like Marr's own discussions, have been focused on the computational and algorithmic levels for vision, what Marr himself characterizes, respectively, as the "what and why" and "how" questions about vision. As we will see, there is particular controversy over what the computational level involves. In addition to the trichotomy of levels at which an informational-processing analysis proceeds, there are two further interesting dimensions to Marr's approach to vision that have not been widely discussed in the philosophical literature. These add some complexity not only to Marr's theory, but also to the issue of how "computation" and "representation" are to be understood in it.

The first is the idea that visual computations are performed sequentially in stages of computational inference. Marr states that the overall goal of the theory of vision is "to understand how descriptions of the world may efficiently and reliably be obtained from images of it." He views the inferences from intensity changes in the retinal image to fullblown three-dimensional descriptions as proceeding via the construction of a series of preliminary representations: the raw primal sketch, the full primal sketch and the 2 1/2-D sketch. Call this the *temporal dimension* to visual computation. The second idea is that visual processing is subject to modular design, and so particular aspects of the construction of 3-D images – stereopsis, depth, motion, and so on – can be investigated in principle independently. Call this the *modular dimension* to visual computation.[12]

A recognition of the temporal and modular dimensions to visual computation complicates any discussion of what "the" computational and algorithmic levels for "the" process of vision are. Minimally, in identifying each of Marr's three levels, we need first to fix at least the modular dimension to vision in order to analyze a given visual process; and to fix at least the temporal dimension in order to analyze a given visual computation. We will see how these points interact with the debate over Marr's theory shortly.

Burge's argument that Marr's theory is not individualistic is explicitly and fully presented in the following extended passage:

(1) The theory is intentional. (2) The intentional primitives of the theory and the information they carry are individuated by reference to contingently existing physical items or conditions by which they are normally caused and to which they normally apply. (3) So if these physical conditions and, possibly, attendant physical laws were regularly different, the information conveyed to the subject and the intentional content of his or her visual representations would be different. (4) It is not incoherent to conceive of relevantly different (say, optical) laws regularly causing the same non-intentionally, individualistically individuated physical regularities in the subject's eyes and nervous system.... (5) In such a case (by (3)) the individual's visual representations would carry different information and have different representational content, though the person's whole non-intentional physical history... might remain the same. (6) Assuming that some perceptual

states are identified in the theory in terms of their informational or intentional content, it follows that individualism is not true for the theory of vision.[13]

The second and third premise make specific claims about Marr's theory of vision, while the first premise, together with (4) and (5), indicate the affinity between this argument and Burge's original argument for individualism, cast in Twin Earth-like terms, that we discussed in Chapter 4.

Burge concentrates on defending (2)–(4), largely by an appeal to the ways in which Marr appears to rely on "the structure of the real world" in articulating both the computational and algorithmic levels for vision. Marr certainly does make a number of appeals to this structure throughout *Vision*. For example, he says

> The purpose of these representations is to provide useful descriptions of aspects of the real world. The structure of the real world therefore plays an important role in determining both the nature of the representations that are used and the nature of the processes that derive and maintain them. An important part of the theoretical analysis is to make explicit the physical constraints and assumptions that have been used in the design of the representations and processes...

And Marr does claim that the representational primitives in early vision (such as "blobs, lines, edges, groups, and so forth") "correspond to real physical changes on the viewed surface." Together these sorts of comments have been taken to support (2) and (3) in particular.[14]

Marr's appeals to the "structure of the real world" do not themselves, however, imply a commitment to externalism. For these remarks can be, and have been, interpreted differently. Consider two alternative interpretations available to individualists.

The first is to see Marr as giving the real world a role to play only in constructing what he calls the computational theory. Since vision is a process for extracting information from the world in order to allow the organism to act effectively in that world, clearly we need to know something of the structure of the world in our account of what vision is for, what it is that vision does, what function vision is designed to perform. If this is correct, then it seems possible to argue that one does not need to look beyond the head in constructing the theory of the representation and algorithm. As it is at this level that visual states are taxonomized qua the objects of computational mechanisms, Marr's references to the "real world" do not commit him to an externalist view of the taxonomy of visual states and processes.

The second is to take these comments to suggest merely a heuristic role for the structure of the real world, not only in developing a computational taxonomy but also in the computational theory of vision more generally.

That is, turning to the beyond-the-head world is a useful shortcut for understanding how vision works and the nature of visual states and computations. This heuristic is effective either in providing important background information that allows us to understand the representational primitives and thus the earliest stages of the visual computation, or by serving as an interpretative lens that allows us to construct a model of computational processes in terms that are meaningful. Again, as with the previous option, the beyond-the-head world plays only a peripheral role within computational vision, even if Marr at times refers to it prominently in outlining his theory.

Beyond questions of how to interpret Marr's own comments, individualists have objected to Burge's argument in two principal ways. First, Gabriel Segal and Robert Matthews have both in effect denied (2), with Segal arguing that these intentional primitives (such as edges and generalized cones) are better interpreted within the context of Marr's theory as individuated by their narrow content. Second, Frances Egan has more strikingly denied (1), arguing that, qua computational theory, Marr's theory is not intentional at all. Both objections are worth exploring in detail, particularly insofar as they highlight issues that remain contentious in contemporary discussions. In fact, Marr's theory raises more foundational questions than it solves about the nature of the mind and how we should investigate it.[15]

5 SEGAL AND EGAN ON COMPUTATION AND REPRESENTATION

Segal points out that there are two general interpretations available when one seeks to ascribe intentional contents to the visual states of two individuals. First, one could follow Burge and interpret the content of a given visual state in terms of what normally causes it. Thus, if it is a crack in a surface that plays this role, then the content of the corresponding visual state is "crack"; if it is a shadow in the environment that does so, then the content of the visual state is "shadow." This could be so even in the case of *doppelgängers*, and so the visual states so individuated are not individualistic. But second and alternatively, one could offer a more liberal interpretation of the content of the visual states in the two cases, one which was neutral as to the cause of the state, and to which we might give the name "crackdow" to indicate this neutrality. This content would be shared by *doppelängers*, and so would be individualistic.

The crucial part of Segal's argument is his case for preferring the second of these interpretations, and it is here that one would expect to find an appeal to the specifics of Marr's theory of vision. While some of Segal's arguments here do so appeal, Segal also introduces a number of quite general considerations that have little to do with Marr's theory in particular. For example, he points to the second interpretation as having economy on its side, thus appealing to considerations of simplicity, and says,

> The best theoretical description will *always* be one in which the representations fail to specify their extensions at a level that distinguishes the two sorts of distal cause. It will *always* be better to suppose that the extension includes both sorts of thing.[16]

Why always? Segal talks generally of the basic canons of good explanation in support of his case against externalism, but as with the appeals to the nature of scientific explanation that turned on the idea that scientific taxonomy individuates by "causal powers" that we discussed in Chapter 4, here we should be suspicious of the level of generality (and corresponding lack of substantive detail) at which scientific practice is depicted. Like Burge's own appeal to the objectivity of perceptual representation in formulating a general argument for externalism, these sorts of *a priori* appeals seem to me to represent gestural lapses entwined with the more interesting, substantive, empirical arguments over individualism in psychology.

When Segal does draw more explicitly on features of Marr's theory, he extracts three general points that are relevant for his argument that the theory is individualistic: Each attribution of a representation requires a bottom-up account, a top-down motivation, and is checked against behavioral evidence. Together these three points imply that positing representations in Marr's theory does not come cheaply, and indeed is tightly constrained by overall task demands and methods. The first suggests that any higher-level representations posited by the theory must be derived from lower-level input representations; the second that all posited representations derive their motivation from their role in the overall perceptual process; and the third that "intentional contents are inferred from discriminative behavior."[17]

Segal uses the first assumption to argue that since the content of the earliest representations ("up to and including zero-crossings") in *doppelgängers* are the same, there is a prima facie case that downstream, higher-level representations must be the same, unless a top-down motivation can be given for positing a difference. But because we are considering *doppelgängers*, there is no behavioral evidence that could be used to diagnose a representational difference between the two (Segal's third point), and so no top-down motivation is available. As he says, "[t]here would just be no theoretical point in invoking the two contents [of the twins], where one would do. For there would be no theoretical purpose served by distinguishing between the contents."[18]

How might an externalist resist this challenging argument? Three different tacks suggest themselves, each of which grants less to Segal than that which precedes it.

First, one could concede the three points that Segal extracts from his reading of Marr, together with his claim that the lowest levels of representation are individualistic, but question the significance of this. Here one could agree that the gray arrays with which Marr's theory begins do, in a sense, represent light intensity values, and that zero-crossings do, in that same sense, represent a sudden change in the light intensity. But these are both merely representations of some state of the retina, not of the world, and it should be no surprise that such intraorganismic representations have narrow content. Moreover, the depth of the intentionality or "aboutness" of such representations might be called into question precisely because they don't involve any causal relation that extends beyond the head. They might be thought to be representational in much the way that my growling stomach represents my current state of hunger. However, once we move to downstream processes, processes that are later on in the temporal dimension to visual processing, genuinely robust representational primitives come in to play, primitives like "edge" and "generalized cone." And the contents of states deploying these primitives, one might claim, as representations of a state of the world, metaphysically depend on what they correspond to in the world, and so are not individualistic. The plausibility of this response to Segal turns on both the strength of the distinction between a weaker and a stronger sense of "representation" in Marr's theory, and the claim that we need the stronger sense to have states that are representational in some philosophically interesting sense.

Second, and more radically, one could allow that all of the representational primitives posited in the theory represent in the same sense, but challenge the claim that the content of any of the corresponding states is narrow: it is wide content all the way out, if you like. The idea that the representational content of states deploying gray arrays and zero-crossings is in fact wide might itself take its cue from Segal's second point – that representations require a top-down motivation – for it is by reflecting on the point of the overall process of constructing reliable, three-dimensional images of a three-dimensional visual world that we can see that even early retinal representations must be representations of states and conditions in the world. This view would of necessity go beyond Marr's theory itself, which is explicitly concerned only with the computational problem of how we infer three-dimensional images from impoverished retinal information. But it would be very much in the spirit of what we can think of as a Gibsonian aspect to Marr's theory.

Third, and least compromisingly, one could reject one or more of Segal's three points about Marr's theory or, rather, the significance that Segal attaches to these points. Temporally later representations are derived from earlier representations, but this itself doesn't tell us anything about how to individuate the contents of either. Likewise, that Marr himself begins with low-level representations of the retinal image tells us little about whether such representations are narrow or wide. Top-down motivations are needed to justify the postulation of representations, but since there are a range of motivations within Marr's theory concerning the overall point of the process of three-dimensional vision, this also gives us little guidance about whether the content of such representations is narrow or wide. Behavioral evidence does play a role in diagnosing the content of particular representations, but since Marr is not a behaviorist, behavioral discrimination does not provide a litmus test for representational difference.[19]

This third response seems the most plausible to develop in detail, but it also seems to me the one that also implies that there is likely to be no definitive answer to the question of whether Marr's theory employs either a narrow or a wide notion of content, or both, or neither. Marr was not concerned at all himself with the issue of the intentional nature of the primitives of this theory. But the depth of his methodological comments and asides has left us with an embarrassment of riches when it comes to possible interpretations of his theory. This is not simply an indeterminacy about what Marr meant or intended, but reflects a similar indeterminacy within the computational approach to vision itself, and, I think, in computational psychology more generally.

With that in mind I shall turn now to Egan's claim that the theory is not intentional at all. This is a minority view of Marr's theory that has brought some incredulous stares in light of the constant references within Marr's work to notions of representation. How could Egan's claim be squared with such appeals? Of interest in this connection is Chomsky's endorsement of Egan's interpretation of Marr. Speaking of Ullman's studies of the determination of structure from motion within a broadly Marrian framework, Chomsky says

The account is completely internalist. There is no meaningful question about the 'content' of the internal representations of a person seeing a cube under the conditions of the experiments, or if the retina is stimulated by a rotating cube, or by a video of a rotating cube; or about the content of a frog's 'representation of' a fly or of a moving dot in the standard experimental studies of frog vision. No notion like 'content' or 'representation of' figures within the theory, so there are no answers to be given as to their nature.[20]

On Chomsky's view, the nonintentional interpretation of such theories of vision is made plausible by the claim that "content" and "representation of" are terms that are neither defined within such theories nor with a clear meaning within them. The theories are nonintentional in the relevant sense, and thus it is a moot question whether they operate with a narrow or a wide notion of content. As Chomsky continues,

> The same is true when Marr writes that he is studying vision as a 'mapping from one representation to another, and in the case of human vision, the initial representation is in no doubt – it consists of arrays of image intensity values as detected by the photoreceptors in the retina' (Marr 1982:31) – where 'representation is not to be understood relationally, as representation of.'

We will return to these points in assessing Egan's interpretation below.[21]

At the heart of Egan's view of Marr is a particular view of the nature of Marr's computational level of description. Commentators on Marr have almost universally taken this to correspond to what others have called the "knowledge level" or the "semantic level" of description, that is, as offering an intentional characterization of the computational mechanisms governing vision and other cognitive processes. Rather than ignoring Marr's computational level, Egan rejects this dominant understanding of the computational level, arguing instead that what makes it a computational level is that it specifies the function to be computed by a given algorithm in precise, mathematical terms. That is, while this level of description is functional, what makes it the first stage in constructing a *computational* theory is that it offers a function-theoretic characterization of the computation, and thus abstracts away from all other functional characterizations.[22]

Thus, while vision might have all sorts of functions that can be specified in language relatively close to that of common sense – it is for extracting information from the world, for perceiving an objective world, for guiding behavior – none of these, in Egan's view, form a part of Marr's computational level of description. Given this view, the case for Marr's theory being individualistic because computational follows readily:

> A computational theory prescinds from the actual environment because it aims to provide an abstract, and hence completely general, description of a mechanism

that affords a basis for predicting and explaining its behavior in any environment, even in environments where what the device is doing cannot comfortably be described as *cognition*. When the computational characterization is accompanied by an appropriate intentional interpretation, we can see how a mechanism that computes a particular mathematical function can, in a particular context, subserve a cognitive function such as vision.[23]

According to Egan, while an intentional interpretation links the computational theory to our common sense–based understanding of cognitive functions, it forms no part of the computational theory itself. Egan's view naturally raises questions not only about what Marr meant by the computational level of description but more generally about the nature of computational approaches to cognition.

There are certainly places in which Marr does talk of the computational level as simply being a high-level functional characterization of what vision is for, and thus as simply orienting the researcher to pose certain general questions. For example, one of his tables offers the following summary questions that the theory answers at the computational level: "What is the goal of the computation, why is it appropriate, and what is the logic of the strategy by which it can be carried out?" Those defending the claim that Marr's theory is externalist have typically rested with this broad and somewhat loose understanding of the computational level of the theory.[24]

The problem with this broad understanding of the computational level, and thus of computational approaches to cognition, is that while it builds a bridge between computational psychology and more folksy ways of thinking about cognition, it also creates a gap within the computational approach between the computation level and the algorithmic level. For example, if we suppose that the computational level specifies simply that some visual states have the function of representing edges, others the function of representing shapes, and so on, there is nothing about such descriptions that guides us in constructing algorithms that generate the state-to-state transitions at the heart of computational approaches to vision. More informal elaboration of what vision is for, or of what it evolved to do, do little by themselves to bridge this gap.

The point here is that computational specifications themselves are a very special kind of functional characterization, at least when they are to be completed or implemented in automatic, algorithmic processes. Minimally, proponents of the broad interpretation of computational approaches to cognition need either to construe the computational level as encompassing but going beyond the function-theoretic characterizations of cognitive capacities that Egan identifies, or they must allocate those characterizations to the algorithmic level. The latter option simply exacerbates the "gap" problem identified above. But the former option lumps together a variety of quite different things under the heading of "the computational level," and subsequently fails to recognize the constraints that computational assumptions bring in their wake. The temporal and modular dimensions to Marr's theory exacerbate the problem here.

There is a large issue lurking here concerning how functionalism should be understood within computational approaches to cognition, and correspondingly how encompassing such approaches really are. Functionalism has usually been understood as offering a way to reconcile our folk psychology, our manifest image of the mind, with the developing sciences of the mind, even if that reconciliation involves revising folk psychology along individualistic lines. And computationalism has been taken to be one way of specifying what the relevant functional roles are: They are "computational roles." But suppose that Egan is right about Marr's understanding of the notion of computation as a function-theoretic notion, and we accept the view that this understanding is shared in computational approaches to cognition more generally. Then the corresponding version of functionalism about the mind must be function-theoretic in Egan's sense: It will "prescind from the actual environment," as must the computational level, but also from the sort of internal causal role that functionalists have often appealed to. Cognitive mechanisms, on this view, take mathematically characterizable inputs to deliver mathematically characterizable outputs, and qua computational devices, that is all. Any prospects for the consilience of our "two images" must lie elsewhere.

In arguing for the nonintentional character of Marr's theory of vision, Egan presents an austere picture of the heart of computational psychology, one which accords with the individualistic orientation of computational cognitive science as it has traditionally been developed, even if computational psychologists have sometimes attempted to place their theories within more encompassing contexts. We have seen that Chomsky shares this austere view of at least Marr's theory in suggesting that "content" and "representation of" are not terms that the theory traffics in. How plausible is such a view of computational psychology?

One general problem, as Larry Shapiro points out, is that a computational theory of X tells us very little about the nature of X, including information sufficient to individuate X as (say) a visual process at all. While Egan seems willing to accept this conclusion, placing this sort of concern outside of computational theory proper, this response highlights a gap between computational theory, austerely construed, and the myriad theories – representational, functional, or ecological in nature – with which such a theory must be integrated for it to constitute a complete, mechanistic account of any given cognitive process. The more austere the account of computation, the larger this gap becomes, and the less a computational theory contributes to our understanding of cognition. One might well think that Egan's view of computational theory in psychology errs on the side of being too austere in this respect.[25]

We can relate this more directly to Chomsky's views here. If Marr's theory in particular, or computational theories of cognitive abilities more generally, do not contain an account of the content of (say) visual states, or what those states are representations of, then surely they leave out something critical about perception or cognition more generally. Part of the promise of computational approaches to cognition has been to show how computation and representation go hand-in-hand. But on the Egan-Chomsky view, it now seems that "representation" itself is a term of art within such theories with no connection either to intentionality or to the underlying, algorithmic level. The promise cannot be fulfilled.

Making Egan's view less austere in a way that would bridge the gap between computational and various other levels (both "higher" and "lower") would likely require one of two things. First, incorporating either part of what Egan thinks of as the intentional interpretation of the computational theory within that theory. Or, second, offering a richer conception of the mechanisms specified at the computational level that compromise their context independence. Either way, addressing this problem removes the bases for Egan's argument for an individualistic view of computational psychology.

As I shall make clear in the next two sections, I share the view that there may be no fact of the matter about whether Marr's theory employs a notion of narrow or wide content. Thus, I am sympathetic to Egan's development of an interpretation of Marr's framework that bypasses this issue. But the austere conception of computation shared by Egan and Chomsky seems to me an interpretation with too high a price.

6 EXPLOITATIVE REPRESENTATION AND WIDE COMPUTATIONALISM

As a beginning on an alternative way of thinking about computation and representation, consider an interesting difference between individualistic and externalist interpretations of Marr's theory that concerns what it is that Marrian computational systems have built into them. Individualists about computation, such as Egan and Segal, hold that they incorporate various innate assumptions about what the world is like. This is because the process of vision involves recovering 3-D information from a 2-D retinal image, a process that without further input would be underdetermined. The only way to solve this underdetermination problem is to make innate assumptions about the world. The best known of these is Ullman's rigidity assumption, which says that "any set of elements undergoing a two-dimensional transformation has a unique interpretation as a rigid body moving in space and hence should be interpreted as such a body in motion." The claim that individualists make is that assumptions like this are part of the computational systems that drive cognitive processing. This is the standard way to understand Marr's approach to vision.[26]

Externalists like Shapiro have construed this matter differently. Although certain assumptions must be true of the world in order for our computational mechanisms to solve the underdetermination problem, these are simply assumptions that are *exploited* by our computational mechanisms, rather than innate in our cognitive architecture. That is, the assumptions concern the relationships between features of the external world, or between properties of the internal, visual array and properties of the external world, but those assumptions are not themselves encoded in the organism. To bring out the contrast between these two views, consider a few simple examples.[27]

An odometer keeps track of how many miles a car has traveled, and it does so by recording the number of wheel rotations and being built to display a number proportional to this number. One way it could do this would be for the assumption that 1 rotation = x meters to be part of its calculational machinery. If it were built this way, then it would plug the value of its recording into an equation representing this assumption, and compute the result. Another way of achieving the end would be to be built simply to record x meters for every rotation, thus exploiting the fact that 1 rotation = x meters. In the first case it encodes a representational assumption, and uses this to compute its output. In the second, it contains no such encoding but instead uses an existing relationship between its structure and the structure of the world, in much the way that a polar planimeter measures the area of closed spaces of arbitrary shapes without doing any representation crunching. Note that however distance traveled is measured, if an odometer finds itself in an environment in which the relationship between rotations to distance traveled is adjusted – larger wheels, or being driven on a treadmill – it will not function as it is supposed to, and misrepresent that distance.[28]

Consider two different (unconscious) strategies for learning how to hit a baseball that is falling vertically to the ground. Since the ball accelerates at 9.8 ms², there is a time lag between swinging and hitting. One could either assume that the ball is falling (say, at a specific rate of acceleration), and then use this assumption to calculate when one should swing; alternatively, one could simply aim a certain distance below where one perceives the ball at the time of swinging (say, two feet). In this latter case one would be exploiting the relationship between acceleration, time, and distance without having to encode that relationship in the assumptions one brings to bear on the task.

Exploitative representation is an efficient form of representation when there is a constant, reliable, causal or informational relationship between what a device does and how the world is. Thus, rather than encode the structure of the world and then manipulate those encodings, "smart mechanisms" can exploit that constancy. As the odometer example suggests, the encoding view also presupposes some mind-world constancy, but this is presumed only for "input" representations to start the computational process on the right track. Exploitative representation makes a deeper use of mind-world constancies.

The fact that there are these two different strategies for accomplishing the same end should, minimally, make us wary of accepting the claim that innate assumptions are the only way that a computational system could solve the underdetermination problem. But I also want to develop the idea that our perceptual system in particular and our cognitive systems more generally typically exploit rather than encode information about the world and our relationship to it, as well as say something about where Marr himself seems to stand on this issue.

An assumption that Egan makes and that is widely shared in the philosophical literatures both on individualism and computation is that at least the algorithmic level of description within computational psychology is individualistic. The idea here has, I think, seemed so obvious that it has seldom been spelled out: Algorithms operate on the syntactic or formal properties of symbols, and these are intrinsic to the organisms instantiating the symbols. We might challenge this neither by disputing how much is built into Marr's computational level, nor by squabbling over the line between Marr's computational and algorithmic levels, but, rather, by arguing that computations themselves can extend beyond the head of the organism and involve the relations between individuals and their environments. This position, wide computationalism, holds that at least some of the computational systems that drive cognition reach beyond the limits of the organismic boundary. Its application to Marr's theory of vision marks a departure from the parameters governing the standard individualist-externalist debate over that theory. Wide computationalism constitutes one way of thinking about the way in which cognition, even considered computationally, is "embedded" or "situated" in its nature, and it provides a framework within which an exploitative conception of representation can be pursued.

The basic idea of wide computationalism is simple. Traditionally, the sorts of computation that govern cognition have been thought to begin and end at the skull. But why think that the skull constitutes a magic boundary beyond which true computation ends and mere causation begins? Given that we are creatures embedded in informationally rich and complex environments, the computations that occur inside the head are an important part but are not exhaustive of the corresponding computational systems. This perspective opens up the possibility of exploring computational units that include the brain as well as aspects of the brain's beyond-the-head environment. Wide computational systems thus involve minds that literally extend beyond the confines of the skull into the world. In the terms introduced earlier in Part Two, they have wide realizations (see Figures 7.1 and 7.2).

Standard Computationalism

An example: Multiplying with only internal symbols

Computational system ends at the skull; computation must be entirely in the head.

1. Code external world.

2. Model computations between internal representations only.

3. Explain behavior, based on outputs from Step 2.

FIGURE 7.1. Standard Computationalism

An example: Multiplying with
internal and external symbols

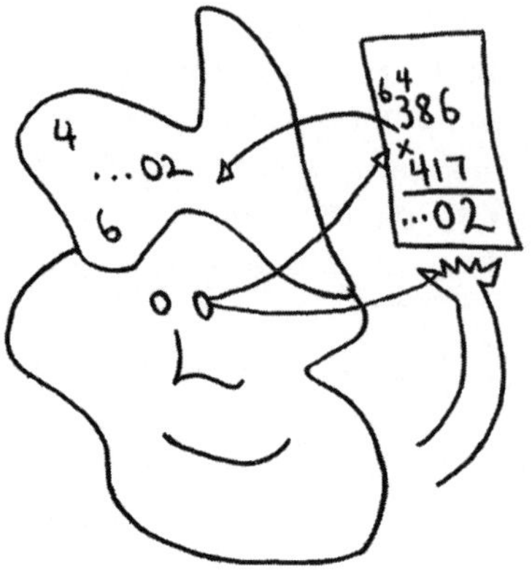

Wide Computationalism

Computational system can extend beyond the skin into the world; computation may not be entirely in the head.

1. Identify representational or informational forms – whether in the head or not – that constitute the relevant computational system.

2. Model computations between these representations.

3. Behavior itself may be part of the wide computational system.

FIGURE 7.2. Wide Computationalism

One way to bring out the nature of the departure made by wide computationalism draws on a distinction between locational and taxonomic conceptions of psychological states. Individualists and externalists are usually presented as disagreeing over how to taxonomize or individuate psychological states, but both typically presume that the relevant states are *locationally individualistic*: They are located within the organismic envelope. What individualists and externalists typically disagree about is whether in addition to being locationally individualistic, psychological states must also be taxonomically individualistic. This is, as we have seen, what is usually at issue in the debate over Marr's theory of vision, where the focus has been on whether Marr uses a wide or a narrow notion of content. Wide computationalism, however, rejects this assumption of locational individualism by claiming that some of the "relevant states" – some of those that constitute the relevant computational system – are located not in the individual's head but in her environment.

If some cognitive systems, wide computational systems, are not locationally individualistic, then they, and thus the states that constitute them, are not taxonomically individualistic. That is, locational width entails taxonomic width. This is because two individuals could instantiate different wide computational systems simply by virtue of differences in the beyond-the-head part of their cognitive systems. Again, the framework introduced in Chapters 5 and 6 makes this claim easy to state: Total realizations of wide computational systems differing only in their noncore parts, or that are radically wide, could instantiate the cognitive systems of individuals who were molecularly identical.

The intuitive idea behind wide computationalism is easy enough to grasp. But there are two controversial claims central to defending wide computationalism as a viable model for thinking about and studying cognitive processing. The first is that it is sometimes appropriate to offer a formal or computational characterization of an organism's environment, and to view parts of the brain of the organism, computationally characterized, together with this environment so characterized, as constituting a unified computational system. Without this being true, it is difficult to see wide computationalism as a coherent view. The second is that this resulting mind-world computational system itself, and not just the part of it inside the head, is genuinely cognitive. Without this second claim, wide computationalism would at best present a zany way of carving up the computational world, one without obvious implications for how we should think about real cognition in real heads. I take each claim in turn, with specific reference to Marr's theory of vision, and with an eye to highlighting some of the broader issues that arise in thinking about individualism and cognitive science, including how we think about mental representation.

Offering a formal or computational characterization of an organism's environment, if it is to capture important aspects of the structure and dynamics of that environment, is neither trivial nor easy. Or, to put it in terms that help to locate whatever mystery there is to this claim within perhaps more familiar territory for those working in the cognitive sciences, doing so is no more and no less trivial than doing so for psychological states themselves.

To construct a computational model of an internal, psychological process, one postulates a set of primitive states, $S_1 \ldots S_n$, and then formulates transition rules that govern changes between these states, as well as some set of initial states to which the transition rules apply in the first instance. The computational model's adequacy is proportional to how closely its primitives, transition rules, and starting point(s) parallel aspects of the corresponding cognitive system being modeled. The crucial assumption in computational modeling is that causal transitions between physical states can be represented as inferential transitions between computational states.

This view of what a computational system of mental states is has been elaborated by Rob Cummins as the "Tower-Bridge" picture of computation, and is, I think, implicit in Egan's function-theoretic conception of computational psychology. It applies not only to "classic" models of computational cognition but also to connectionist models, at least those in which there remain notions of computation and representation. In the former, the computational states tend themselves to be rich in structure and thus differentiated, and often approximate everyday concepts, with the transition rules serving both to "unpack" the complexity to these symbols (for example, "dog" –> "animal") and the relations between symbols more generally. In the latter, the computational states tend to be simple, relatively undifferentiated, and they do not correspond readily to everyday concepts. Here the transition rules are connection strengths between the computational states (the "nodes"), and the dynamics of the system is governed primarily by these together with the initial layering of the nodes.[29]

This basic idea is general enough that it applies not only to cognitive systems but in principle to any type of system whose structure and dynamics we wish to model. This is a desirable feature, since (i) the notion of computation that we apply to cognitive systems should not be *sui generis*, applying only to such systems; and (ii) a variety of noncognitive systems – from planets to ecosystems to colonies of social insects to intracellular biological systems – can be computationally modeled. What marks off cognition as special here is not the notion of computation that it employs but the idea that the cognitive system itself and its components are themselves computing, and not just being computationally modeled. This idea itself is manifest in talk of "neural computing," "single-neuron computation," and the "computational brain," as well as in the Ur-idea of the "mind as computer." If I have correctly characterized the idea of computational modeling, however, then the idea that cognition is different in kind from all (or even the vast majority of) other domains to which computational modeling is applied – that is, that, plus or minus a bit, only cognition is genuinely computational in and of itself – involves mistaking the features of the model for the features of what is modeled.

Given this notion of computation, the idea that there can be computational systems that involve the nonbrain part of the world is trivial. Less trivial is the claim that the brain *plus parts of the nonbrain part of the world* together can constitute a computational system, a locationally wide computational system, since that rests on there being a robust, structured causal relationship between what is in the head and what is outside of it that can be adequately captured by transition rules. As I said in Chapter 5, processes in the brain, or more generally the organism, that have evolved via world-mind dependencies are good candidates for forming parts of wide computational systems, assuming that the nonorganismic contribution to this system itself has some sort of rich causal structure that admits a formal characterization.

Perceptual processing is a good candidate place to look for such systems. A lens that transforms the light that passes through it does not itself compute the spatial frequency of the resulting image. But there is a causal process whose inputs (target object) and outputs (resulting image) we can characterize formally (in terms of spatial frequency), and that involves the lens as a mediating causal mechanism. This mechanism, and those that it feeds, can exploitatively represent objects in the world. The computation here is wide, since the computational states extend beyond the boundary of the relevant individual, the lens. The same is true if that lens happens to form part of somebody's eye.

How might this apply to Marr's theory of vision? As we have seen, Marr himself construes the task of a theory of vision to show how we extract visual information from "arrays of image intensity values as detected by the photoreceptors in the retina." Thus, as we have already noted, for Marr the problem of vision begins with retinal images, not with properties of the world beyond those images, and "the true heart of visual perception is the inference from the structure of an image about the structure of the real world outside." Marr goes on to characterize a range of physical constraints that hold true of the world that make this inference possible, but he makes it clear that "the constraints are used by turning them into an assumption that may or may not be internally verifiable." For all of Marr's talk of the importance of facts about the beyond-the-head world for constructing the computational level in a theory of vision, this is representative of how Marr conceives of that relevance. It seems to me clear that, in terms that I introduced in the previous sections, Marr himself adopts an encoding view of computation and representation, rather than an exploitative view of the two. The visual system is, according to Marr, a locationally individualistic system.[30]

Whatever Marr's own views here, the obvious way to defend a wide computational interpretation of his theory is to resist his inference from "x is a physical constraint holding in the world" to "x is an assumption that is encoded in the brain." This is, in essence, what I have previously proposed one should do in the case of the multiple spatial channels theory of form perception. Like Marr's theory of vision, which in part builds on this work, this theory has usually been understood as postulating a locationally individualistic computational system, one that begins with channels early in the visual pathway that are differentially sensitive to four parameters: orientation, spatial frequency, contrast, and spatial phase. My suggestion was to take seriously the claim that any visual scene (in the world) can be decomposed into these four properties, and so see the computational system itself as extending into the world, with the causal relationship between stimulus and visual channels itself modeled by transition rules. Rather than simply having these properties encoded in distinct visual channels in the nervous system, view the in-the-head part of the form perception system as exploiting formal properties in the world beyond the head. In Marr's theory, there is one respect in which this wide computational interpretation is easy to defend, and one respect in which it is difficult to defend.[31]

The first of these is that Marr's "assumptions," such as the spatial coincidence assumption and the "fundamental assumption of stereopsis," typically begin as physical constraints that reflect the structure of the world; in the above examples, they begin as the constraint of spatial localization and three matching constraints. Thus, the strategy is to argue that the constraints themselves, rather than their derivative encoding, play a role in defining the computational system, rather than simply filling a heuristic role in allowing us to offer a computational characterization of a locationally individualistic cognitive system.[32]

The corresponding respect in which a wide computational interpretation of Marr's theory is difficult to defend is that these constraints themselves do not specify what the computational primitives are. One possibility would simply be to attribute the primitives that Marr ascribes to the image to features of the scenes perceived themselves, but this would be too quick. For example, Marr considers zero-crossings to be steps in a computation that represent sharp charges in intensity in the image, and while we could take them to represent intensity changes in the stimuli in the world, zero-crossings themselves are located somewhere early in the in-the-head part of the visual system, probably close to the retina.

A better strategy, I think, would be to deflate the interpretation of the retinal image and look "upstream" from it to identify richer external structures in the world, structures, which satisfy the physical constraints that Marr postulates. That is, one should extend the temporal dimension to Marr's theory so that the earliest stages in basic visual processes begin in the world, not in the head. Since the study of vision has been largely conducted within an overarching individualistic framework, this strategy would require recasting the theory of vision itself so that it ranges over a process that causally extends beyond the retinal image.

Mark Rowlands has contrasted Marr's approach, beginning with the retinal image in his analysis of vision with the ecological approach of James J. Gibson, which begins with information contained in what Gibson called the ambient optical array. Although Rowlands locates the views of Marr and Gibson on a continuum, his chief aim in drawing the contrast is to argue for a view of perception closer to Gibson's end of the continuum than to Marr's. My idea here is somewhat different: to suggest that we can augment Marr's computational view by something like Gibson's view of what the starting point for an analysis of vision is. The importance of the idea that the ambient optical array, as an information-bearing structure external to the organism, is the appropriate starting point for this analysis, as Rowlands argues, is that doing so enriches the information available for visual processing and so reduces the complexity that we need to attribute to the organism itself in accounts of vision. The ambient optical array is part of the locationally wide computational system that vision takes place in.[33]

Once we take this step, then the interaction between information-processing structures inside organisms and information-bearing states outside of them becomes central to a computational account of vision. And there seems a second Gibsonian insight that can direct research here: The idea that vision is exploratory, and that crucial to that exploration is the movement of the organism through the ambient optical array. Rowlands thinks of this as an organism manipulating its environment to extract information from it, whereas it seems to me that it is more fruitful to see this as an organism manipulating itself, in particular, its body and its parts, for this purpose. This is one way in which vision is animate, a point to which I shall return and a second respect in which we can extend the temporal dimension to Marr's theory. Visual inputs are not simply snapshots that are then (somehow) assembled and bound together within the organism in generating complete visual scenes. Rather, they are complete visual scenes sampled through bodily explorations of the ambient optical array over time.

If we view these aspects of Gibson's views as extensions of the temporal dimension to Marr's theory, then we might also wonder about revisiting the modular dimension to Marr's theory. As I said near the beginning of section 4, a second complexity to Marr's theory is the assumption that visual computations are highly modular, such that features of the final visual image, such as depth and motion, are computed independently. But if the information available to wide computational systems is enriched in the temporal dimension relative to that available to their narrow counterparts, then surely there is less need to assume that vision is as modular as Marr himself assumed.

The second task – of showing that wide computational systems themselves, and not just their in-the-head components, are cognitive systems – is perhaps better undertaken once we have a range of examples of wide computational systems before us. Before turning to that, now that we have explored the debate over Marr's theory in some detail, I want to return to the idea of narrow content.

7 NARROW CONTENT AND MARR'S THEORY

Consider the very first move in Segal's argument for the conclusion that Marr's theory of vision is individualistic: The innocuous-looking claim that there are two general interpretations available when one seeks to ascribe intentional contents to the visual states of two individuals, one "restrictive" (Burge's) and one "liberal" (Segal's). In introducing the distinction between narrow content and wide content in Chapter 4, I indicated that something like these two general alternatives were implicit in the basic Twin Earth cases with which we – and the debate over individualism – began. There I also said that the resulting idea – that twins must share some intentional state about watery substances (or about arthritislike diseases, in Burge's standard case) – is the basis for attempts to articulate a notion of narrow content, that is, intentional content that does supervene on the intrinsic, physical properties of the individual.

The presupposition of a liberal interpretation for Marr's theory, and a corresponding view of the original Twin Earth cases in general, are themselves questionable. Note first that the representations that we might, in order to make their disjunctive content perspicuous, label "crackdow" or "water or twater," do represent their reliable, environmental causes: "Crackdow" is reliably caused by cracks or shadows and has the content crack or shadow; similarly for "water or twater." But then this disjunctive content is a species of wide, not narrow, content. In short, although being shared by twins is necessary for mental content to be narrow, this is not sufficient for narrow content.

To press further, if the content of one's visual state is to be individualistic, it must be shared by *doppelgängers* no matter how different their environments. Thus, the case of twins is merely a heuristic for thinking about a potentially infinite number of individuals. But then the focus on a content shared by two individuals, and thus on a content that is neutral between two environmental causes, represents a misleading simplification insofar as the content needed won't simply be "crackdow" but something more wildly disjunctive. This is because there is a potentially infinite number of environments that might produce the same intrinsic, physical state of the individual's visual system as (say) cracks do in the actual world. It is not that we can't simply make up a name for the content of such a state – we can: Call it "X." – Rather, it is that it is difficult to view a state so individuated as being about anything. And if being about something is at the heart of being intentional, then this calls into question the status of such narrowly individuated states as intentional states.

Segal has claimed that the narrow content of "crackdow," or by implication "water or twater," need not be disjunctive, just simply more encompassing than, respectively, crack or water. But casting the above points in terms of disjunctive content simply makes vivid the general problems that (1) individuation of states in terms of their content still proceeds via reference to what does or would cause them to be tokened; and (2) once one prescinds from a conception of the cognitive system as embedded in and interacting with the actual world in thinking about how to taxonomize its states, it becomes difficult to delineate clearly those states as intentional states with some definite content. As it is sometimes put, narrow content becomes inexpressible. Two responses might be made to this second objection.[34]

First, one might concede that, strictly speaking, narrow content is inexpressible, but then point out ways of sneaking up on it. One might do so by talking of how one can "anchor" narrow content to wide content, or of how to specify the realization conditions for a proposition. But these suggestions, despite their currency, seem to me little more than whistling in the dark, and the concession on which they rest, fatal. All of the ways of "sneaking up on" narrow content involve using wide contents in some way. Yet if wide content is such a problematic notion (because it is not individualistic), then surely the problem spreads to any notion, such as snuck-up-on narrow content, for whose intelligibility it is crucial.[35]

If narrow content really is inexpressible, then the idea that it is this notion that is central to psychological explanation as it is actually practiced, and this notion that does or will feature in the natural kinds and laws of the cognitive sciences, cannot reasonably be sustained. Except in Douglas Adamesque spoofs of science, there are no sciences whose central explanatory constructs are inexpressible. Moreover, this view would make the claim that one arrives at the notion of narrow content via an examination of actual explanatory practice in the cognitive sciences extremely implausible, since if narrow content is inexpressible, then one won't be able to find it expressed in any existing psychological theory. In short, the idea that snuck-up-on narrow content is what cognitive science needs or uses constitutes a lapse back to the cognitive science gesture.

Second, it could be claimed that although it is true that it is difficult for common-sense folk to come up with labels for intentional contents, those in the relevant cognitive sciences can and do all the time, and we should defer to them. For example, one might claim that many if not all of the representational primitives in Marr's theory, such as blob, edge, and line, have narrow contents. These concepts, like many scientific terms, are technical and, as such, may bear no obvious relationship to the concepts and terms of common sense, but they still allow us to see how narrow content can be expressed. One might think that this response has the same question-begging feel to it as does the claim that our folk psychological states are themselves narrow. However, the underdetermination of philosophical views by the data of the scientific theories, such as Marr's, that they interpret remains a problem for both individualists and externalists alike here. As my discussion of exploitative representation and wide computation perhaps suggests, my own view is that we need to reinvigorate the ways in which the computational and representational theories of mind have usually been construed within cognitive science. If this can be done in more than a gestural manner, then the issue of the (in)expressibility of narrow content will be largely moot.

8 LOCATIONAL VERSUS TAXONOMIC EXTERNALISM

Our extended treatment of Marr's theory of vision over the previous four sections has distinguished two different ways of thinking about externalism that make contact with computational approaches to cognition: taxonomic and locational externalism. A cluster of related paradigms within contemporary cognitive science that go under various names – situated cognition, embedded cognition, distributed cognition, improvisational cognition – can be understood in terms of either or both of these forms of externalism. In this section, I shall discuss specific accounts of cognition within these paradigms. In particular, I shall argue that these accounts of cognition posit wide computational systems, and thus are committed to the idea that cognition is locationally externalist.

Coincident with the rise of connectionist models of cognition in which cognitive representations can be distributed across a number of nodes, rather than localized in few of them, has been the idea that cognition can be distributed either across agents or across agents and their environments. In the former case, cognition is construed interpersonally, involving social and cultural mediation between people, while in the latter case it is seen as extending beyond the individual into her environment. Since the relevant part of the environment is often a cultural product, such as a cognitive artifact, in both cases there are cultural and social dimensions that are integral to cognition.

Edwin Hutchins's work on seafaring navigation exemplifies a distributed approach to cognition of both types. While Hutchins is critical of some of the turns that the computational approach to cognition has taken, his own views have been developed within an overarching computational framework, one with interpersonal and artifactual dimensions. Hutchins argues that technology should not be thought of simply as a way of augmenting individual cognitive capacities, however, but as a means of changing the nature of the representational spaces or media in which computations are performed. Computations are not simply performed in the head of the individual, but occur in interpersonal activities that make use of recent (and not so recent) cognitive artifacts. The locus of computation is in the beyond-the-individual world.

Hutchins brings out the interpersonal dimension to the cognitive processes involved in navigation when he says, in his introduction to *Cognition in the Wild*, that he hopes

> to show that human cognition is not just influenced by culture and society, but that it is in a very fundamental sense a cultural and social process. To do this I will move the boundaries of the cognitive unit of analysis out beyond the skin of the individual person and treat the navigation team as a cognitive and computational system.[36]

Here the unit of cognition is not the individual but the navigation team of which the individual is a part. Both the social organization of the team and the relationships between its members and various cognitive artifacts (for example, the alidade, the bearing log, phone circuits, the hoey, the chart, the fathometer) serve as the cognitive architecture of this larger cognitive unit. For example, consider the fix cycle, which plots the position of the ship (output) given two lines of position as inputs. This cycle involves the generation, transformation, and utilization of representations in many individuals and many cognitive artifacts. There is no one place, in particular, no one individual, in which this process is implemented.

Hutchins's study of navigation does tell us a lot about the social distribution of cognitive tasks in navigation, showing how modern navigation involves exploiting social structures and relationships. And perhaps this even provides an example of group-level cognition, a topic we will discuss in detail in Part Four. But Hutchins also highlights how individual-level cognition often extends beyond the boundary of the individual. For example, he does so in emphasizing how integral the use of representational artifacts – from calendars, to maps, to alidades, to hoeys – are not only to team-level cognitive tasks, but also to component, individual-level cognitive tasks – such as reading the alidade, writing an entry on the bearing log, or adjusting the hoey. Such cognitive processes are realized in wide computational systems, where these systems constitute ways of extending an individual's cognitive capacities. Navigation involves a wide range of locationally externalist cognitive processes.

A second example that exemplifies this perspective is Dana Ballard's work on animate vision in general and on the role of deictic coding in cognition in particular. Here the central idea is that cognition often involves rapid updates of information from one's environment in an unplanned or improvised manner, where this improvisational cognition involves adjusting one's body, particularly through eye movements, to take advantage of information that is stored in the environment and thus does not need to be computed or stored by the individual. For example, vision involves intense, repeated causal interactions with an environment via rapid saccadic fixations, which shifts the burden of the computational load from inside the head to the head-world. Inside-the-head representations are computationally expensive to construct and to maintain, hence the idea that they are constructed only when necessary, with the representational slack being taken up by interaction with the world.

Here we have an instance of Andy Clark's previously discussed 007 principle – know only as much as you need to know to get the job done, where to "know" something is to internally represent it in some way. Exploitative representation reduces the informational load that individuals bear through exploiting world-mind regularities, and in the case of animate vision this exploitation makes use not simply of such regularities but the ease with which our bodies can be adjusted to utilize them.

The central novelty of Ballard's approach to vision is to combine the concept that looking is a form of doing with the claim that vision is computation. As we saw in section 6, although the ecological psychologist James J. Gibson recognized the first of these points long ago, his views were developed in direct opposition to the idea that vision was inferential or computational. Ballard integrates these two points by introducing the idea that eye movements constitute a form of deictic coding in that they are behaviors that orient visual attention and fixation in ways that allow perceivers to exploit the world as an external storage device. Eye movements are a type of pointing device, a type of "doing-it-where-I'm-looking" strategy, that means that perceivers need not copy all of the information in a scene in order to use it in guiding further action.[37]

Ballard applies this model of cognition not only to vision but also to attention, memory, and action. Despite the name "deictic *coding*," it represents an approach to cognition that involves exploiting rather than encoding beyond-the-head environments. Here too, I would suggest that the idea of a wide computational system, and of cognitive processes being locationally wide, aptly describe the program of animate vision.

To take a third and final example, consider the "intelligence without reason" approach to robotics championed by Rodney Brooks that concentrates on developing behavior-based systems. Brooks makes no attempt to model how people behave, and is explicit that he is not engaged in any sort of "cognitive modeling" that captures (even in part) the cognitive processes that people instantiate. His goal is to build robots, which he calls "Creatures," that behave intelligently in natural environments, rather than to mimic human performance on some specific task (for example, decision making, block moving, question answering) in an artificially restricted domain (for example, chess, the blocks worlds, restaurant scripts, respectively). This is achieved by a subsumption architecture that builds more complex layers on top of less complex layers, with each layer achieving some action-centered goal, such as avoiding objects, wandering, or exploring its environment. Layers themselves are composed of simple finite-state machines that achieve specific behaviors.

One of Brooks' recurring themes is that *the world is its own best model*, and so there is no attempt to have his Creatures encode features of the world in order to negotiate successfully in it; rather, they exploit those features. But his Creatures are not relying on symbolic aspects of their environments or external representations. Indeed, Brooks has claimed that the notion of representation itself is not readily applicable to his Creatures at all. Like the claim of some early connectionists that they had bypassed the notion of representation altogether, this claim seems to me hyperbole, especially given that these are computationally driven processes. Given that we admit no computation without representation, then either his creatures are a combination of reaction (world-creature) and computation (in the creature), or they are computational all the way out. Either way, we have a locationally externalist view of processing, and in the latter case, wide computationalism.[38]

We can now, finally, return to the question of whether wide computational systems themselves – as opposed simply to their in-the-head realizations – are properly thought of as cognitive. Suppose that one grants locational externalism about computational systems. Might one still maintain individualism about cognition by conceptualizing the cognitive system proper as ending at the boundary of the individual? Such a view rests on the plausibility of identifying what is internally realized as always being the appropriate cognitive system. Consider the navigation systems that Hutchins discusses. For individualism to be defensible as a view of cognition in this case, there must be an in-the-head account of all of the relevant cognitive processes; likewise for the case of animate vision and behavior-based robotic systems. But recall that all three cases share the strategy of pointing to individual-world interactions as a way to avoid positing more costly internal representations and computations. Thus, this individualistic move is unmotivated and runs counter to the reconceptualization of representation within these paradigms. Representation is not something implanted in individuals but something that individuals do by exploiting the rich structures of their environments in cycles of perception and action.

9 HAVING IT BOTH WAYS?

Two strands to the discussion in this chapter might be thought to be in tension with one another. The first stems from my claim that once one moves from the cognitive science gesture to a more full-fledged examination of theory and explanation in the cognitive sciences themselves, the debate between individualists and externalists becomes tougher to resolve. This debate turns not only on how one interprets particular theories but also on broader issues in the philosophy of mind and science. Thus, I have been arguing not only for the conclusion that the cognitive science gesture itself far from resolves the debate in favor of individualism, but that going beyond that gesture leaves much in the debate open. But, second, in offering a wide computational interpretation of Marr's theory of vision in section 6 and in linking this interpretation to a range of contemporary work on situated and embedded cognition, I have argued for a resolution of the debate in favor of externalism. Surely, one might think, one can't have it both ways.

The way to resolve this tension, I think, is to lean on the distinction between taxonomic and locational externalism. The debate over Marr's theory of vision has been conducted almost exclusively in terms of taxonomic externalism and individualism, as manifest in the focus on narrow and wide content. But whether any theory employs a notion of narrow or wide content will turn largely on what it says about wildly counterfactual cases, cases in which there are molecular (or even narrowly functional) duplicates in minimally different environments. And Marr's own discussion of the computational level supports both a narrow and a wide interpretation of it. Thus, the nonconclusiveness of individualism construed as a taxonomic thesis about the cognitive sciences.

Once we shift to the locational conception of externalism, matters become somewhat clearer. The issue here is whether the cognitive sciences sometimes investigate wide computational systems, and at least on the surface this seems easier to resolve, since it is a matter primarily of identifying whether the computational system of interest literally extends beyond the boundary of the individual. Locational externalism is a stronger view than taxonomic externalism and entails it unless there is a plausible, nonquestion-begging way to individuate mental states independent of their total realizations. We have seen several failed attempts to do this, in this chapter and the last, and the emphasis on locational externalism clearly places an onus on individualists.

This is a matter of whether the cognitive sciences do or should employ a strategy of integrative synthesis, as well as constitutive decomposition, whereby individuals and their cognitive processes are located in systems larger than those individuals. My argument in this chapter has been that by refashioning the notions of representation and computation, there is no reason why not, and much reason to be bipartisan or pluralistic about these notions, as we should be about realization.

10 BEYOND COMPUTATION

In introducing and defending wide computationalism as a perspective on research in cognitive science, I have appealed to existing work that I think can plausibly be seen as exemplifying a wide computational view of cognitive processing. Such a view shifts the attentional focus of the field in such a way as to suggest new directions for the computational theory of mind. I have suggested that one could interpret Marr's theory of vision as invoking the strategy of integrative synthesis (over Marr's own likely objections), and identified three paradigms of research in cognitive science – in distributed cognition (Hutchins), in animate vision (Ballard), and in reactive robotics (Brooks) – that prima facie also exemplify this strategy.

Individualists, of course, are unlikely to take the appearances here at face value. Indeed, they should challenge the claim that any of these research paradigms are properly construed as invoking wide computational systems. There are various ways of doing so: from questioning the conceptual integrity of the very idea of wide computationalism, to arguing that individuals are the largest cognitive units to which we in fact apply computational analyses of cognition, to pointing to inadequacies of wide computational interpretations of specific theories and explanations.[39]

Rather than continuing this debate here, I want to move on beyond computation. In the next chapter, I shall generalize from wide computationalism to wide psychology, and in so doing discuss a variety of noncomputational approaches to cognition that also exemplify the externalist view of cognition that I have been developing in Part Two. The argument will be that both taxonomic and locational externalism, whether computational or not, provide the theoretical backdrop for redirecting some of the individualistic research on core topics, such as memory and cognitive development, in ways that draw on some largely forgotten insights within psychology itself.

PART THREE

THINKING THROUGH AND BEYOND THE BODY

8

The Embedded Mind and Cognition

1 REPRESENTATION AND PSYCHOLOGY

The chapters in Part Two have laid the foundation for a general externalist view of cognition and the mind. For the most part, the concepts of realization, representation, and computation have been developed and deployed within individualistic frameworks. In broadening and refashioning these concepts I have created a space for an externalist psychology, and illustrated ways in which some of that space has already been occupied within computational cognitive science.

The most important of these concepts for cognitive psychology is that of representation. Cognitive psychology explores the nature and structure of mental representations and how they are processed: how they are stored, retrieved, transformed, and related to one another. In the last chapter we saw that a representational view of cognition need not be individualistic. In this chapter, I move beyond foundation laying for externalism to show the place that exploitative representation has within cognitive psychology. I shall focus on areas of psychology in which representation has played a central role – on memory (section 4), developmental psychology (section 5), and folk psychology and the theory of mind (section 6).

Representation is not simply a form of encoding but more generally a form of informational exploitation of which encoding is a special case. Representations need not be thought of as internal copies of or codes for worldly structures. Rather, representation is an activity that individuals perform in extracting and deploying information that is used in their further actions. It involves an agent enmeshed with the world not prior to or following but in the very act of representing. On the traditional view of representation, cognition is wedged between perception and action, implying, in the philosopher Susan Hurley's words, that "[t]he mind is a kind of sandwich, and cognition is the filling." The shift in perspective that the concept of exploitative representation introduces opens the way to developing a view of cognition that treats what is inside the head and what is beyond it in a symmetrical fashion.[1]

One form that this symmetry takes, exemplified by the multiplication example depicted in Figures 7.1 and 7.2, is the identification of explicit symbolic structures in a cognizer's environment that, together with explicit symbolic structures in its head, constitute the cognitive system relevant for performing some given task. Such symbols in the world can be exploited rather than encoded by individuals and their in-the-head computational systems.

The same is true when information in the world does not take this explicit symbolic form. To use the psychologist George Miller's apt term, we are *informavores* and can exploit causal and probabilistic dependencies in the world in generating in-the-head structures that, in part, direct our behavior. In the previous chapter, I developed this idea in terms of computation extending beyond the boundary of the individual, but we could express this more generally in terms of information-processing systems that do so. Again, there is no metaphysical significance to the boundary of an individual's body for individuating where that individual's mind begins and ends.[2]

The exploitative view of representation is one way to develop the idea that cognition is situated, embedded, and embodied. One often-expressed concern about this idea is that it involves an unacceptably deflationary understanding of what cognition is. Proponents of the embedded mind either focus exclusively on aspects of cognition that involve the organism's interface with the world (for example, perceptual and motor capacities), or they offer thin behavioristic or functionalist construals of intuitively more central cognitive abilities. By showing how exploitative representation applies to paradigm cases of core cognitive capacities, and by suggesting extensions of existing paradigms of representational psychology, I hope to preempt both of these standard criticisms.

2 LIFE AND MIND: FROM REACTION TO THOUGHT

Our paradigm of a living thing is an organism. All organisms react to occurrences in their environments, and use those reactions to control their bodies in some way. I am inclined to think that much the same is true of all thinking things, and to view this as a simple but deep fact about the nature of cognition, certainly for animate creatures like us whose cognitive functioning is tied to their continued existence as living beings. Cognition allows us to register what is in our here and now environment, and to adapt our bodies to what we register. But this truth should not overshadow the fact that cognition also allows us to do much more, to go beyond our immediate environments, back to the experienced past through memory and ingrained habit, and forward to the distant future through planning and imagination.

The parallel between the embeddedness and embodiment of living and thinking things, together with the recognition that thought is something more than stimulus driven and response driving, invite the question of what more there is to cognition than simple registration and reaction. We can think of this in terms of types of representational system that organisms possess, each with a distinctive locus of control for action.

First, a creature might simply have a *reactive* representational system. It registers one or more states of its environment, and this guides its behavior, but the connections between registration and reaction are simple and fixed. The behavior of the creature is thus effectively under control of the stimuli in its environment. Vary the position of an intense light source within certain parameters and you cause a variation in the position of the flower in a sunflower. Adjust the magnetic field in the liquid in which a paramecium floats, and you change the direction in which it moves. Although the internal structure of creatures with such representational systems plays a crucial role in the mediation of stimuli and response, the locus of control for their behavior is external. It is environmental.

But many creatures are able to exercise more control over their own reactions. Their representational systems are not simply reactive but *enactive*: They endow those blessed with them some control over the nature and strength of the behaviors enacted. A registration of the environment is made, but it does not automatically generate a reaction in what is done. Rather, it is combined with other registrations, and these together generate a bodily output. I would hazard the guess that most of the animal behavior that we are familiar with in everyday life – from interacting with our domestic pets, to watching squirrels or birds in a yard, or vertebrates in general in the wild – involves enactive representational systems. It is with enactive representational systems that most of us are first comfortable in invoking distinctly psychological language in a literal sense. Enactive creatures perceive and decide, whereas reactive creatures do so only in a metaphorical sense.

Yet even if an enactive representational system introduces a distinctive psychology, it does not suffice, one might think, for the full range of psychological capacities that creatures have. In particular, enactive representation is still closely tied to what one does with one's body, and does not yet give us higher cognition with a completely internal locus of control. That is what a *symbolic* representational system creates, the capacity to divorce cognition from its bodily origins. It creates genuine thinkers, creatures who can use representations to generate other representations, and so whose cognition may have a high level of autonomy from the here and now. It is here that we have thought, inference, reasoning, planning, wishful thinking, and reflection. The heart of cognition. Cognition central.

Human beings manifest all three types of representational systems: in reflexes (reactive), in bodily skills (enactive), and in higher cognition (symbolic). Although there is a physiology that underlies any biological reflex, it is plausible to view the reflex as under the control of the stimulus that elicits it. (In fact, if standard stimuli do not elicit the reflexive reaction, then we view the system responsible for it as having broken down, as not functioning as it is supposed to.) It is in this sense that reflexes are not options exercised by the organisms that have them, and the external locus of their control is one reason not to think of them as cognitive in nature. Bodily skills, such as riding a bicycle, introduce such options, given any particular stimulus. Yet there remains a sense in which they are not controlled "centrally" in that they do not require – in fact, often require the absence of – direction from paradigmatic cognitive states, such as beliefs and desires. This is why they can be performed, as we say, "without thinking about it," and why thinking about it sometimes gets in the way of successful performance. Once you know how to perform a bodily skill – from bicycle riding, to typing, to gymnastics, to catching a ball – it is a matter of letting the body do what it knows to do, rather than trying to control what it does consciously. One may concentrate in performing a bodily skill, but not on the sequence of actions that constitute that performance.

In contrast to both these cases, higher cognitive capacities have an internal locus of control: What directs them, what leads them ultimately to govern our behavior, lies within us. This control is not always conscious but it is neither environmental nor bodily and is mediated by "what is in

TABLE 8.1 *Locus of Control and Representational Type*

Locus of Control	Type of Organism/ Representational System	Example in Humans
environmental	reactive	reflexes
bodily	enactive	mimetic skills
cranial	symbolic	beliefs, desires

the head." Symbolic representation makes for truly voluntary behavior, free action, genuine choice, and rationality. Some would say that symbolic representational systems are uniquely human, or that they are what make language, a social life structured by conventions, and rich cultural traditions possible. Table 8.1 summarizes the three types of representational system, and some of what I have said about them.

Given this view of higher cognition, it is perhaps natural to think that cognitive psychology, at least in studying such higher cognitive capacities, should focus just on what is in the head. It should bracket off what the head is in and maintain that the symbols that it is concerned to understand supervene on the intrinsic, physical states of not just the individual, but of the brain. In short, higher cognition is individualistic.

The plaint of this chapter is that this final conclusion is mistaken. Before turning to the substantive discussion, a brief sketch of why I think even so-called higher cognition is best viewed from the externalist point of view.

3 THE EMBEDDEDNESS AND EMBODIMENT OF HIGHER COGNITION

All three areas of cognitive psychology that I will discuss in this chapter – memory, cognitive development, and folk psychology – have a *metarepresentational edge* to them in that they involve the representation of representations. Metarepresentation epitomizes much of what I have said about symbolic representational systems. It has been thought to characterize, and even to uniquely identify, human intelligence. It is a prerequisite for rational reflection and deliberation, and for living one's life in accord with goals, plans, values, and ideals. If the locus of control as we move from reflexes to bodily skills to symbolic abilities becomes increasingly internal, increasingly removed from the direct effects of what the head is in, then surely metarepresentational abilities go one further step in this direction. And if an internal locus of control makes for an individualistic perspective, then externalists should face an uphill battle in trying to make sense of these abilities.

What I would like to do is grant the symbolic nature of cognition and the idea that cognition has an internal locus of control, and show how neither leads to individualism. But this simply makes space for an externalist view of cognitive psychology. What reason is there to occupy this space? A common theme in my discussion will be that as we move from simpler to more complicated cognitive processing in our accounts of memory, cognitive development and folk psychology, the pressure to move from an individualistic to an externalist psychology increases. Far from being the province of an inwardly withdrawn mind, metarepresentation and the levels of cognitive performance that it facilitates belong to the mind as it is located in the social and physical world.

The strategy of argument that I shall use in making a case for an externalist cognitive psychology will draw on an inversion of the trichotomy between reactive, enactive, and symbolic representational capacities. I will argue that many cognitive capacities in symbol-using creatures, far from being purely internal, are either *enactive bodily* capacities, or *world-involving* capacities. These capacities are not realized by some internal arrangement of the brain or central nervous system, but by embodied states of the whole person, or by the wide system that includes (parts of) the brain as a proper part (Table 8.2).

Enactive bodily capacities and the embodiment of cognition more generally are important in their own right. But understanding the ways in which higher cognition is embodied can also be instrumental in showing the limitations to encoding views of representation that underpin individualistic views of the mind. Since my primary concern is to show how higher cognition extends beyond the boundary of the individual, I

TABLE 8.2 *Higher Cognition and Its Realizations*

Cognitive Capacities in Symbol-Using Creatures	Realization of the Capacity
purely internal	internal cognitive arrangement of the brain
enactive bodily	cerebral + bodily configuration
world involving	cerebral arrangement + environmental configuration

shall concentrate more heavily on world-involving capacities than enactive bodily capacities in what follows.

4 MEMORY

Memory has been one of the most active areas of research in psychology. From Ebbinghaus's initial experiments on the recall of lists of nonsense syllables to current PET or fMRI research on the localization of particular memory systems, the bulk of this research has reflected psychology's origin as an experimental discipline. There is, I think, a real question about the scope and ultimate value of the bulk of the work in this tradition, a question that has manifested itself in recent debates over the relationship that such research bears to everyday memory, memory in the wild, memory as it is used in our day-to-day lives. Over two decades ago the cognitive psychologist Ulric Neisser provocatively stated that

> the results of a hundred years of the psychological study of memory are somewhat discouraging. We have established firm empirical generalizations, but most of them are so obvious that every ten-year-old knows them anyway. We have made discoveries, but they are only marginally about memory; in many cases we don't know what to do with them, and wear them out with endless experimental variations. We have an intellectually impressive group of theories, but history offers little confidence that they will provide any meaningful insight into natural behavior. Of course, I could be wrong: perhaps this is the exceptional case where the lessons of history do not apply, and the new theories will stand the test of time better than the old ones did.... But because they say so little about the everyday uses of memory, they seem ripe for the same fate that overtook learning theory not long ago.[3]

For all that has happened in the past twenty-five years, there is little reason to revise this general judgment.

Part of Neisser's point about the experimental tradition in memory research is that the near exclusive concentration on controlled conditions under which one finds a significant difference between two groups of subjects has led many investigators to forget what memory is for, what it does for us as individuals in our day-to-day lives. "Real life" memory – for example, on memory as it is used by an individual to construct narratives about what has happened to her, or on memory as it operates in eyewitness testimony in the legal system – is the remedy that Neisser himself sees as a way of rectifying this deficit in traditional memory research. Reconceptualizing the study of memory as a form of externalist psychology can contribute to this general project of reenvisioning memory.[4]

Consider first the classic "storehouse" model of memory that derives from a brief comment that the philosopher John Locke made in *An Essay Concerning Human Understanding*. Locke says

> ... This is *Memory*, which is as it were the Store-house of our *Ideas*. For the narrow Mind of Man, not being capable of having many *Ideas* under View and Consideration at once, it was necessary to have a Repository, to lay up those *Ideas*, which at another time it might have use of.

While Locke is often credited with originating the storehouse metaphor in this passage, what is seldom noted is that having introduced it, Locke immediately warns us about how it should be understood, with this passage continuing:

> But our *Ideas* being nothing, but actual Perceptions in the Mind, which cease to be anything, when there is no perception of them, this *laying up* of our *Ideas* in the Repository of the Memory, signifies no more but this, that the Mind has a Power, in many cases, to revive Perceptions, which it has once had, with this additional Perception annexed to them, that it has had them before. And in this Sense it is, that our *Ideas* are said to be in our Memories, when indeed, they are actually no where, but only there is an ability in the Mind, when it will, to revive them again.[5]

Here Locke is cautioning against reifying memory as a storehouse for ideas, since he holds that ideas, including memories, exist only when they are perceived consciously by the mind. Locke is also, I think, suggesting a more dispositional conception of memory. If this is correct, then there is a question whether this ability is, as Locke says, "in the Mind," or whether it is what, in Chapter 6, I called an extrinsic or wide disposition.

The psychologists Asher Koriat and Morris Goldsmith have noted that while the storehouse metaphor has structured the bulk of memory research over the last hundred years, this metaphor has been challenged in recent years by research that draws on what they call the "correspondence metaphor." According to this view, it is of the essence of memory to correspond to some past state of affairs, rather than simply to act as a storehouse for readily identifiable chunks of data. The storehouse metaphor facilitates a conception of memory as a place that can contain a finite number of individual units, and thus suggests that memory be studied in terms of the quantification of those units. By contrast, the correspondence metaphor is conducive to investigating and assessing memory in terms of the accuracy of the contents of the memory, where this is not simply a matter of how much is remembered.[6]

The correspondence metaphor of memory invokes a taxonomically externalist conception of memory. That is, what individuates memory so conceived in general from other types of cognitive processes (for example, imagination, fantasy, wishes) is the relationship memory bears to past, experienced states of affairs in the world. Moreover, there is a second way in which the correspondence metaphor relies on a taxonomically externalist conception of memory. What individuates particular memories from one another, on this view, is at least in part what they are memories of or about, that is, their intentionality. On this view, that memories are not simply self-standing encodings but records of events and episodes in the world is an important fact about them, not something that should be factored out or bracketed off in investigating them.

Ulric Neisser has recently claimed that the underlying metaphor structuring real-life memory research is not that of correspondence, but one of "remembering as doing," with emphasis given to the activity of remembering, rather than the results of that activity, things called "memories." In this respect, Neisser's view is interestingly like Locke's dispositional gloss on his storehouse metaphor. This conception of memory in the experimental literature has its *locus classicus* in Frederic C. Bartlett's influential *Remembering*. Bartlett criticized the view of memory as a storehouse of images or traces, arguing instead that remembering, as an activity, was essentially a constructive process. With some misgivings, Bartlett adapted the term "schema" in referring to the "active organization of past reactions." In tracing parallels between the role of schemata in individual memory and conventionalization in societal remembering, Bartlett laid the foundation for the development of the theory of cultural models that has been influential in recent anthropology and ethnology.[7]

This performative, enactive, or constructive model of memory opens the way for a locationally externalist conception of memory, where what is enacted does not simply stop at the skin but involves engaging with the world through cognitively significant, embodied action. On this view, internally stored memories lose the privileged role that they play on the storehouse conception and become simply one resource that is used in the act of remembering. What are critical are acts of remembering, and while these use internal resources they also take place through bodily activity that is, in turn, engaged with the world. Remembering, on this view, involves exploiting internal, bodily, and environmental resources in order to produce some sort of action, often social in nature.

To adapt Neisser's own examples, to tell a joke or recite an epic tale is not simply to make certain mouth and body movements, any more than it is to produce a certain number of items from memory or to recall something accurately. Rather, it is to make a suitable impact on an audience there and then through one's memory-driven actions. We can conceive of such memory itself as extending into the world through the way in which it engages with and appropriates external symbols, treating them in just the way that internal symbols are treated, and thus giving us a locationally externalist cognitive system.

Such a conception of memory is operative in the psychologist Merlin Donald's view of the evolution of human cognition. On Donald's view, the critical transition that makes distinctively human cognition possible is the formation of what Donald calls the *external memory field*. This is constituted chiefly by visual symbols and the devices that generate them, and these derive from human cultural achievements, such as the development of writing systems and pictorial conventions. We have hybrid minds that combine internal with external symbolic processing. In more recent work, Donald has developed this view with an eye to highlighting its implications for consciousness and its evolution.[8]

Following Donald, the philosopher Mark Rowlands points not only to visuographic representations but also to the role of bodily grounded mimesis as providing a new form of semantic memory, one that allowed for the development of both learned bodily skills and the communication of sophisticated information about one's self and its environment. Rowlands also identifies sound, including spoken language, as an external medium for memory, one whose patterns of repetition, such as rhyme and rhythm, allow complicated sound patterns to be remembered. Sound and the patterns that can be generated through it constitute external memory resources, and provide the basis of the oral traditions that characterize all human cultures even in the absence of permanent external vehicles of representation, such as writing systems.[9]

Thus, this general idea can be developed without restricting oneself to thinking of memory exclusively in terms of symbols. For example, we can think of the performative memory system that extends beyond the head of the individual as incorporating aspects of an individual's environment that are nonsymbolic, including the agent's bodily orientation and actual objects in her environment. So conceived, enactive, procedural memory that is locationally wide is an extension of traditionally conceived procedural memory. The idea that procedural memory may involve doing things with one's body, while itself old hat, does suggest an idea that seems more novel: that one may remember by doing things with one's environment. Perhaps even this idea is old hat; after all, we all know that we can use environmentally cued mnemonics, such as tying a piece of string around one's finger, or leaving a note on the refrigerator.

The Embedded Mind and Cognition

My suggestion is that these need not simply be prompts to remember but themselves are ways of remembering – ways that involve a sustained, reliable causal interaction between an organism and its environment. The magnitude of our symbol-laden environments should be taken seriously, and to do so is to see the mind as extending beyond itself, that is, as being constituted by such symbols and thus as locationally wide.

To make this graphic, consider the popular problem-solving game for children, Rush Hour. The game is played on a square board made up of thirty-six small squares that are snuggly fitted by cars (of length two squares) and trucks (of length three squares). The cars and trucks are of various colors and can be placed either horizontally or vertically on the board. The aim of the game is to move the cars and trucks sequentially so that a designated car (the red car) can proceed to the sole exit on the board. The game comes with two-sided cards. One side depicts the way in which the cars should be set up initially (Figure 8.1, with the red car designated "X"). The other side provides a code for completing a series of sequential moves sufficient to solve that particular "rush hour" problem (Figure 8.2). The first move, in the game depicted, is to move car G one square to the right. And in just forty-six more simple moves you're home free!

The way in which most of us go about solving even a relatively simple Rush Hour problem involves a sustained perceptual and cognitive

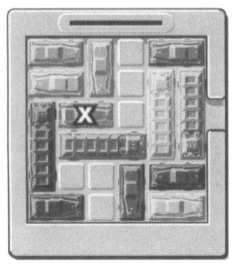

FIGURE 8.1. Rush Hour: A Problem

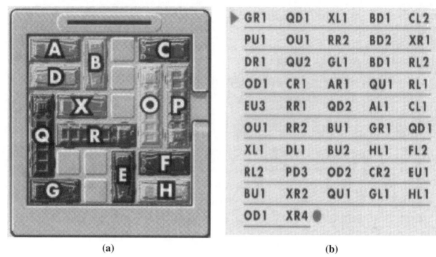

FIGURE 8.2. (a) and (b): Rush Hour: A Solution

interaction with a highly structured environment. The board of fixed dimensions, the rules for the movement of the cars and trucks, and the objective of the game all structure and constrain what we can do in playing the game. But in playing it we do not simply encode all of this and then solve the problem. (Go ahead, be my guest!) For most of us, at any rate, that is not possible. Rather, we solve the problem by continually looking back to the board and trying to figure out sequences of moves that will get us closer to our goal, all the time exploiting the structure of the environment through continual interaction with it. We look, we think, we move. But the thinking, the cognitive part of solving the problem, is not squirreled away inside us, wedged between the looking and the moving, but developed and made possible through these interactions with the board.

Of course, there is a second way to solve any given Rush Hour problem (apart from asking your kids to do it). Set up the board, flip to the "answer side" of the card, read off the code, and move each car or truck accordingly. (This was actually my six-year-old daughter's preferred solution, after she got the hang of the game – and perhaps after she tired of my fumbling around with standard ways of proceeding.) While we might be tempted to think of this as involving a problem-solving technique that is more purely internal, note how much external structure even it exploits. There is the labeling of the cars and trucks on the board, the code for each move, the convention that we read from left to right, both in each encoded move and from one instruction to the next. We then have to put all of that together with finding the car corresponding to the symbol, moving it in accord with the instruction, then finding our place back on the solution card (lose your place there and you're dead).

The first point to make is that much of our everyday cognition is more like the first way than the second way of solving a Rush Hour problem. We are geared to interact cognitively with external structures. This is not simply because of our memory and reasoning limitations. Rather, it is because in doing so we can take advantage of both the rich environmental structures that we find all around us and our ability to control ourselves and adjust our relationship to those structures. The second point is that even the second way of solving the problem, the method intuitively that relies on encoding rather than exploitation, is still not purely internal and symbolic in how it proceeds. With both methods the mind extends itself beyond the purely internal capacities of the brain by engaging with, exploiting, and manipulating parts of its structured environment.

The externalist perspective is most compelling in cases in which systems of external symbols come to change, in significant ways, the cognitive capacities of individuals who interact with them. Most striking here are cases in which an organism, such as a bonobo, develops in a symbol-enriched environment and subsequently displays massively increased symbolic capacities. Consider Kanzi, the human-raised bonobo that has been central to both the life and research of the primatologist Sue Savage-Rumbaugh. Kanzi has been thoroughly enculturated, and engages in sophisticated linguistic communication through a 256-symbol keyboard that he can carry with him. Given Kanzi's actual developmental environment, Kanzi plus a 256-symbol keyboard forms a cognitive system with memory and other cognitive capacities that far exceed those of just Kanzi. (Much the same holds true of Alex, Irene Pepperberg's African grey parrot.) My point here is not the trivial one that enriched environments can causally produce smarter critters; rather, it is that what metaphysically determines the smartness of at least some critters is their being part of wide cognitive systems. To connect this back to Donald's views of human cognition and its evolution, we are certainly such creatures, although I also think that we are not alone in this respect.[10]

In illustrating this perspective by invoking these exotic sorts of case I do not mean to imply that memory is locationally wide only in extreme or unusual circumstances. Rather, the real import of these cases is that they provide us with a way to think about everyday, human memory. Both our day-to-day acts of remembering and the systems that those acts involve are locationally wide, drawing as they do on the symbolic and nonsymbolic environments that we individually and collectively create, from the more obvious forms of external storage devices, such as notepads, diaries, books and memos, to the daily routines we form to structure our lives and the habits that form us and the structure of our lives. This is not chiefly a developmental point, one about how we come to possess rich internal structures, memories, that guide our lives; nor is it a claim about what stimulates or prompts particular acts of memory. Rather, it is a claim about what is at the heart of memory, what memory is: It is a locationally wide ability that creatures like us have that allows us to make use of the past in acting for the future.

Such a conception of memory is implicit in the work of those, such as Michael Cole and Paul Connerton, whose interests lie in the use of memory in specific, culturally mediated activities. Cole develops the Vygotskyan view of cognition as a mediated activity that relies as much on external as on internal symbols, symbols that are culturally loaded and are to be understood in terms of their location in a broader cultural system; I shall discuss it further in section 5 on cognitive development. Connerton, by contrast, views what he calls *social memory* not primarily as inscriptional, a matter of encoding, but as performative in nature, where the performances establish both individual and cultural habits. The idea of procedural memory, and the distinction between procedural and episodic memory, is well entrenched in individualistic paradigms within psychology, and Connerton extends this idea to apply collectively to whole groups of individuals. Performances, things that we do, are both bodily and ritualistic in that they involve the repetition of the performance in light of the perceived past, where this past is the shared past of a group of individuals. Performance is memory in action.[11]

Although Connerton's argument concentrates on specific rituals in just one historical episode – the ritualistic beheading of Louis XVI in the French Revolution and the ways in which body posture and dress style changed in the aftermath of the revolution – the conception of social memory that he develops is quite general. It applies to how we greet others; how we lock and unlock our doors; how and what we eat; the forms of entertainment we engage in; our work styles and choices; how and where we sleep; what, whether, and how we drive; how we read; how we die. Individual habits and civil rituals act reciprocally to influence both how individual and collective memories are constructed, and the way in which they direct individuals and societies. This perspective on memory invokes a wide psychological conception, a reason why it marks a somewhat radical departure from memory as it has been chiefly studied within departments of psychology. In the little that I have said here, I hope it is clear that such a conception of memory was not, however, always alien to the experimental tradition of research. In fact, it would have been quite familiar to someone like Bartlett.

There are two caveats to enter about the externalist conception of memory that I have advocated here. First, "memory," like "cognition," is something of a catch-all term, and the phenomena it refers to are ubiquitous in our mental lives: in language acquisition and use, the performance and learning of skills, communication and socializing, daily routines, and any form of employment. If memory is externalized, then so too is much of our cognitive life. While I have suggested that we think of memory as a locationally wide system that constitutes a form of exploitative representation, I don't mean to impose a false unity on the diverse manifestations of memory in everyday life. Second, both Donald and Rowlands present externalist conceptions of memory in terms of the distinctions traditionally used to characterize memory – between episodic and semantic memory, implicit and explicit memory, procedural and declarative memory, and long-term and short-term memory. This enhances the impression that the externalist view of cognition is primarily an extension of existing ways of thinking about memory beyond the head, and I have followed Donald and Rowlands part of the way here. Yet if truth be told, many of these distinctions seem to me to be problematic, to be understood and used inconsistently by psychologists, and to stand in need of radical rethinking. Insofar as the externalist view suggests further modifications to how memory is conceptualized, some of these may be extensions of existing dichotomies, others replacements for them.

In concluding my discussion of the claim that mental states have a wide realization in Chapter 6, I raised and addressed the question of whether this entailed that the subjects or bearers of such states were themselves locationally wide, that is, larger than individual organisms. A similar question can be raised about conceptualizing memory in terms of a locationally externalist system: Does this mean that memories do not belong to individual people, but are only a part of some collective memory or (worse) free-floating cognitive flotsam?

The short answer is "No," and we can explain why by returning again to the idea of a locus of control. We do have forms of external memory storage, such as diaries, which while deriving from (and often recording, in part) an individual's mental life, can exist beyond the life of their author and come to be accessed as a form of memory by others. And each time any one of us speaks or writes in a communicative context, we create tokens of external symbols that constitute a common symbol pool from which each of us draws. But these become integrated into an overall cognitive system that we control, and that control is critical to cognition being ours and to bearing on our lives as agents. Each of us forms a core part of a specific wide memory system, one in which we serve as a locus of control. And that is why the individual remains the entity that has memories, even if memory is neither taxonomically nor locationally individualistic.

5 COGNITIVE DEVELOPMENT

Research on cognitive development, particularly that of the last twenty-five years, has produced fascinating data about the richness of the internal representational structures that infants and young children have. This research has brought with it a sea change in how children's minds are conceived, implying that children have significantly richer, more specific knowledge than previous research had indicated. Moreover, the corresponding cognitive mechanisms that generate such knowledge appear to be in place at significantly younger ages than previously thought, in some cases at birth or shortly thereafter. The common-sense view, largely accepted and developed within earlier developmental psychology, holds that cognitive abilities are heterogeneous across children, they develop largely through learning and other forms of environmental interaction, and anything like adult abilities only begin to appear after the age at which children begin formal schooling in Western societies. By contrast, the picture of cognitive development painted by the research of the last twenty-five years is one of widely-shared abilities with a rich, innate component that differ from those of adults in many ways but not in terms of their basic cognitive nature.

To make this contrast more concrete, consider two striking examples. One might think that the concept of number is acquired by the child through long-term exposure to basic mathematics at a relatively late age, or that the concept of a physical object is acquired through an inductive extrapolation from exposure to instances of physical objects over time. Intuitively, the possession of each of these concepts seems to be a cognitive achievement gained through relevant experience with the world. Indeed, this was the accepted view of these concepts within developmental psychology, largely due to the influence of Piaget's interactionist view of cognitive development. By contrast, the "new" developmental psychology, using sensitive (and clever) experimental techniques, indicates that infants early in their first year of life are already acting in ways that imply that they have some concept of number and of physical object. At this age, there has been no instruction in mathematics, and little opportunity to extrapolate from instances. Thus, the data suggest that these concepts are not acquired in these ways. Moreover, these concepts seem present so early in the life of the child that there is reason to think that they are not acquired at all, but instead are innate, part of the intrinsic endowment of the child, with the development of these concepts merely triggered by environmental stimulation, not shaped by it.[12]

This strand of nativism in recent work on cognitive development corresponds to one dimension I identified in strong nativist views of cognition in Part One, the external minimalism thesis. This work also manifests the other dimension to strong nativist views, the internal richness thesis, through its commitment to the modularity of cognitive development. Characterized generally, the new developmental psychology views preschool children and infants as having domain-specific, cognitive modules, including not only modules for knowledge about physical objects and number during infancy, but also those for biology, social relationships, and minds. This knowledge ascribed to infants and young children has typically been understood in terms of the child's possessing a theory about the relevant domains, which lead her to expect certain outcomes rather than others, and thus behave in some ways rather than others.[13]

Like the maturational view of cognitive development that endorses the external minimalism thesis, this aspect to contemporary research puts it at odds not just with common sense but with much of the earlier work on development. Included here is work conducted within an associationist framework, and that within the Piagetian, constructivist tradition, each of which has claimed that cognitive development is driven by domain-general processes, such as learning and stage-relative décalage. These recent claims not only fly in the face of the received folk and scientific wisdom of the past, but also raise many questions about the nature of development that were previously unasked.

For example, if infants begin with at least some domain-specific, conceptual knowledge, how is that knowledge used in the development of children's later domain-specific knowledge? Given the characterization of the child as a theorist of sorts, what is the relationship between the child and the scientist, or between development and scientific change?

What place is there in this conception for more global, domain-general processes in cognitive development of the sort that had, until recently, populated the mind of the child within developmental psychology? Such questions are the subject of ongoing research within the "new" developmental psychology.

There are limitations to the individualistic way in which this paradigm thus far has been and can be developed. There are also natural ways to extend the gambit of a developmental psychology that incorporates and builds on its insights. The massive redescription of the nature and extent of the child's knowledge at the heart of recent developmental psychology has concentrated on the structure of an infant or child's internal mental representations, on what it is that the child does or doesn't know about the relevant domain at a given age, and on the internal mechanisms governing developmental change. If infants and children know significantly more than theorists had previously ascribed to them, as they appear to, then views of later developmental changes will also almost certainly need to be modified, since the state of (say) the three-year-old from which the (say) eight-year-old develops is not what it was once thought to be. But ultimately this nativist emphasis within the new developmental psychology must move beyond the head in making sense of the full pattern of cognitive development.

One way of augmenting the resulting individualistic perspective on development is via the mediational approach that derives from the work of the Russian psychologists Lev Vygotsky and Alexander Luria in the 1930s that I mentioned in passing in the previous section. Motivating this view was, in Luria's words, the idea that

> [t]he chasm between natural scientific explanations of elementary processes and mentalist descriptions of complex processes could not be bridged until we could discover the way natural processes such as physical maturation and sensory mechanisms become intertwined with culturally determined processes to produce the psychological functions of adults.[14]

This idea carried with it the working assumption that individual mental abilities are significantly modified by the various mediational tools that they employ. Such mediators include maps, tools, and other artifacts, numerical systems, memory aids, and, most importantly for Vygotsky, spoken and written language. All such mediators are not only cultural products in that they are the products of particular cultural histories and thus are available to an individual only within the corresponding cultural contexts. In addition, they are employed primarily within the context of social interaction and facilitation, typically in small groups or dyads. As Vygotsky says, "The path from object to child and from child to object passes through another person. This complex human structure is the product of a developmental process deeply rooted in the links between individual and social history."[15]

For Vygotsky, higher thought processes, such as remembering, attending, and speaking, are essentially mediational, and thus social. Cognitive development is not, like embryological development, primarily the unfolding of existing structures in accord with some fixed, biological program, but a dialectical or dynamic process that involves in-the-head mental structures, beyond-the-head mediators, and the social context in which they interact. Humans have natural psychological abilities, but these are qualitatively different from the forms they take when they are deployed together with specific mediators. Attention or memory without the direction provided by spoken, communicative language or other external aids, are fleeting and transient. Literacy, the ability to read and write, itself augments and restructures significant aspects of thought and language, including the size of the lexicon, the development of metalinguistic abilities, and strategies of memorization. Employed with the mediators of spoken and written language, these core cognitive capacities become enhanced abilities that make a qualitative difference to the sorts of things that children can do.[16]

If this is true of higher cognitive processes in general, then it is true of the domain-specific processes postulated by the new developmental psychology. This suggests a natural direction in which to shift the focus of research attention: In investigating the conceptual changes within a given domain, or the emergence of new domains of thought in childhood, look outside of the head and consider how it modifies the child's "theory of X." In emphasizing both the social and dynamic nature of cognitive processing and its development, it should be clear how the Vygotskyan perspective ascribes an integral role to the environment outside the individual. The external environment does not simply cause changes in a child's internal mental structures but in a literal sense comes to constitute part of new, locationally wide abilities, which are then used to interact with other mediators, leading to further locationally wide abilities, and so on. Once such wide, mediational abilities are established, they can then be transformed and internalized by individuals and thus decontextualized from their initial context of acquisition.

A part of the child's environment that serves as a crucial mediator across a range of contexts is other people. Here some work within the recent "theory of X" approach in developmental psychology shows convergence with some of the conclusions central to the Vygotskyan tradition. The work that I shall discuss in the remainder of this section focuses on *explanation* and how a grasp of it develops in children over time.[17]

A striking paradox about explanation and how it is cognized turns on a triad of claims. First, explanation is a crucial and ubiquitous part of much of everyday life. Second, its ubiquity in our day-to-day dealings with one another seems to reflect the various "theories" that we have about corresponding domains. Yet third, the level at which we are able both to offer and understand explanations is amazingly *shallow*. What I mean by this is simply that explanations typically stop or bottom out surprisingly early on. Consider an intuitive example from the developmental psychologist Frank Keil. Although almost everyone who owns a car can give some sort of explanation as to why their car starts (or doesn't) when the key is placed in the ignition and turned, few of us are able to respond with any depth to even the next few follow-up "why" or "how" questions. The shallowness in this case is the norm: We rarely have ready access to explanations of any depth for all sorts of phenomena for which we are able to offer some sort of explanation. Indeed, we often carry with us what Leonid Rozenblit and Frank Keil call an *illusion of explanatory depth* until we are faced with the actual task of explanation. Thus, people frequently seem to think they have vivid, fully mechanistic models of how something works or how it got the way it did. But when forced to state explicitly that mechanism as an explanation, their own intuitions of explanatory competence are shattered.

Although overconfidence in one's abilities has been experimentally documented in a range of domains, this experimental work has been recently extended to apply to explanation in particular. For example, Rozenblit and Keil asked college students whether they know how various familiar devices work, such as flush toilets, piano keys, and zippers. Subjects were given a list of forty-eight familiar items, and asked to use a seven-point scale to rate how well they understood how each worked. They were then given a four-item test list and asked to produce a detailed, step-by-step causal explanation for each item on this list, followed by a series of reratings of how well they understood each of these items following their attempt to produce an explanation and their reading of expert explanations. Rozenblit and Keil also ran versions of the experiment with facts, narratives, procedures, and natural phenomena.

The general finding of this series of studies supports the original claim that Keil and I made about the shallows of explanation: There is an illusion of explanatory depth in ordinary people's understanding of familiar devices. Many subjects assert or imply that they have a complete and fully worked out understanding such that they could explain all the necessary steps in any process involving the object. Yet when asked for such explanations, a large percentage of these participants show striking inabilities to put together a coherent explanation, missing not just a few arbitrary details, but critical causal mechanisms. Until they attempt such explanations, they are often under the illusion that they have a complete, "clockworks," vivid understanding. The inflation in the estimate of one's ability to provide an explanation seems not to be general, since the pattern of results obtained in the device condition was not repeated in the facts, narrative, or procedure conditions, although it was repeated in the natural phenomena condition. Thus, there seems to be something about our knowledge of both familiar devices and natural phenomena that creates this illusion of explanatory depth. Rozenblit and Keil suggest that these domains promote the illusion because they are sites at which a series of confusions converge. For example, people mistake something like the perceptual vividness of familiar devices and natural phenomena for the richness of their own internal representations, and their understanding of these at their functional or operational levels with that of the underlying causal mechanics.

Apart from the illusion of explanatory depth itself, what is striking and relevant for developmental paradigms that emphasize the nature of underlying theories that guide cognitive development is the lack of explanatory depth itself. Missing in adults seems to be the sort of theoretical understanding of the underlying workings that anything more than a skeletal theory in the relevant domain would surely provide. The shallows of explanation thus pose a prima facie problem for accounts of cognitive development that ascribe to the child a theory of the relevant domain, which is then added to or modified through development. This is because the shallows of explanation we observe in adults seem to be due to a sort of theoretical abyss. The gap between the limited explanations that suffice for the purposes at hand and those that we believe ourselves to have is due ultimately to a lack of corresponding, detailed theoretical knowledge that would allow us to provide more satisfying, detailed explanations. That is, the shallows of explanation are not due simply to contextual or pragmatic features of the practice of explanation, or from general processing or access abilities. Rather, it stems from something central to the "child's theory" view of cognitive development: the absence of detailed theories themselves.

This implies that the problem in the triad generating the paradox concerning explanation (ubiquity-theory-shallows) lies in the second claim: that explanations reflect an individual's theoretically rich understanding of some phenomenon. If explanation is theoretically shallow, how can it also be ubiquitous?

What makes explanation both ubiquitous and shallow is a certain division of cognitive labor, one which makes detailed theoretical knowledge in the head of each individual unnecessary, a division akin to the social division of labor central to Putnam-Burge externalism. We rely on knowledge in others extensively in our explanatory endeavors (as in our linguistic endeavors), and we rely on the assumption of knowledge in others to give us a sense of explanatory insight. Everyday folk know enough about the "nominal essences" of the things that they interact with on a regular basis in order to be able to offer relatively shallow explanations for their behavior. But there are also experts who have either the within-level or across-levels knowledge that the folk typically lack, and who are in a position to offer explanations with more depth. Although we are faced, as individuals, with the theoretical abyss as the norm, the theoretical knowledge that we lack exists somewhere, just not in our heads. This is to say that explanation and the theories that underwrite their depth, are wide: They do not supervene on an individual's intrinsic, physical properties.

This reliance on the knowledge of others in explanatory understanding seems to be a feature that even preschool children recognize, as recent work by Donna Lutz and Frank Keil indicates. For example, a child might be told that Bill knows all about why two magnets, if turned the right way, stick together; and that John knows all about why a china lamp breaks into pieces if it falls off a table. The child is then asked who knows more about why television screens get all fuzzy sometimes during thunderstorms. Even preschoolers will cluster explanations about electricity and magnetism together to a greater extent than either of those explanation types with mechanics. There is no doubt that they are in nearly full ignorance of any specific mechanisms, yet as early as the age of four they have some sense of how some explanations are more likely to be related in the minds of experts. (In this particular example they may be keying into notions of invisible forces and action at a distance.) As Lutz and Keil report, even three-year-olds have a sense of how knowledge is clustered in the minds of at least some experts, although this knowledge is far less secure and extensive than it is at four.

One may think that this simply reveals more complexity to the internal structure of the child's mind than one might have initially suspected. Indeed, die-hard strong nativists and individualists might even see it as providing evidence for another module, that for deference to others! But this would be to miss the significance of these results and their relationship to those concerning the illusion of explanatory depth. What both indicate is not simply that children are built to rely on the knowledge of others but that their actual reliance limits the extent to which internally rich structures are needed to understand the world around them and function effectively in it.

This cognitive division of labor and the relatively impoverished, internal cognitive structures that go with it are instances of more general features of cognition. As we rely on other people, so too do we make use of information in the world more generally. And as we have at least a skeletal understanding of what others know (and who knows what), so too do we have "modes of construal" of the various ways in which properties are causally clustered and distributed in the world. The general point is that throughout much of development, and long before formal schooling, a set of framework explanatory schemata are at work and seem to be essential for further theory growth and conceptual change. This suggests the basis for some rapprochement between the new developmental psychology and a mediational view of cognitive development.[18]

Vygotsky himself had a specific, general view of how mediational abilities developed. All mediational abilities, he claimed, begin as *interpsychological* abilities, abilities that are developed and manifested only in social relations between individuals. This is a consequence of his "general genetic law of cultural development," which says that

> [a]ny function in the child's cultural development appears twice, or on two planes. First it appears on the social plane, and then on the psychological plane. First it appears between people as an interpsychological category, and then within the child as an intrapsychological category.[19]

While this is plausible for communicative abilities, it is incompatible with the results of what I have been calling the new developmental psychology, even Keil's modes of construal tempering of it, which indicate the theoretical and conceptual sophistication that young children bring to cognitive tasks "on the social plane."

In any case, Vygotsky's "law" seems less plausible for the full range of forms that even mediational cognition can take. For example, many of the mediational devices that have been culturally developed, such as maps, signs, and even numerical systems, can be and are used by individuals from the outset, particularly once they have already acquired related psychological abilities. They do not appear to have the double existence – first on the social plane, and then on the psychological plane – that Vygotsky's law attributes to all higher cognitive processes. But even if some mediational cognition is not tied as closely to a social context as Vygotsky seems to have thought, the more general point that mediational cognition is ontogenetically prior to many of the higher cognitive processes that are the focus of individualistic psychology gives us reason to rethink what it is we are discovering when we discover the "child's theory of X."

I want to conclude the substantive part of this chapter with some thoughts about externalist psychology and the child's theory of X, where X = mind.

6 FOLK PSYCHOLOGY AND THE THEORY OF MIND

We are mindreaders. The explosion of work over the last twenty years in both cognitive development and primatology exploring the developmental and evolutionary origins of this ability has largely construed the capacity itself as a theory of mind, a theory that attributes folk psychological states to agents, and that allows one to predict and explain an agent's behavior in terms of the relationships between those states, perception, and behavior. Folk psychology itself forms a core part of many areas within psychology, including work in social cognition, group dynamics, and decision making. Philosophical discussions of folk psychology concentrating on the relationship between folk psychology and a truly scientific psychology have sometimes implied that such work could form no part of a truly scientific psychology. Yet such eliminativist conclusions have not garnered widespread support either in philosophy or in psychology. Here I focus my discussion of our mindreading abilities on the end state of these ontogenetic and phylogenetic processes, the folk psychology that we end up sharing and relying on in everyday life.

Folk psychology has played a prominent role in the debate between individualists and externalists, beginning with the widely accepted conclusion from the Putnam-Burge thought experiments that folk psychological states are externalist. In Part Two, I argued that we should see this as entailing that folk psychological states have a wide realization, and that folk psychology itself is a type of wide cognitive system, one that involves the social relations between individuals. But what of the underlying capacity or disposition that each of us has to construct such a folk psychology? The large and growing literature on our mindreading abilities has, by and large, treated this capacity as an individualistic system, a theory of mind module or the child's theory of mind.[20]

In discussing this I want to distinguish between our *bare-bones* folk psychology, belief-desire psychology, and a richer conception of folk psychology, one that includes the full range of psychological states, such as emotions (anger, elation, fear), moods (restless, horny, inattentive), and sensations (pain, experiencing red, tickling). I shall refer to this richer conception as *full-blown* folk psychology. Consider, first, bare-bones folk psychology.

It is plausible to think that the capacity that normal human adults have to ascribe belief and desire to one another is both locationally narrow and taxonomically wide. It is locationally narrow because the realization of the capacity is purely internal to the individual who has the capacity. But it is taxonomically wide because beliefs and desires are individuated, in part, by their intentional content, that is, what they are about, and such content is wide. This is so whether one thinks that this ascriptive ability operates via a theory or via acts of imaginative simulation. Bare-bones folk psychology admits of what we might think of as an intellectualist construal, one in which the ability itself is purely internal to the cognitive agent, and I suspect that this is one reason why it has been the focus of attention in the "theory of X" tradition. Matters are less straightforward, however, when one considers both the full-blown capacities we have for folk psychological explanation and some of our more advanced deployments of folk psychology.

Consider full-blown folk psychology, which augments bare-bones folk psychology with a heterogeneous bunch of further states and capacities that the folk readily attribute to one another. Apart from the various emotions, moods, and sensations already mentioned, full-blown folk psychology includes character and temperament states (sturdy, reliable, happy-go-lucky) and global cognitive assessments (rational, intelligent, scatty). It is much less plausible to think that the realization of the capacity to ascribe full-blown folk psychology is purely internal than to think so in the case of bare-bones psychology. That is because these states have a felt component, whether it be experiential or bodily (or both), and it is difficult to see how one could accurately and reliably ascribe such states to others without knowing what they were like in one's own case. Such knowledge itself is procedural and has a bodily realization in that it involves not simply having one's brain in some internal state but, at least, having one's brain and body in a certain state.

The most obvious ploys for proponents of the theory view of folk psychology would be to argue that (i) full-blown folk psychology can be reduced to bare-bones psychology, or (ii) however important experiential and bodily aspects are to the acquisition of folk psychology, they do not form part of its realization, which is purely internal. Both of these options seem unpromising, however, in light of the argument of Chapters 5 and 6: (i) would seem to involve either the sort of hierarchical decomposition that I argued against there, while (ii) overlooks or simply dismisses wide realizations as a species of total realization. Even if we conceded that bare-bones folk psychology was individualistic, both strategies would seem to manifest one or another form of smallism.

My claim, then, is that the move from bare-bones to full-blown folk psychology involves a shift from a purely internal mental capacity to a bodily enactive skill. But I also want to suggest that some of our most sophisticated deployments of folk psychology, such as understanding a complicated narrative about the mental lives of others, and manipulating another's full-blown folk psychology – involve a symbolic capacity that is world-involving. In such cases, folk psychology starts to look not just taxonomically but locationally externalist.

Consider narrative engagement that involves understanding the full-blown folk psychology of characters in a literary, dramatic, or cinematic genre. To understand, say, a certain kind of novel one must not only ascribe full-blown folk psychological states to the characters in the novel but also understand those characters' (partial) views of the world, a world which naturally includes other people. (Very effective novels in this respect include Vladimir Nabokov's *Lolita* or Ian McEwan's *The Innocent* or *Amsterdam*.) As you read deeper into the novel, you must modify your representations of the folk psychological representations that each character has. But since the metarepresentational load here increases dramatically with the complexity of the portrayal of the characters and their relationships to one another, it is no surprise that even partial expertise typically involves *knowing how to find one's way about in the novel*. It involves knowing how to locate and identify the folk psychological representations that respective characters have, and the signs of these in the novel itself. Here the representations that are the object of your own representations are located somewhere other than in your own head. In short, this understanding involves constructing a representational loop that extends beyond the head and into the minds of the fictional characters – and perhaps the narrator or even the author – with which you are engaged.

Perhaps unsurprisingly, much the same is true of appreciating the full-blown folk psychology of real people, especially those to whom you are close. Our representations of the mental lives of companions and friends are more sophisticated not simply because of the added internal complexity such representations have in our own heads, but because they index richer mental representations in the minds of one's companions than those in the minds of strangers. Rather than simply encoding information about these mental representations, we engage and interact with them, and in so doing extend the system of mental representations to which we have access beyond the boundary of our own skins. As with our reliance on cognitive artifacts to bear some of the representational load borne during a complicated cognitive task, here we exploit rather than replicate the representational complexity of our environments. But unlike at least the cases of distributed cognition that I have discussed in which the individual is displaced as the unit of cognition, here individuals remain both the unit of cognition and the locus of representational control, with interactions with external representations augmenting the internal representational systems of individuals.

Both the case of narrative engagement and that of locationally wide, full-blown folk psychology involve representational capacities whose locus of control is still, by and large, internal. But such a locus of control is not an essential feature of full-blown folk psychology, even if in many instances it is a pervasive feature. We can come to rely on others psychologically in such deep ways that we in effect surrender our autonomy, or have it stripped away from us. A final word about such cases and their relationship to the externalist mind.

Consider, first, cases in which one person has blind devotion to and trust in another, which can involve close kin relations, lovers and life partners, or religious believers and their spiritual leaders. In extreme cases, the thought and action of the trusting party is given over to that of the trustee. When that happens we have a locus of control that is external to the trusting party. As with these other forms of locationally wide folk psychological systems, the cognitive capacity here is world involving, with the relevant folk psychological representations being located both inside and outside of a given individual's head. The folk psychological states of the trustee come to form a part, a controlling part, of the folk psychology of the trusting party, effectively displacing some of the higher cognitive capacities of that individual.

Suppose, for example, that sustained deception arises in a relation of blind devotion or trust. From the perspective of the deceiver, manipulator, or person trusted, the locus of control here remains internal. But from the perspective of the deceived, the manipulated, or the person trusting, their representational folk psychological states are controlled by folk psychological states beyond what we usually think of as their own mind.

There are, I think, many real-life situations that approximate this sort of case, even if few involve pure blind devotion and trust. They may involve partial or mutual psychological dependencies, enhanced or diminished cognitive functioning, asymmetrical power relations, or sustained deception and manipulation. It would be close to the pure cases that the self, the subject of psychological states, is no longer bounded by the body, and where separation of the trusting and trustee approximates the severing of a physical boundary within an organism. This provides one way to think about the death of a loved one, or the betrayal of personal trust, and the psychology of the grief or anger that follows in its wake.

7 THE MIND BEYOND ITSELF

Dissatisfaction with the idea that mental representation is simply a form of encoding has motivated a variety of ways of thinking of the mind as embodied and embedded. While some radical, early expressions of both connectionism and dynamic approaches to cognition implied that the notion of representation itself was the source of the problem, for the most part representation has been reconceptualized rather than consigned to the dustbin of Bad Ideas. By articulating the notion of exploitative representation in Part Two and showing how it applies to higher cognition in this chapter, I have presented a view of the mind as encultured, as embedded in social and technical networks, and as constructed through its extension beyond the boundary of the individual.

Although I have said less about the mind as embodied, I think that the exploitative view of representation can be applied to make sense of the embodiment of cognition as well, where the body becomes another resource that cognitive systems use to work their magic no different in kind from cognitive resources in the environment to which the individual is coupled. Recent representational views that emphasize the embodiment of cognition, such as those of George Lakoff and Mark Johnson, Arthur Glenberg, and Rick Grush, different as they are from one another, maintain what we might think of as a Cartesian bias in thinking of cognition in terms of what lies within the skin – ideally, the brain – of the individual cognizer. In this respect, I suspect that they remain too closely wedded to encoding views of representation. Mental representation is metaphorically structured (Lakoff and Johnson), operates through embodied encodings (Glenberg), or is processed through inner dynamical models or emulators (Grush). We need to look beyond both head and body in thinking about mental representation.[21]

There is a more conservative and a more radical strand to the argument in this chapter for adopting an externalist perspective of cognition.

The more conservative strand works with the idea that a range of psychological states and processes are taxonomically wide in that how they are individuated, classified, or taxonomized relies on factors outside of the head, and thus do not supervene on the intrinsic, physical properties of individuals. By applying this strategy not just to the philosopher's favorite, (bare-bones) folk psychology, but to a range of states and processes in subpersonal psychological theories, one can see that taxonomic externalism is not simply an implication of the Putnam-Burge thought experiments but implicit in much existing psychological explanatory practice. I suspect that, particularly amongst cognitive psychologists, there will remain the feeling that taxonomic externalism remains merely a metapsychological perspective on explanatory practice, rather than a view that actively guides the research that is done within the cognitive sciences themselves.

Such a view is more difficult to sustain with respect to the more radical strand to the argument, the one that holds that at least some psychological states and processes are locationally externalist: In a literal sense, they physically extend beyond the head of the individual who has them. Much of the discussion in this and in the previous chapter has focused on both examples of this type of externalist psychology and general features of cognition that make this a viable paradigm for structuring psychological research. Locational externalism appeals to the ways in which cognition, particularly human cognition, relies on and incorporates physical and social aspects of the environment of the individual. The context-sensitive view of realization in Part Two provides a missing link between a materialist metaphysics and such an externalist psychology. The accompanying explanatory strategy of integrative synthesis does the same for methodology and explanatory practice in the cognitive sciences and externalism.

While psychologists and cognitive scientists themselves may not view the metaphysics here as all that relevant to the type of science they develop, the direction of research in the cognitive sciences has been shaped by the general sense that a properly scientific psychology should be individualistic, in much the way that it was shaped in a previous generation by the sense that such a psychology should be behavioristic. Views of what there is, of what you need to postulate (or can't postulate) in your ontology to do your science, are often not far beneath the surface of explanatory practice. In particular, by scratching a little at the idea that cognition is a form of symbol-processing, we were able to identify why individualism might seem a necessary or desirable view of cognitive psychology, whether or not one held in addition a computational view of the nature of these symbols and how they are processed.

The externalism developed in this chapter takes the symbolic nature of thought seriously but suggests that internal symbols are simply one kind of cognitive resource used in memory (as constituents of a storehouse), cognitive development (as constituents of theories), and in folk psychology (as constituents of beliefs). External symbols are an obvious second kind of cognitive resource used in cognitive processing, but simply to see externalists as adding external to internal symbols would be to mischaracterize the shift in perspective implied by the externalism I have defended. For the central notion becomes that of a cognitive system. Some cognitive systems are wide, and some contain both internal and external symbols. But enactive, bodily cognitive systems, such as wide procedural memory systems, may be conceptualized in terms of explicit symbols only with some strain, as may wide perceptual systems that involve the extraction of information that is usually thought of as non- or subsymbolic.

A large part of the significance of mind-world coupling lies in its iterative nature. We take part of the world, and learn how to incorporate and use it as part of our cognitive processing. That, in turn, allows us to integrate other parts of the world that, in turn, both boost our cognitive capacities and allow us to cognitively integrate further parts of the world. And so on. Although some recent discussions of the embeddedness of cognition have focused on novel and future technologies – from cell phones, to electronic implants, to telerobotics – the two most significant forms of iterative scaffolding are older than the human species: the advent of spoken language (itself a scaffold for much higher cognition and written symbol systems), and the *cognitive* dependence of infants on their parents (the mother of all inventions?).[22]

With that in mind, we can see how externalism departs from the smallist views that typically drive researchers to look further "into" the brain in search of cognitive systems. Rather, the externalism I have defended suggests that in order to understand central aspects of cognition we look not to what's in the brain but what the brain is in. Memory, the process of cognitive development, and folk psychology as they actually guide the cognitive lives we lead are embedded cognitive systems, and much of the embedding framework is social.

In both species of externalism that I have discussed, the individual remains the subject or bearer of psychological states, even if she no longer serves as a boundary demarcating the entities of a respectable psychological science. At the end of the previous section I introduced cases in which this is no longer true – where subjectivity itself may no longer be easy to locate – but I think that such cases, unlike those involving locationally externalist cognitive systems, are rare. Thus, this is one respect in which the individual is a focal point even for an externalist psychology.

To further extend this externalist account of the mind, I turn in the next chapter to consciousness.

9

Expanding Consciousness

1 THE RETURN OF THE CONSCIOUS

Although the 1990s was officially the "Decade of the Brain" in the cognitive sciences, judging by the volume and range of literature, for philosophers of mind it was, rather, the "Return of the Conscious." From early in the decade, works such as John Searle's *The Rediscovery of the Mind* and Owen Flanagan's *Consciousness Reconsidered* aimed to restore consciousness to center stage in the philosophy of mind. This restoration was in part a way of correcting a distortion that the cognitive revolution's emphasis on unconscious mental processing had wrought, initiating a culture in which discussions of consciousness could be held without philosophical embarrassment. By the end of the decade, the philosophical literature on consciousness had outstripped that on any other topic in the philosophy of mind.

To be sure, much of the work on consciousness in the last dozen years or so has attended to or even stemmed from research in the neurosciences. But perhaps the issue that has most engaged philosophers of mind has been whether the neurosciences or indeed any physical science could reveal all there is to know about consciousness. Claims that there would, of necessity, remain some sort of explanatory or ontological gap between the world revealed by science and conscious phenomena themselves had been articulated and defended by Thomas Nagel, Frank Jackson, and Joseph Levine more than twenty years ago. Such claims have more recently received a reinvigorated examination by David Chalmers in his *The Conscious Mind*, the most widely discussed book by philosophers of mind since at least D. M. Armstrong's *A Materialist Theory of the Mind*.[1]

In this chapter, I do not attempt to review this literature, nor develop a substantive, comprehensive theory of consciousness, which I suspect is an illusive goal (more on which in a moment). Rather, I shall focus specifically on the implications that the sort of externalist psychology that I have defended in the last two chapters has for consciousness and its study. Consciousness has seemed especially problematic for externalists, involving mental phenomena – ranging from pain to visual experience to self-knowledge – for which internalist accounts have seemed prima facie inescapable. In general terms, conscious mental phenomena have appeared to be so intimately or immediately related to facts about the conscious subject, their bearer, and so distantly or mediately related to facts about the world of that subject, that externalism has faced an uphill battle in presenting itself as even a possible view of consciousness.

2 PROCESSES OF AWARENESS AND PHENOMENAL STATES

To begin, let's review a sampling of the range of mental phenomena that have been regarded as either being conscious or being at the heart of consciousness:

- bodily sensations, particularly the feelings one has through the senses of touch or proprioception
- pain, particularly intense or acute pain
- visual experience, particularly that of color
- higher-order cognition, cognition with mental states as their objects
- attention, particularly that directed at aspects of one's experience of the world
- introspection, being reflection of some sort on one's own mental life, including one's self

Specifying the relationship between any of these phenomena, as well as how each is to be understood – what processes each involves, what is essential to each, what role each plays in consciousness – takes one immediately into the various debates over consciousness. I shall proceed by focusing on particular phenomena on this list and considering externalism with respect to each.

At the core of the chapter is articulation of what I call the *TESEE* conception of consciousness: consciousness as *T*emporally *E*xtended, *S*caffolded, and *E*mbodied and *E*mbedded. In sections 3 and 4, I focus on aspects of consciousness that are thoughtlike – higher-order cognition, attention, and introspection – what collectively I shall call *processes of* *awareness*, first introducing (section 3) and then arguing for (section 4) an expanded view of such processes. The TESEE conception of processes of awareness complements the externalist view of cognition already developed in Chapters 7 and 8.

One further question is how well such an externalist view accounts for the "other half" of consciousness – pain, bodily states, and visual experience – what I group together as *phenomenal states*. Recently, philosophers such as Michael Tye, Bill Lycan, and Fred Dretske have attempted to extend externalism from its roots in characterizing intentional phenomena to consciousness by adopting a representational view of the phenomenal. Although the relationship between the intentional and the phenomenal will be the focus of the next chapter, in section 5, I consider Dretske's view, in particular a dilemma argument that he uses to motivate his externalism about the phenomenal. The discussion here will set the scene for considering views of at least some phenomenal states that exemplify the TESEE conception of them in section 6. Here I will draw on recent work by the philosopher Alva Noë and the psychologist J. Kevin O'Regan on the sensorimotor contingency theory of visual experience, and by Susan Hurley on consciousness more generally. As we will see, especially in sections 7 and 8, there are limitations to how extensive an externalist account of the phenomenal can be, and there are various ways in which the TESEE view stops short of the sort of "global externalism" defended by Dretske and other representationalists.

There are two reasons for suspecting that a general theory of consciousness will prove illusive. The first is that, as suggested by the six kinds of mental phenomena I presented as examples of conscious states, such a theory either has to lump quite diverse phenomena together, or explain why some are more fundamental as conscious mental phenomena than others (and treat just those). Even though I view as useful the categorization of these into processes of awareness and phenomenal states, and shall show ways in which the TESEE conception of consciousness illuminates both, there are important differences between these two categories of mental phenomena, as well as within them.

The second is that the demands placed on a theory of consciousness come from many different quarters. These range from demands peculiar to philosophers – such as those sometimes imposed within the explanatory gap literature or work on the problem of self-knowledge – to constraints that derive from facts about neural processing or limits to the accessibility of consciousness within cognitive neuroscience and psychology. These demands sometimes apply primarily or paradigmatically to just some subset of mental phenomena considered conscious (thus interacting with my first reservation), or address distinct aspects to consciousness. We are in the same position as are those searching for a theory of matter that satisfies both idealists, who think that all there is are ideas – and so need to be shown how any theory of matter is compatible with that – and physicists who are convinced that matter exists but disagree about whether it is continuous or particulate, uniform or differentiated.

3 EXPANDING THE CONSCIOUS MIND: PROCESSES OF AWARENESS

Consider higher-order cognition, attention, and introspection, what I refer to collectively as processes of awareness. These are sometimes thought of as processes that characterize distinctively human consciousness, though they have also been deployed in offering a general account of consciousness. For example, so-called higher-order theories of consciousness (HOT) propose that consciousness is awareness. On HOT views, what makes a given state conscious is that it is the object of some other, higher-order mental state of a particular kind. In keeping with my general aims in this chapter, here my focus will not be on such theories but instead on the processes that they posit and their relation to externalism about the mind.

I shall suggest that processes of awareness call out for a radical rethinking along externalist lines, one that turns on taking locational externalism about consciousness more seriously than it has been taken, even by externalists about the phenomenal. The argument here will be similar to that given in the previous chapter, where I argued that our conceptions of memory, cognitive development, and folk psychology should be explicitly refashioned along externalist lines. At the heart of this reconceptualization are three features of processes of awareness that have usually been ignored or downplayed: They are *temporally extended*; they are typically *scaffolded* on environmental and cultural tools; and they are both *embodied* and *embedded*. Let me take each of these in turn.

In both the philosophical and psychological literature, processes of awareness are usually thought of as enduring for very short periods of time. The sort of first-person introspection or retrospection on one's own mental states that allows philosophers to conceptually analyze the products of processes of awareness, and psychologists to experimentally record reports and other outcomes of those processes, are temporally quite limited. (Seconds for philosophers, milliseconds for psychologists.)

It is this conception of processes of awareness that generates many of the classical problems in each field. For example, consider Hume's problem of the elusive I, the self being difficult to detect amongst the fleeting objects of awareness that one finds in introspection. Or consider a version of the binding problem, the problem of accounting for how various aspects of experience are bound together to form a seamless, relatively unified experience, given the different ways in which these aspects are processed within the organism. Both problems presuppose a conception of consciousness as operating on a timescale of seconds or less.

These forms of the processes of awareness are important to cognition, but they are not the only form that such processes take. As Merlin Donald has pointed out, there is a more temporally extended form that awareness takes, enduring minutes or hours, that is especially significant for human cognition, for without it we would not be able to perform many of the tasks that are uniquely human (or near enough so). Constructing or following a narrative, planning to and then acquiring a given motor skill, and negotiating a crowded street in a car or on foot are three examples of such actions that we perform not simply while conscious but at least sometimes consciously. The processes that generate such actions temporally extend beyond the limit of the second hand. Such tasks are ubiquitous in everyday, waking life, and they involve an agent who is both embodied and embedded (more of which in a moment). Both culturally and individually we construct perception-action cycles that involve attuning ourselves to the world, and the world to ourselves. Many such cycles are constituted primarily by conscious experiences and acts, and their temporal extension, over minutes or hours, goes hand-in-hand with their spatial extension beyond the brain of individual cognizers.[2]

This temporally and spatially extended conception of processes of awareness makes it easy to understand consciousness as environmentally and culturally scaffolded. Although the cognitive role of specific, culturally developed tools, such as navigation equipment, clocks, and maps, has been acknowledged in some recent thinking about cognition, there has been a concentration in this literature on symbolic devices and symbols and their interaction with the cognizers. One consequence of this is that the scaffolded nature of processes of awareness and cognition more generally appears as a relatively esoteric, specialized addition to biological cognitive systems, the cream on the cake of cognition. This is seriously misleading, for the range of scaffolding involved in processes of awareness is extensive, and includes, in addition to humanly constructed

Expanding Consciousness 219

symbolic devices, a range of environmental and social structures that are appropriated by cognizers.

For example, systematic, reflective, cognitive use has been made of natural features of land and sky, such as the position of the sun or the shift in position of landmarks, in traditional seafaring navigation. These natural features play similar roles that cognitive devices play in modern, Western navigational practices. To take another example, a road itself can be as important a cognitive resource as signs along that road or a map that shows where the road leads. Individuals who want to get from A to B need to be appropriately coupled to all of these resources. In neither case need we think of such cognitive resources as symbolic in nature, or as carrying meaning in and of themselves. What is crucial about such scaffolds for processes of awareness is that they are causally integrated into what can be complicated, conscious actions that simply could be not performed without them. Their integration, their causal connectedness to the individual cognizer, extends not only that individual's cognitive abilities, but expands her consciousness.[3]

Once we adopt a temporally extended view of consciousness and start looking for scaffolding in everyday mental life, examples multiply. In the section on cognitive development in the previous chapter, I said that both spoken language and written language serve as mediators that change the structure of cognition, and much the same is true of consciousness. The written word and all that invokes it – from labels on packages, to directional signs, letters, books, advertisements, addresses, codes, computers, scientific instruments, t-shirts, shopping lists, cards, the Western educational system, law courts, and bureaucracy in general – is ubiquitous in how we cognitively negotiate our daily lives. Its integration into our conscious mental lives alters the very processes of awareness that constitute consciousness through augmentation. Those processes now incorporate the written word, just as the processes of awareness of speaking creatures came to incorporate speech into consciousness. In each case, consciousness itself relies on, and expands to include, such external scaffolds.

A second, general source for mental scaffolding, as suggested in my discussion of folk psychology in the previous chapter, is other people. Others often constitute cognitive resources that can be accessed fairly directly (for example, by asking them) or in more indirect ways. If we accept that cognitive resources can be distributed across individuals, then it is a small step to viewing these resources as sometimes the object of processes of awareness. But let me say something more about the third plank to the view of processes of awareness I am articulating – the embodiment and embeddedness of such processes – before taking this small step.

Temporally extended processes of awareness are embodied and embedded not simply in that they are processes of organisms, organisms have bodies, and these bodies exist in environments. This is banal. Rather, conscious life in the time scale of minutes or hours involves agency, and agents, at least the sorts of agents that we are and all examples that we know of are, exercise their agency in the physical world through their bodies over extended periods of time. There is no other way to act. And bodies gain traction with the physical world through reliably causing changes in, and in turn by being changed by, that world. No environment, no bodily action; no bodily action, no agency; no agency, no temporally extended processes or awareness.

The countervailing claim that we can imagine such processes taking place without either or both a body and an environment at all has a rich philosophical history, from Descartes's evil demon hypothesis to brain-in-a-vat thought experiments. Does the TESEE view of consciousness deny that we can imagine processes of awareness going on in such cases? Two related replies. First, such imaginings are typically radically underdescribed, and they often become incoherent once they are imagined more fully. More pointedly, I have found that disagreements over what can and can't be imagined here often turn on one's prior philosophical commitments, and that makes me suspicious about relying too heavily on such claims of imaginability. So, while I am sympathetic to arguments that point to problems with completely disembodied and disembedded consciousness, I think that methodologically it is wise to steer clear of a commitment here. Second, the TESEE view of consciousness is a view of how consciousness works as a matter of fact for creatures like us with bodies in environments, rather than of necessity for any creature that we think we can imagine having the very same processes of awareness as us. One need not have a view of the latter in order to have a theory of the former.[4]

The previous paragraphs suggest an argument from the temporally extended nature of processes of awareness to the further claims about agency, embodiment, and embeddedness. But in fact I think that the argument could be run from any one of these claims to the others, since they form a cluster of claims that stand or fall together. If this is right, then the position I am advocating needs to be either accepted or resisted *in toto*; there is no middle-ground position. Thus, the tempting idea of viewing processes of awareness as temporally extended, but taking the somatic and extrasomatic resources they involve as simply inputs for internal processes of awareness to operate on, is a nonstarter. If one expands consciousness on one of these dimensions, one expands it on all.

There are thus two sorts of arguments that need to be made to support this position. The first should make a prima facie case for this conception of processes of awareness being a package deal, so that the only live options are two: Accept or reject the package. The second should then argue for the former over the latter of these options.

4 ARGUING FOR EXPANDED CONSCIOUSNESS

Much of the previous section constitutes a start on the first of these tasks, since it proceeded by showing how accepting one part of the package leads to accepting other parts. But we need also to show that there is no privileged starting point in this conception of processes of awareness. Consider the following argument, which draws on the premise that processes of awareness – higher-order cognition, attention, and introspection – involve not only various sorts of access to environmental resources, but that these are properly thought of as cognitive resources for the individual.

(1) Processes of awareness involve accessing and using cognitive resources.
(2) Some of these resources lie beyond the head of the individual.
(3) Accessing and using these resources requires acting on the world in specific ways (through eye and head movement, bodily orientation and motion, manipulation). [2]
(4) Even accessing and using internal cognitive resources (for example, memories, perceptions) often requires such bodily action and worldly engagement through mnemonics, intention-fixing, rehearsal and repetition, talking to oneself, doodling. [1]
(5) The actions required in 3 and 4 typically occupy a temporal extension of many seconds, minutes, or even hours. [3 and 4]
(6) So processes of awareness are temporally extended [5], scaffolded [2], and embodied and embedded. [3 and 4]

I take it that the most controversial premise in this argument is 2. An individualist who concedes 1 (and perhaps 4), might think that she can embrace the temporally extended nature of processes of awareness, and so a version of 5 and 6, without going externalist. This would be to concede that consciousness is temporally extended and relies on environmental resources, while denying that the beyond-the-head resources are cognitive.

Traditionally, the mind is conceptualized as beginning and ending inside the head. We can express this view as implying two things about the cognitive resources that constitute the mind. First, they lie exclusively in the head, and second, they do not strictly require bodily action and worldly engagement to be cognitive resources. Suppose that we attempt to tackle the argument for the TESEE view of consciousness by staying close to the traditional view of the mind on just the first of these two points. That is, suppose that we have a head-bound view of cognitive resources, but acknowledge that these internal cognitive resources often interact with the world beyond the head in their role as cognitive resources. In effect, this would be to attempt to cede both 1 and 4 in the argument, while denying 2 and 3. Can such a position be defended?

One reason to think not is that the very processes that 4 concedes involve resources in the world. If these processes are cognitive, then these are cognitive resources. Hence 2. What is needed is a way to deny that processes of awareness are ever themselves embedded. But the range of examples that we have seen in the previous section makes this prima facie implausible. The chief and perhaps most obvious way to deny this is to defend the view that there is a fundamental asymmetry between what is inside the head and what is outside of it, such that only the former can constitute processes of awareness. An encoding view of representation would fit the bill, since then resources inside the head would code for those outside the head, but not vice-versa. But I have argued against such a view in the previous chapter. I see no nonquestion-begging way to defend the requisite asymmetry.

On the view of representation as exploitative that I have developed, individuals or cognitive systems exploitatively represent objects, properties, events, and propositions, but the boundary of the individual does not mark the place where representation begins or ends. States inside the head are caused by, and carry information about, states outside the head, but they bear this same relation to states of the body. Applying this view to processes of awareness that operate through bodily action beyond the limit of milliseconds and seconds makes it particularly difficult to distinguish as cognitive resources just those states that occur inside the head. As we manipulate our relationship to the world through action – whether it be through physical grasping with the hands in object manipulation, head or eye movements in visual attention, talking to oneself or another in figuring out what one thinks – we also cause changes in those parts of the beyond-the-head world. Processes of awareness often have a phenomenology, can be "directly" activated or accessed, and have a subjective dimension. But as both internal and external resources play crucial and similar roles in these processes, it is arbitrary to assume that the processes themselves begin and end in the head.

Thus, it is problematic to maintain just the first part of the traditional view of the mind, that it lies in the head, while granting that there is some sense in which processes of awareness are temporally extended and embodied and embedded. So suppose instead that we look to uphold both parts of the traditional view, conceding less to the TESEE conception. In terms of the argument for TESEE that I have given this would be to reject 4 as well as 2 (and so 3), or maintain only an individualistic version of 4. This is clearly possible, but at a price. For then the only processes of awareness the argument encompasses are those that don't involve bodily action and worldly engagement. This would exclude much higher-order cognition and attention, both of which are often worldly directed, and even large parts of introspective practice, which is seldom purely internally directed. More importantly, there seems little sense in which processes of awareness could be temporally extended, rather than a series of temporal snapshots strung together, since nearly any such process that endures many seconds, minutes, or hours does involve precisely what the denial (or modification) of 4 prohibits. Thus, removing 2, 3, and 4 from the argument also removes, or at least radically attenuates 5. This reinforces the sense that the elements of the TESEE conception form a package, and need to be accepted or rejected as a whole.

Now, to stage 2: The defense of the conception of processes of awareness as temporally extended, scaffolded, and embodied and embedded, the TESEE conception. There are two chief, general, related desiderata to appeal to here: to account for as full a range of mental life involving processes of awareness as possible; and to deliver a conception of such processes that shows how such processes are integrated with the rest of cognition.

On the question of integration, how well one thinks this view integrates with the rest of cognition will depend on what one thinks the rest of cognition is like. In Chapters 7 and 8, I sketched a view of both computational and noncomputational psychology in which the ideas of exploitative representation and locational externalism played a central role. Most relevant for thinking about processes of awareness, is the externalist conceptualization of memory, cognitive development, and folk psychology, introduced in the previous chapter. For a strong case can be made for seeing continuity between such processes and processes of awareness. This is in part due to the metarepresentational edge to these processes, involving, as they often do, the representation of representations. The theme developed in that chapter was that there was some pressure to move from individualistic to externalist conceptions of such processes as we considered the more sophisticated and real-life forms that they took. The same is true of processes of awareness.

Take higher-order cognition, which is typically characterized as explicitly metarepresentational. The simplest example of such a process is that of occurrently thinking about or entertaining some other thought you have. Suppose that you are sitting in a room, eyes closed, and you do this. The second-order thought is distinct from simply having the corresponding occurrent first-order thought. Intuitively, it involves something in addition, some further mental process whose object is that first-order thought. There is some pressure to construe such examples of higher-order cognition as taxonomically externalist; in effect, this was argued with respect to memory in the previous chapter. But consider more complicated instances of this sort of process that are common in everyday life.

Consider, for example, the higher-order cognition you have when you are reflecting on whether your beliefs or opinions are consistent, or engaging in a conversation whose aim is to resolve a dispute over values, or writing up a sort of informal balance sheet that allows you to explore how realistic it would be for you to seek to achieve a long-time goal. These are all processes that typically involve you *doing something*, from reading over your diaries, to drawing on the views of others, to making diagrams, notes, and annotations. As such, they depart from the simple case we began with and its Cartesian overtones. The something that is done involves the body's interactions with cognitive resources beyond the head. This action is possible only because the relationship between what is inside the agent and what is beyond its boundary is reliable and systematic, and thus can be taken for granted in acting.

If this is correct, then this is one respect in which the externalist views defended here offer a way to treat conscious and unconscious cognitive processing in an integrated or unified way. The traditional division between higher cognition (what I have referred to as "cognition central"), and processes of awareness as forms that consciousness takes is an artifact of history and of disciplinary orientation. If we adopt this view of how processes of awareness are integrated with other parts of cognition, then the issue of integration becomes entwined with that of its range of application. For now there is no firm boundary between what I have been calling processes of awareness and "merely" cognitive processes. The TESEE conception applies to both.

To avoid one misunderstanding of what the TESEE conception of processes of awareness aims to do, let me locate that view within the pluralistic view of cognition that is part of my externalism. The aim in the preceding and present section is not to tout the behavioristic view that there is nothing internal to conscious experience, or that processes of awareness are never purely internal. Sometimes they are. Rather, it is to suggest that such processes more often involve what is inside the individual insofar as they are integrated into what that individual does as an embodied and embedded agent. And more so, rather than less, as we move to more sophisticated, realistic examples of consciousness in action. What the TESEE conception challenges is the near exclusive focus on what is inside the head in thinking about processes of awareness, just as the introduction of context-sensitive realization, wide computationalism, and exploitative representation challenge a similar focus elsewhere in our conception of the mind and cognition.

5 GLOBAL EXTERNALISM AND PHENOMENAL STATES

Let us now turn from processes of awareness to the other half of consciousness, what I have referred to collectively as phenomenal states: bodily sensations, pain, and visual experience. However plausible the TESEE conception of processes of awareness is, it is prima facie more problematic to apply it to phenomenal states. Individualism is a natural starting point for accounts of such states, even if we grant the externalist view of computation and cognition developed thus far.

Philosophers have generated a variety of views from the idea that phenomenal states are a sort of internal event, one wedged at the interface between mind proper and world. These include classic empiricist accounts of the mind, Cartesian skepticism within epistemology, twentieth-century sense-data theory, and versions of the explanatory gap problem for contemporary materialism. Phenomenal states have typically been regarded as occurring completely within the boundary of the individual. Thus, they are at least locationally individualistic. Why think that they are also taxonomically individualistic?

In the case of at least bodily sensations and pain, this has been taken to imply taxonomic individualism about those conscious states, typically because such states are not thought to have propositional content (though see below). Thus, the chief feature of mental states that has been seized on as the basis for viewing them as taxonomically externalist – their intentionality – is absent in the case of bodily sensations and pain. What is pain about? Nothing: It just hurts. There is a distinct feel to pain, and a distinct phenomenology more generally to bodily sensations, such as those gained through touch. This "raw feel" to pain and bodily sensations supervenes on just the intrinsic, physical states of the body.

In the case of visual experience, a version of this view has been thought to hold because there is an aspect of visual experience – how things seem to me visually here and now – that can be distinguished from the conceptual or propositional content that such experience possesses. Even if we concede that visual experiences have a (wide) propositional content, there is some further content they have that is narrow, and so taxonomic individualism is plausible for at least this aspect of visual experience. This has sometimes been expressed by saying that visual experience has *nonconceptual* content, that this content is individualistic, and that it plays an important role in understanding the phenomenology of visual experience.[5]

Fred Dretske has defended the view, by contrast, that if one is an externalist about conceptual or propositional representational mental states, then one should also be an externalist about all experiential mental states, particularly those of visual experience. Dretske holds this view as a function of his endorsement of a representational account of experience that entails that consciousness is a species of representation. If one is an externalist about representation, then one should also be externalist about consciousness. Dretske recognizes, however, that to avoid the conditional being run as part of a *reductio* against externalism, externalism about conscious experience requires defense. Why should one be an externalist about phenomenal states?

Dretske's chief positive argument for this position can be put succinctly. What are phenomenal states? If they are or involve conceptual or thoughtlike entities, then those states are taxonomically externalist: They inherit their width from that of the concepts or thoughts they involve. In this respect, phenomenal states are like metarepresentational or higher-order mental states, and the sorts of arguments that motivated externalism about intentional states apply to them. If, on the other hand, phenomenal states are completely divorced from conceptual or thoughtlike entities, such that (for example) it is possible for two individuals to have distinct phenomenal states despite their sharing all their intentional states (including beliefs about those states), then phenomenal states are unknowable – even from the first-person point of view. Apart from running counter to how phenomenal states are usually thought of, this in turn removes whatever ground we might have for thinking that such states of intentional *doppelgängers* are different.[6]

Although Dretske is concerned in the first instance with sensory states, he views this argument as applying more generally to phenomenal states, including pains, emotions, and motivational states. The position that Dretske defends, or sees as defensible, thus might be termed *global externalism*, since it holds, or suggests, that externalism is true of all mental states. William Lycan and Michael Tye have also defended such a view. All three philosophers reach their global externalist conclusion via their endorsement of a representational account of the phenomenal. Here I want to focus on global externalism itself, and why it should be resisted.[7]

At the end of the previous section, I located the TESEE conception of processes of awareness within the over-arching pluralistic view of the mind and cognition that I have been developing in Parts Two and Three. I have also said that I am skeptical about the prospects for a general theory of consciousness in part because of the diversity among the phenomena and processes considered as falling within its ambit. These points together suggest at least a certain caution in speaking of "the mental" or "the conscious" as categories that we might systematically and globally theorize about. I suspect that the same is true of "the phenomenal" as a category about which we might have substantive, interesting, true generalizations. I include here Dretske's complex generalization that all such states are externalist if language- or thoughtlike, and unknowable if not. While Dretske's dilemma argument identifies something correct about some phenomenal states, both horns of the dilemma can be resisted.

This resistance is easiest to mount in the case of pain and bodily sensations, and it turns on the fact that we have ways to individuate such states other than via whatever propositional content they have. (This is the truth, I assume, in the idea that such states are not representational at all.) Intuitively, one such way is by an appeal to their phenomenal character, but Dretske's dilemma argument suggests that there is a fragility to this sort of appeal, one that may well call into question our first-person knowledge of the nature of our phenomenal states. Let me explain.

Some phenomenal states, such as pain or the bodily sensations generated by touch, have an aspect to them that is only distantly related to language. Subsequently, they have a content that may be difficult to express in language. I feel a dull, gnawing pain in my knee area, and thus come to believe that I have a pain here, but what has been called the phenomenological content of the former state is not identical to the propositional content of the latter. If I have another such recurrent pain in my thigh, it will likely cause me to believe that I have a pain in my thigh. But if I am attentionally preoccupied, or tired or drowsy, then I could experience one or the other of these pains without forming the corresponding belief. I know that I was or even am in pain, but I don't know just which pain it is. (Suppose that I have both kinds of pain regularly, and they are not all that phenomenologically distinctive from one another.) Maybe I form the wrong belief here, or maybe I form no belief at all. Perhaps I simply register discomfort somewhere or other in my leg, not conceptualizing it as pain, or as more specifically located.

Suppose we view such states as being thoughtlike, so that they fall under the first horn of Dretske's dilemma. Then according to Dretske they should inherit their content from that of the thoughtlike states to which they are causally related. Yet it is difficult to say just what this content is precisely because the relevant thoughtlike states are fleeting and poorly attended. But insofar as they have a clearly specifiable content, it seems doubtful that their content is itself wide. Since it both pertains to a condition of one's body and is suitably general (because vague), it seems that at least embodied *doppelgängers* must share it. There can be no externalist contagion from the thoughtlike to the phenomenal if the thoughtlike states themselves are individualistic. When I think "I feel discomfort around here," thinking of what is in fact a pain in my knee but which I do not conceptualize as such, then this is a state that would seem to be shared by anything identical to me in its intrinsic, physical properties. Even though there will be representational descriptions that *doppelgängers* need not share (for example, I have a pain *in my knee*), they are not the only thoughtlike states that one may have. Thus, the thoughtlike character of some phenomenal states does not entail externalism.

Suppose, on the other hand, that we view such phenomenal states as not being thoughtlike (and so as falling under Dretske's second horn). Then Dretske's claim is, in effect, that such states are unknowable, and that we thereby lack a way of individuating them. Thus, we have no basis for saying that they are either identical or different in any particular pair of cases. This claim is effective in undermining the use to which individualists put Twin Earth–styled arguments about phenomenal states, which is part of Dretske's own aim here, and it rightly flags some significant limits to the first-person knowledge we have of phenomenal states. Yet we should reject the more general claim that phenomenal states lack criteria of individuation unless they are thoughtlike.

Again, consider the state of pain that leads me to be in the thoughtlike state "I feel discomfort around here." Even if we think of this state of pain itself as not thoughtlike, we can still individuate it by reference to the relevant entity-bounded system. As I said in Chapter 5, states of pain form part of the nociceptive system, and at least as conceptualized within the medical communities that focus on pain (anesthetists, oncologists, pharmacologists, for example), this is an entity-bounded system, and so locationally individualistic. On this view, the principal entity that is individuated is the nociceptive *system*, and the individuation of various states of the system is derivative. The nociceptive system in a particular overall state will instantiate pain (say, rather than a tickle, or nothing at all), or pain with a certain representational content, because that is the kind of state that this type of system instantiates. In short, the systemic conception of individuation and realization defended in Part Two provides a way of resisting the second horn of Dretske's argument. Thus, denying the thoughtlike character of at least some phenomenal states does not entail externalism.

Consider a variant on the simple example I have focused on so far to clarify what this response to Dretske's argument for global externalism shows, and what it suggests is special about at least some phenomenal states. I have granted that states of pain have a representational content but argued that this concession does not entail externalism about phenomenal states via either horn of Dretske's dilemma. There is something right about Dretske's argument, but it does not lead to global externalism about phenomenal states.

Suppose I feel a strong pain in my left knee for about five seconds, and so come to think

There is a strong pain in my left knee that has lasted about five seconds.

The representational content of this thought concerns the strength, location, and duration of the pain. We might think, first, that we can thus individuate the pain itself by the corresponding content: It is strong, occurs in my left knee, has lasted about five seconds, and so on. And, second, that such content must be narrow. But consider a *doppelgänger* who lives in a community that uses "knee" to refer to what we call the knee, plus the lower half of the thigh area, yet (here, mistakenly) thinks that knees are just the joint area in the middle of the leg. Suppose that we each undergo two pains, the first in what we each think of as our knees, the second in the lower part of our thighs. Then the thought that we each report by saying

The first pain was in my knee but the second was not

is one whose wide content differs, depending on which one of us has the thought. And so if the pain itself is individuated in terms of this content, then it is an externalist state.

About this, I think Dretske is correct. Since the sort of content that is typically ascribed to phenomenal states is subject to precisely this sort of point, the representational nature of phenomenal states often entails taxonomic externalism about the phenomenal. Perhaps paradoxically, this is particularly true of the content typically reported in the first-person, since we tend to characterize our own states more precisely and in richer detail using a public language than we do the states of others.

Yet *contra* Dretske, representationalism doesn't strictly entail such externalism because one can individuate pains other than via this kind of propositional content. In fact, there are at least two different ways to do so that together constitute the basis for a response to both halves of the dilemma that Dretske poses. First, there remains a kind of content – vaguer, more general, and bodily – for which such Twin Earth examples cannot be constructed (and so the first horn can be grasped). Second, we can individuate states in terms of the systems of which they are a part, and in the case of at least pain and presumably bodily sensations these are entity bounded. Thus, there remains a path to an individualistic scheme of individuation for those phenomenal states that is independent of their connections to any kind of content (and so the second horn can be grasped).

One might wonder whether this response can be generalized in either of two ways. First, might a version of the first half of the response also apply to the propositional attitudes themselves? Some proponents of narrow content, such as Frank Jackson, Philip Pettit, and Stephen White, have suggested that there is such a notion of general narrow content. Second, could a version of the second half of the response be generalized to apply to phenomenal states in general? This would support the idea that externalism stops with the intentional and need not infect all of the phenomenal, some of which remains individualistic. I think that neither of these generalizations is warranted.[8]

Consider the first. The notion of narrow content for states of pain and bodily sensations is quite restricted; I have suggested that it does not apply even to such states in general. Our usual ways of thinking about our own states of pain and bodily sensations make use of the same public language used to specify the content of our intentional states. Dretske is correct in holding that phenomenal states, so specified, inherit the wide content of the concepts employed here. Yet there remains something it is like to have such states that can be specified more vaguely and generally that does not make implicit commitments to how the physical or social world beyond the individual is. Once we move to other sorts of phenomenal states, such as those of visual experience, or to the propositional attitudes more generally, the Dretskean point about the inheritance of width applies so broadly that talk of narrow content appears misplaced. To claim, for example, that the narrow content of

> *There is some water*

is

> *There is some drinkable, transparent, sailable-on kind of stuff*

or

> *There is some stuff that I use "water" to refer to*

is perhaps to identify a psychological state that some *doppelgängers* share.[9] Yet that state and the content it has is not truly narrow, since *doppelgängers* may differ with respect to them. The sort of content that we would need would be something like

> *There is some watery stuff*

where "watery stuff" refers to stuff that has the appearance of water. Even were this individualistic (which I doubt), note that it is not something that has the same meaning as "There is some water," and not something that individuals need be in position to use to characterize their own internal states, since they may lack the concept "watery stuff."

Whether phenomenology itself might be sufficient to determine an individualistic form of content is something I shall discuss in more detail in the next chapter. My point here is simply that the relationship between phenomenology and content may vary depending on the type of phenomenal state we are considering, and this is sufficient to undermine the first generalization, from the restricted appeal to narrow content that I have made.

Consider, more briefly, the second generalization, one that generalizes on the appeal to entity-bounded systems. Why not simply appeal to a range of entity-bounded cognitive systems – for belief, for perception, for motivation, for emotion – in order to undermine Dretske's claim that phenomenal states divorced from language cannot be individuated? There are two reasons, perhaps familiar in light of our discussion in earlier chapters. First, whether any given system is entity bounded or wide is not simply "up to us," something we invent rather than discover. Rather, that is resolved by looking to the relevant sciences: How do they taxonomize the kinds of system that they traffic in? Second, perception, memory, and folk psychology all constitute cognitive systems that are not just taxonomically but locationally externalist, at least when we consider the full range of their uses. Given this, it is plausible to view entity-bounded cognitive systems as the exception, rather than the rule. Pain and bodily sensations are two paradigms of mental states realized by such systems. Their assimilation to physiological systems together with the limited role that wide content plays in their individuation, is a rare combination for cognitive states. For example, I doubt that either feature holds of all forms of visual experience, particularly given the locationally externalist view of perception.

6 TESEE AND SENSORY EXPERIENCE

In the previous section I intimated that even though Dretske's dilemma could be averted in at least the case of pain and bodily sensations, it retained its force against its primary target, individualistic accounts of visual experience. I have also suggested that the TESEE view of consciousness applies both to processes of awareness and to phenomenal states. In light of this pair of points it seems most plausible to develop the TESEE view for visual experience in the first instance, stepping back to consider sensory experience and then phenomenal states more generally once we gain a foot in the door of the phenomenal.

The TESEE view holds that consciousness is temporally extended, scaffolded, and embodied and embedded. As with the application of the TESEE view to processes of awareness in sections 3 and 4, developing a TESEE view of visual experience requires reconceptualizing the relevant aspect of consciousness. Such a rethinking of the nature of visual experience is underway and has been facilitated by a more general paradigm change in the study of vision. Traditional views of vision conceive of it as passive, internal, and essentially in the business of generating rich internal representations. Much of the work on vision over the last dozen years or so within the cognitive sciences has self-consciously contrasted itself with tradition here and conceptualized vision as active, embedded, and essentially in the business of generating action. For example, the psychologists

David Milner and Mel Goodale have reinterpreted the original distinction between the "what" and "where" visual systems in the ventral and dorsal processing streams as systems for, respectively, perception and the guidance of action. They remind us that "vision evolves in the first place, not to provide perception of the world per se, but to provide distal sensory control of the many different movements that organisms make." The computer scientist Dana Ballard distinguishes literalist from functionalist views of visual representations, the latter of whose "principal tenet is that the machinery of the brain has to be accountable to the observed external behaviour, and that there are many ways to do this other than positing literal data structures."[10]

While such views stop short of reenvisioning the kind of experience that we call visual, they make that reenvisioning possible. I shall discuss two ways in which a rethinking of visual experience has been undertaken recently. Both involve abandoning or significantly tempering the idea that particular states of an individual are phenomenal in and of themselves, and in light of this they have broader implications for our thinking about consciousness.

The first of these views has been articulated by the philosopher Alva Noë and the psychologist J. Kevin O'Regan as part of their *sensorimotor theory of visual consciousness*. Developed specifically as a view of visual experience, their view holds that seeing is a kind of act that involves an organism actively exploring its environment. This activity is governed by what they call sensorimotor contingencies, knowledge of which is gained through active exploration of an environment, and whose acquisition constitutes the acquisition of the relevant experience. As O'Regan and Noë say, visual experience is "the activity of exploring the environment in ways mediated by knowledge of the relevant sensorimotor contingencies."[11]

To explain this view on an intuitive level and how it departs from traditional views of visual experience, Noë draws an illuminating parallel between visual and tactile perception. Seeing a bottle is more like tactilely feeling a bottle than it is like taking a picture of a bottle. When you have a bottle in your hands, you only make contact with a part of the bottle at any given time, although you feel the bottle, not simply some part of it. That is how you would naturally report your experience, in any case. Moreover, you move your hands to actively explore the bottle in perceiving it, and this interaction with the sensed environment is not a way of taking a series of tactile snapshots of the bottle in order to infer that it is a bottle. Rather, the active manipulation of the bottle is your tactile experience of it. The tactile experience is not some event inside you, but something that you undergo in your active exploration. Tactile experience, so conceived, is temporally and spatially extended, and since it requires active exploration of objects in the environment, it is also clearly embodied and embedded.[12]

One of the strengths of the sensorimotor view is that it provides a natural way to interpret a range of cases where sensory experience itself seems to involve or require either explicit or intended action on the part of the subject. These include well-known cases of adaptation to left-right inverting lenses and those involving change blindness, where scene change during saccadic motion creates stunning lapses in change detection. On the sensorimotor contingency theory, the first of these involves a subject learning a new sensorimotor contingency within one sense modality that comes to be integrated with the rest of the adapting subject's behavior. The second not only reveals the limitations to the information that is encoded in visual representation, but also shows the importance of active and continual sensory interaction of the surrounding environment for complete perception.[13]

The account also predicts that there should be cases of sensory substitution just when alternative sensory channels preserve the same structure of sensorimotor contingencies. This arguably is precisely what is found in cases of remote tactile sensing and the visual experiences of users of tactile visual sensory systems (TVSSs). The best known of the latter of these use a camera to feed visual information into cutaneous stimulators attached to the subject's back, stomach, finger, or forehead, where tactilely stimulated, blind subjects report visual experiences. On the sensorimotor contingency theory, what subjects learn through their practice in actively controlling the camera is a new series of tactile sensorimotor contingencies, each of which is isomorphic to those that the visual system would normally develop.[14]

Critical to the success of these TVSSs in producing reported visual experiences in the blind is the ability of the subject to control the movement of the camera. This makes sense on the sensorimotor theory, since what is critical to visual experience, on that view, is establishing or maintaining sensorimotor contingencies, not simply receiving or preserving information that codes for visual features of the environment. (For readers with CAVE – Complete Audio-Visual Environment – virtual reality experience, note the experiential difference when you walk with the virtual reality control, as opposed to when you follow someone else holding the control.) The reported visual experience of TVSSs are, to be sure, incomplete vis-à-vis those using their eyes – for example, subjects do not report seeing objects as colored. But this can be explained in terms of the limitations to the isomorphism between the visual information that the subject controls through the camera and the tactile stimulation she receives on her skin. The relevant (and surprising) finding is that there is any visual experience reported at all.

Likewise with remote tactile sensing more generally. Tactile sensations are usually felt at a particular location in the body, but they can also be felt in objects that extend beyond the body, typically in (and through) objects to which the body is reliably connected. On the sensorimotor theory, and TESEE views more generally, this is because sensory experience is a matter of embodied and embedded agents doing things in the world. While the body is a critical mediator for that action, it forms only a contingent boundary for sensory experience that can be, and sometimes is, transcended. Blind people experienced in the use of a cane feel objects in the world, not the cane itself; bicyclists can feel the flatness of a tire while they are riding, and without touching the tire itself; and experienced hockey players feel the puck on the end of their sticks. Just as in the case of visual perception using TVSSs, here the phenomenology requires practice and is more restricted than that gained through regular tactile stimulation. In particular, the sensations are chiefly spatial in character, concerning the position, size, texture, and shape of what is felt, and do not correspond to the full range of bodily sensations. On this view, tactile sensations are felt at a bodily location, despite the fact that it is in the brain where they are registered, because the body is systematically linked to the brain as a matter of physiology. When the body becomes similarly causally linked to other features of the world, at least some of those felt sensations can be transferred to those features.

A second way in which a TESEE view has been developed is through Susan Hurley's attack on the equation of perception with input (and action with output). Hurley's alternative idea is that perception and action are better thought of as forming dynamic cycles. In the previous chapter, I briefly mentioned Hurley's caricature of the traditional view of the mind as a sort of sandwich in which cognition is the filling. Her conception of perception and action as forming dynamic cycles is one of her positive alternatives to the traditional characterization of perception and action.

On the traditional view, perception is input and the cognitive architecture of the organism is conceptualized in terms of what Hurley calls *vertical modules*:

Each vertical module performs a broad function, then passes the resulting representations on to the next. Within the perceptual module, information about locations, color, motion, and so on, is extracted from inputs by various parallel streams of domain-specific perceptual processing. The representations produced by the various streams of input processing converge and are combined by perception. The unified result is sent on to cognition, the central module that interfaces between perception and action.[15]

The chief puzzles about visual experience on this view are where, when, and how consciousness of particular features of visual experience come about, and how they are unified to form the seamless experience that, for the most part, we typically have.

In challenging this division between perception and action, Hurley introduces the idea of *horizontal modules*, content-specific layers that loop "dynamically through internal sensory and motor processes [as] well as through the environment." There are several features of horizontal modules relevant to our rethinking of visual experience. The first is that they are dynamic in that they involve continuous feedback between perception and action. For this reason, any process that they govern must be conceptualized as temporally extended, rather than simply as a series of temporal atoms, and as embodied in an active agent. The second is that there is no barrier to such modules being either entirely internal to an organism, or extending partly into that organism's environment and so locationally externalist. Perception does not occur first, with perceptual inputs being processed "centrally" and then instructions passed off to action. Rather, the many systems engaged in perceptual processing, including in the production of the "what it's like" of experience, cut across the input-output distinction. As with the sensorimotor contingency theory, on Hurley's view there is a "there" to sensory experience, but sensory experience is not properly conceived as a type of internal event, or a property of some specific internal state of the organism.[16]

There are several affinities between the views of O'Regan and Noë and Hurley. Like the sensorimotor contingency theory, Hurley's views of perception challenge the traditional way of thinking of visual experience as some sort of internal event that is then monitored or processed further in some other way to give rise to consciousness. And like that theory again, Hurley's view makes sense of a range of actual and hypothetical experiments concerning perceptual experience, intention, and action. These include some of those that O'Regan and Noë also discuss – including the left-right inversion prism experiments of James Taylor and Paul Bach-y-Rita's studies of TVSSs – but also Ivo Kohler's classic experiments with blue-left and yellow-right inverting goggles and cases of asymmetrical bodily and perceptual neglect.[17]

Particularly relevant is Hurley's discussion of the spatial inversion experiments that Taylor performed using the mathematician Seymour Papert as a subject. In these experiments, the subject was fitted with left-right inverting goggles, and then trained on a range of tasks that required moving his body through the actual (uninverted) world. Taylor reports that Papert adapted to the wearing of goggles fitted with these lenses to the extent of being able to ride a bicycle with them on, and also readily adapted to the removal of the goggles, even while riding the bicycle. As Hurley recounts, to perform successfully on such tasks requires calibrating a range of cognitive systems, including those for motor control and for proprioceptive feedback and control, for now we have a subject acting in the world, not simply recording information about it visually and reporting that information. To the adapted subject, the world appears the right way up, and his performance on a range of spatial tasks suggests that his adaptation is not simply visual but proprioceptive, intentional (in the sense of concerning the agent's intentions), and motoric as well. What has adapted is the whole subject, or at least cognitive systems that cut across the perception-action divide of traditional views of perception.

One conclusion that Dretske draws in his argument for global externalism can be expressed by saying that visual experience is metaphysically determined in part by the thoughts and language that we use to describe it. We can draw a similar moral from left-right inversion cases: that visual experience itself is metaphysically determined in part by what one does. More specifically, visual experience lies in the coordination and integration of visual "input" and behavioral "output," in the exploitation of visual information for the direction of behavior.

Hurley also considers a variation on the actual experiment that places the subject in "Mirror Earth," an environment in which there are left-right reversals that are, in turn, reversed by the left-right inverting lenses. This scenario parallels Ned Block's "Inverted Earth" thought experiment, which Block uses to challenge externalist, particularly representationalist, views of the nature of consciousness. What is distinctive about the Mirror Earth thought experiment that Hurley offers is that, as with the left-right inversion experiments, it involves a "where" rather than a "what" system. As such, this Mirror Earth case allows – indeed, requires – the subject's exploration and manipulation of things in his environment. Since the inverting lenses correct only for the worldly inversion as it is perceived visually, further inversions are required if the subjects are to remain in the same internal states through their respective explorations of their environments. As Hurley points out, unlike the case of Inverted Earth, which involves color inversion, this case of Mirror Earth requires that motor output and proprioceptive feedback be systematically inverted in order to have a case in which we have two individuals identical in their intrinsic, physical properties.[18]

Part of Hurley's point in considering the dynamic Mirror Earth variant on the actual left-right inversion experiments is that it is in fact very difficult to satisfy the assumption that an individual on Mirror Earth can be a duplicate of an individual on Earth, since the requisite inversions in the dynamic case seem to require changes in the subject himself. It is with this Mirror Earth case in mind that I want to turn to some broader issues about consciousness and philosophical methodology in the next section.

7 INDIVIDUALISTIC RESIDUES

Both the sensorimotor contingency theory and the dynamic cycles conception of perception-action exemplify how a TESEE view of visual experience can be developed. Together with representationalist views of visual experience of the type that we have seen Dretske articulate, they provide a basis for challenging much that has been taken for granted about visual experience that has suggested an individualistic view of it. I think that the root assumptions at play here are that (a) particular states are phenomenal in and of themselves, and that (b) phenomenal properties are intrinsic properties of the states that have them, or (at least) of the subjects of those states. The systemic view of taxonomy and individuation introduced in Part Two provides a general reason to be skeptical about (a), and the TESEE view of visual experience provides a positive framework that either does without (a) or tempers its significance. Insofar as particular visual experiences are phenomenal, they are part of an embodied system as it dynamically interacts with a particular environment and with an agent's motoric, intentional, and proprioceptive capacities. The rejection of (b) amounts to a rejection of "qualia," or at least one significant feature of them, but it is important to see that the TESEE view is not eliminativist about visual experience, as other critiques of qualia have been. There are phenomenal properties that we have first-person access to, just as there is mental content to which we have such access. But in both cases, these properties are not what we might have thought they were. In particular, they are externalist, rather than individualistic, in nature.[19]

What of the TESEE view and the remaining phenomenal states, those of pain and bodily sensations? In section 5, I argued that Dretske's argument for global externalism was weakest in the case of these sorts of phenomenal states. This is because their representational character is not necessarily subject to the social deference that drives taxonomic externalism, and they are realized in entity-bounded systems, and so are not subject to an appeal to locational externalism. These points make it prima facie less plausible to apply the TESEE view to these kinds of phenomenal states across the board, even if many of the representational characterizations we naturally give to such states follow those of visual experience and are at least taxonomically externalist. This constitutes one sort of individualistic residue that remains once we push externalism as far as we can, from the intentional to the phenomenal.

But there is another such residue, a sort of legacy of a philosophical tradition heavily populated by thought experiments. Philosophical discussions have made it seem easier than it in fact is to articulate an individualistic conception of phenomenal states, and correspondingly made externalist views appear more counterintuitive than they in fact are in part because of their reliance on what can and can't be conceived in certain kinds of thought experiments. For example, a familiar line of attack on materialist views of phenomenal states is based on the claim that one can conceive of the possibility of either inverted or absent qualia. The paradigm case discussed in such arguments is that of color experience and its inversion or absence. If all the physical facts about an individual are fixed yet it is possible for that individual to have either inverted or no color experience at all, then physical facts (about an individual) do not determine all of that individual's properties.

The classic response to such arguments, one that opens the door to the sort of argument that we have seen Dretske make for global externalism, is to point to the causal (and so, for some, conceptual) connection between qualitative states and the first-person beliefs that they generate. As the philosopher Sydney Shoemaker says, to think that absent qualia are possible is "to make qualitative character irrelevant both to what we can take ourselves to know in knowing about the mental states of others and also to what we can take ourselves to know in knowing about our own mental states." And qualia severed from first-person knowledge of them under at least some circumstances constitute some sort of conceptual confusion. It is for this reason that Shoemaker holds that while inverted qualia are possible, absent qualia are not. There are two points to make about the dialectic here.[20]

First, such thought experiments themselves presuppose both (a) and (b), for they hold, *ex hypothesi*, all physical or functional facts true of an individual while varying the phenomenal state: Green is the phenomenal state present rather than red (or nothing), or pain is present in one individual while it is absent in her physical or functional duplicate. Insofar as the response that Shoemaker offers shares these assumptions, it too is problematic.

Second, it is easy to radically underdescribe inverted and absent qualia thought experiments, and misleading to draw substantive conclusions on the basis of such underdescription. This is just the sort of point that Hurley has made against Block's appeal to Inverted Earth. Such thought experiments take us so far, but only so far, in telling us something about the nature of the mind and how we can and should conceive of it.[21]

8 GLOBAL EXTERNALISM AND THE TESEE VIEW

While I have rejected Dretske's argument for global externalism, one might wonder about how great the distance is between global externalism and the TESEE conception of consciousness. The appeal to the preceding arguments of Parts Two and Three shares in Dretske's representationalist strategy of arguing for an externalist view of consciousness on the basis of assuming externalism about intentionality.

There are several important differences between global externalism and the TESEE view of consciousness, the most significant of which concerns the universality and necessity of the corresponding externalist claims. If I understand the representationalist position shared by Dretske, Lycan and Tye correctly, it holds that global externalism is necessitated once one allows an externalist foot in the door of intentionality. As we will see in the next chapter, such a view seems to be shared by recent defenders of "narrow phenomenology," such as Terence Horgan and Brian Loar, for they aim to defuse such a global externalism with a "global internalism" about both phenomenology and intentionality.

By contrast, the TESEE view of consciousness, like the more general externalism of which it is a part, is inherently pluralistic. It allows that individualism may be correct about certain mental states – pain is an example that I have given at several points – but that there is no barrier to developing an externalist view of the full range of mental states, intentional or conscious. The TESEE view aims primarily to show that individualism is not a constraint on a psychology or cognitive science of any specific sort of mental state, including the full range of conscious states. As a consequence, it implies that a certain internalist picture of cognition has held us captive, and seeks to replace that picture with another vision of cognitive processing. Given this, the chief ground for preferring the externalist vision is not any *a priori* argument from the purported nature of intentionality to claims about the nature of consciousness. Rather, these concern the general theoretical virtues of the externalist view, such as its range of application and overall integral character.

A second difference between global externalism and the TESEE view is that the TESEE conception takes locational externalism much more seriously than does global externalism. In part because of this, the TESEE conception is more vulnerable to the charge of "changing the subject" in offering an expanded view of consciousness. But rather than allow this to lapse into a dispute over what words mean I have suggested that the attractiveness of the TESEE view lies primarily in its scope and integrity. If the charge against the TESEE view is that it changes the subject by adopting a distinct view of what consciousness is, then the corresponding charge against existing views is that they offer a contracted view of consciousness. Such a view makes it difficult to see how entwined consciousness is with our everyday mental life and the range of cognitive processing that we engage in.

A third and subsequent difference between global externalism and the TESEE view of consciousness is the sensitivity of the latter to developments within the cognitive sciences themselves. There has been a recurring tension throughout Parts Two and Three between the naturalistic deference to explanatory practice in the cognitive sciences that is central to the argument of the book as a whole and the recognition that the cognitive sciences are predominantly individualistic in methodology and outlook. It is in part because of this pair of points that I remain skeptical of global externalism. Its endorsement requires viewing current and past explanatory practice in the cognitive sciences as quite radically mistaken, even about consciousness. Instead, on the TESEE view, there are what we might call philosophical excesses that fall out of the individualistic mindset of tradition – we might include here the belief in qualia, as well as the idea that the first-person perspective is necessarily internalist – but there is no compunction in giving individualism its due. It is just a more limited due than one might have thought.

In the next chapter, I explore some recent philosophical views that attempt to present a robust articulation of consciousness within an individualistic framework. At their heart are some proposals about the relationship between intentionality and phenomenology.

10

Intentionality and Phenomenology

1 THE RELATIONSHIP BETWEEN INTENTIONALITY AND PHENOMENOLOGY

Traditionally, postbehaviorist philosophy of mind and cognitive science has proceeded on the assumption that intentionality and phenomenology can most profitably be treated independently or separately from one another. The intentional and the phenomenal have often been viewed as the two fundamental categories of the mental within the philosophy of mind, with more specific types of mental states falling under one or the other of them. Belief and thought are paradigms of the intentional, and pain and bodily sensations paradigms of the phenomenal. Even when this divide between the intentional and the phenomenal has not been treated as a mutually exclusive categorization scheme for thinking about the constituents of the mind, intentional and phenomenal aspects of particular types of mental states remain distinguished and treated in separation from one another. Terence Horgan and John Tienson have called this general position *separatism*.[1]

These forms of separatism about the intentional and the phenomenal likely have no single cause. For some, separatism makes sense because intentionality is thought to be significantly more tractable than phenomenology. "Look," one might say, "figuring out how the mind works is really hard. Consciousness is a complete mystery. But we at least have some inkling about intentionality and mental representation. So let's work that out first." Hence, the explosion of work on causal, informational, and teleological theories of content, and the attention that the individualism-externalism debate has generated over the past twenty-five years. For others, separatism is driven by viewing a divide-and-conquer strategy in general as more efficient in dealing with difficult-to-understand phenomena. And there are some who simply think that the intentional and the phenomenal are metaphysically quite distinct, so that carving the mind at its joints requires separatism. Extending the Horgan and Tienson terminology, I shall call the former type of positions *pragmatic separatism* and the latter-most position *metaphysical separatism*.

Pragmatic separatism implies a research strategy that can be adopted independent of one's stance on metaphysical separatism. Pragmatic separatism amounts to a two-part gamble. The first assumes that parsing the mental world into the intentional and the phenomenal provides the basis for conceptual and empirical advances in what we know about the mind. This gamble has gone hand-in-hand with the classic computational theory of mind and traditional artificial intelligence, which have chiefly modeled intentionality independent of considerations of phenomenology. The second gamble is to assume that treating intentionality as a unified phenomenon, such that one can theorize about it and explore it in both the mental and nonmental realms, will turn out to have much the same benefits. This gamble has generated informational and teleological accounts of intentionality, which have assimilated mental states to mechanistic detectors (such as thermostats) and bodily organs (such as hearts and kidneys).

Over the last decade or so, these gambles have been challenged, and separatism of both kinds has been rejected. As consciousness of consciousness has increased, a number of philosophers have advocated a central role for the phenomenal in our conception of the mental, and in our conception of the intentional in particular. For example, John Searle has defended the *connection principle*, the principle that "all unconscious intentional states are in principle accessible to consciousness," using this principle to cast doubt on the existence of many of the kinds of mental representations posited within cognitive science. Galen Strawson has claimed allegiance to the widespread view that what he calls "behavioral intentionality can never amount to true intentionality, however complex the behavior, and that one cannot have intentionality unless one is an experiencing being." That is, no matter how complicated a creature's behavior, unless that creature has consciousness all attributions of intentionality are mere "as if" intentionality, not the real McCoy. Both of these views appear to make the existence of phenomenology in a creature a prerequisite for intentionality, at least "original" or "real" intentionality. In doing so, they have brought a focus on human minds – rather than, say, animal minds or computers – as the paradigmatic loci of intentional states.[2]

Some recent views go further than this in suggesting more specific and foundational roles for phenomenology vis-à-vis intentionality. Brian Loar has argued that there is a form that intentionality takes – subjective intentionality, psychological content, intentional qualia, or phenomenal intentionality – that is psychologically pervasive. It is distinct from, and in certain respects more primitive than, the kind or kinds of intentionality that have been discussed in light (or perhaps the shadow?) of the externalist arguments of Putnam and Burge. Loar thinks that such intentionality is narrow, and in part this is because he views phenomenology as being individualistic. In Loar's view, there is not simply a general, presuppositional connection between intentionality and phenomenology. Rather, there is a form of intentionality that is thoroughly phenomenal and that is manifest as the phenomenal content of a range of particular mental states.[3]

Terence Horgan and John Tienson have taken a similar path. They argue for a two-way *inseparability thesis*: that intentional content is inseparable from the phenomenal character of paradigmatic phenomenal states (for example, pain, visual experience), and that phenomenal character is inseparable from the intentional content of paradigmatic intentional states (for example, propositional attitudes). In addition, they defend what they call the *phenomenal intentionality* thesis: that there is a pervasive kind of intentionality determined by phenomenology alone. Like Loar, Horgan and Tienson argue that this intentionality is narrow, and that in important respects it is more fundamental than wide content.[4]

While I think that even the general views typified by Searle and Strawson are problematic, in this chapter, I shall focus on the more specific proposals made by Horgan and Tienson and the views of Loar. Both the inseparability and phenomenal intentionality theses seem to me false, and even were the latter true, the significance that Horgan and Tienson attach to it misplaced. Pinpointing the problems with the Horgan-Tienson position will shed some light, I hope, on the limitations of Loar's more wide-ranging discussion and the broader issues that their shared position and its defense raise.

2 DIMENSIONS OF THE INSEPARABILITY THESIS

One legacy of the attachment that many philosophers of mind had to the intentional during the 1980s was the articulation of various forms of representationalism with respect to the phenomenal through the 1990s. Fred Dretske, William Lycan, Michael Tye, and Gilbert Harman have all defended versions of representationalism about phenomenal states. The basic idea of such representationalist views is to treat phenomenal states as a type of intentional state, to analyze or understand the experiential in terms of the representational. A key strategy of representationalists has been to point to the transparent or diaphanous character of experience and our reflection on it. When I engage in introspection on the character of my experience, I find that it is thoroughly intentional, so thoroughly so that it is hard to distinguish any purely qualitative, nonintentional remainder of the experience. My on-line reflection on my current visual experience, for example, seems to me to yield only what it is I am currently looking at (books, a computer screen, a telephone, a coffee cup, and so on), what is usually taken to be the content of my visual experience. Thus, representationalism serves as a basis for either the rejection of qualia, or their subsumption under the putatively better understood notion of intentionality.[5]

Because of the recent prominence of the intentional and of representationalist views of experience, the two halves to Horgan and Tienson's inseparability thesis are not viewed as equally in need of justification. Representationalists accept, of course, the idea that the intentional pervades the phenomenal, as Horgan and Tienson acknowledge. Thus, it is the second half of their thesis – the claim that paradigmatic intentional states, such as beliefs and desires, have an inseparable phenomenal character – that requires more by way of justification, at least in the dialectical tenor of the times.

There are several dimensions along which versions of the inseparability thesis can vary that make for stronger and weaker views about the relationship between intentionality and phenomenology. Consider three:

(a) *quantificational range*: Are there just some mental states of which the thesis is true, or is it true of all mental states?.
(b) *modal intensity*: Are the intentional and phenomenal merely coincident, physically necessitated or nomologically linked, or conceptually related?
(c) *grain of determinateness*: At how specific a "level" does the inseparability thesis hold? At the least specific level, it would apply to the properties being intentional and being phenomenal. At the most specific level, the thesis would apply to specific mental states (for example, attitude plus content), such as believing that there is a

red tomato in front of me here and now, and having a specific, reddish, roundish visual sensation.

Versions of the inseparability thesis that are strong on any of these dimensions are implausible, not only given Horgan and Tienson's other commitments, but independently. Or so I shall argue.

3 DEFLATING THE INSEPARABILITY THESIS

Consider these dimensions to the inseparability thesis in reverse order.

Grain of determinateness. Few (if any) intentional states have a specific phenomenal character without which they cannot have the wide content that they have. This is clearest in the interpersonal case, and one reason for finding both pragmatic and metaphysical separability theses plausible is that we can generalize about, for example, propositional attitudes in a robust manner without delving into the phenomenology that (let us suppose) accompanies those states. What an arbitrarily chosen pair of people share phenomenologically when they both entertain the thought that George Bush is president of the United States is anyone's guess. Intuitively, people can share this thought despite speaking different languages, and sharing few substantive beliefs or opinions about either George Bush, the presidency, or the United States, all of which typically impact on how it feels, what it is like, to entertain that thought. Phenomenological comparisons are also notoriously under strain in crossspecies cases. Moreover, the finer the grain of determinateness to the experiences, the greater the problem here.

But this is also true *intrapersonally*, even if the variation here is not in general as great, in part because we are creatures of habit mentally as well as behaviorally. It is easy to fall into mental ruts, where the recurring phenomenology is part of the rut. Most pointedly, one and the same intentional state can be realized by a person on two distinct occasions and *have a phenomenology at all on only one of them*. Clearly this is true in cases where an intentional state is *conscious* on only one of those two occasions, assuming (for now) that phenomenology presupposes consciousness. But it is also true even when both occurrences are conscious occurrences.

Horgan and Tienson claim that "[i]f you pay attention to your own experience, we think you will come to appreciate" the truth of their claims about the intentional and the phenomenal. In the spirit of Horgan and Tienson's appeal for a reader to "pay attention to your own experience," I have just done the decisive experiment: I thought first that George

Bush is president of the United States, and had CNN-mediated auditory and visual phenomenology that focused on one of his speeches. I then took a short break, doodled a little, wandered around the room, and then had a thought with that very same content and... nothing. Or at least nothing distinctly Bush-like, as in the first case. I just drew a blank, realized my coffee was finished, and moved on. To be honest, I am not sure whether the drawing a blank or the phenomenal feel of realizing my coffee was finished was the phenomenology that accompanied the thought that George Bush is president of the United States, or whether I was mistaken in some more basic way about what my phenomenology was, or about what thought I was entertaining.[6]

However, there is nothing unusual or weird about this, although I don't claim that everyone will find that they have the same results when they attempt their own replication of the experiment. (It is instructive, however, to try this out on a class of students, and note some of the wild variation in what is reported. And things would no doubt get worse in this respect were we to leave the sanctity of the philosophy classroom.) Some, no doubt, will report as Horgan and Tienson themselves do. But do these different results show that I was mistaken about what I thought, or that I mischaracterized or just missed my phenomenology? No, phenomenology is sometimes like that, tricky to coax out, difficult to map to states with specific content, even fickle or uncooperative. And not attached unwaveringly to intentionality, at least at a relatively fine-grained level of determinacy.

Thus far, I have considered just one specific thought in arguing for the perhaps extreme-sounding conclusion that a thought can be entertained by one person on two occasions and have a distinct phenomenology on only one of them. Part of what I want to suggest is that even when mental states have a certain prominence to them – their content is easily articulated, they are "before the mind's eye," they are shared by lots of people – the notion of their having a distinctive phenomenology can remain puzzling. When we turn to mental states that are less prominent in these respects, the puzzlement deepens. I remember from time to time that I need to repair the spare tire for the car, continue to plan to spend less time watching the seemingly endless hockey finals, and occasionally wonder about stranger things, such as whether metabolic systems must be living. Sometimes these mental episodes are separated by months or by years, and I am certainly in no position now to make a comparison of their phenomenology. But even as mental life is going by it is difficult to know just what the phenomenology of these mundane and not so mundane mental states is. This is not skepticism about phenomenology *per se*, but about the idea that our first-person access to it can be unproblematically granted at a relatively fine grain of determinateness.

Modal intensity. This brings me to the sense in which phenomenology cannot be separated from intentionality. As I have intimated, Horgan and Tienson employ what I shall call the methodology of *imaginative evocation* in motivating and discussing the inseparability thesis. That is, they provide possible scenarios that we are invited to imagine in order to convey some idea of the sort of thing that the phenomenology of intentional states is, why it exists, and why the inseparability thesis is true. But this methodology, despite its increasing deployment in thinking about consciousness, is inherently unsuited to making even a prima facie case for anything but the modally weakest versions of the inseparability thesis, namely, that phenomenology and intentionality are coincident in some range of cases. To establish modally stronger theses through thought experiment they would have to show that we cannot have intentional states without a corresponding phenomenology (or phenomenology without intentionality).

We might think, however, that their methodology issues at least a challenge to those who would deny the inseparability thesis: Given that their examples putatively point to a general feature of intentional states, and their inability to conjure mental states without a corresponding phenomenology, the separatist must describe an occurrent, intentional state that has no phenomenology at all. Yet even if we accept these discussions as shifting the burden of proof in this way (and I am not sure that we should), there are several problems here.

First, as I have indicated, I do not myself seem to have any problem in identifying occurrent intentional states that lack a phenomenology (distinct from their intentionality) or, more accurately, whose phenomenology I feel in no special position to identify with any degree of certainty. Representationalists such as Gilbert Harman report similar abilities. Since I trust the reports of Horgan and Tienson (or rather, I trust them no less than I trust my own erstwhile introspective attempts), it seems that the right conclusion to draw is that there can be differences in what the methodology of imaginative evocation produces in this particular case. The further conclusion that these differences are a result of one's different theoretical starting points is tempting. Such a conclusion would be devastating for the methodology that Horgan and Tienson use.[7]

Loar employs the same methodology, putting particular emphasis on a thought experiment that involves thinking of an isolated brain that has just the same phenomenal experience as you when you are having a particular visual experience (say, seeing a lemon). Loar says he "will be content if you grant at least a superficial coherence to the thought that my isolated twin-in-a-vat has visual experiences exactly like mine." But it is difficult to grant even this if your view is that brains need to be both embodied and environmentally embedded, and actively so, in order to provide the basis for any visual experience at all.[8]

As we saw in the previous chapter, precisely such a view has been recently articulated and defended by Susan Hurley, J. Kevin O'Regan, and Alva Noë. For someone who thinks that embodiment and embeddedness are essential features of visual experience, the thought Loar invites us to entertain is no more and no less conceivable than is the thought that there is a box filled only with air that has just the visual experience that I am having at a particular moment. Those who think that mere air could instantiate mentality – call them *airheads* – are able to conceive something that those with this view of the relationship between experience, embodiment, and embeddedness cannot.

Loar himself considers a version of the objection that phenomenology is a product of theory rather than a reflection of the underlying mental reality (of intentional qualia). He says:

Theory does have a bearing, it is true. But theory does not create the phenomenology. From a neutral position there is a certain phenomenology of perceptual experience. What is missing from the neutral position is a conception of the nature of what is thereby presented.[9]

The bearing that theory has, on Loar's view, concerns how the phenomenology is interpreted, not whether there is a phenomenology there to be interpreted. Yet precisely this latter issue is raised by the version of the objection being pressed.

Second, the modally strongest version of the inseparability thesis is vulnerable to the conceivability of *momentary zombies*, individuals who nearly always have a phenomenology that accompanies their intentionality, but who sometimes (perhaps due to hardware noise) fail to have a phenomenology. Momentary zombies have a phenomenology just like ours, except occasionally there is a gap in it, and they are momentarily zombies. Momentary zombiehood is much easier to concede than full-blooded zombiehood, and surely it is plausible with respect to at least some intentional states (consider, again, the propositional attitudes). If momentary zombies are possible, then it is possible for there to be particular intentional states without an accompanying phenomenology. But I also think it is plausible that *we* are momentary zombies, perhaps due to information-processing bottlenecks and other limitations of our consciousness, with respect to at least some of the intentional states that Horgan and Tienson appeal to. I find this particularly plausible with regard to the example of what Galen Strawson has called *understanding-experience*, being the experience of hearing "someone speaking non-technically in a language one understands," and which I sometimes find I have, and sometimes not. In any case, the general point is that modally strong versions of the inseparability thesis are particularly vulnerable to relatively tame versions of some standard thought experiments.[10]

Quantificational range. If one concedes that there are dispositional intentional states, such as belief and desire, then the scope of the inseparability thesis needs to be restricted at least to occurrent intentional states, or to dispositional states when they are occurrent. But does the thesis need to be restricted further, not just to occurrent states but to those occurrent states of which one is conscious? One reason to think so is that if we think of occurrent states at a given time as those that govern our behavior at that time, those of which we are conscious at that time will be a proper subset of our occurrent states. But it is not clear that occurrent states of which we are not conscious at a given time have any more of a phenomenology than do nonoccurrent states. I noticed a short while ago that the room was getting dark and that I should turn on a desk lamp; I noticed more recently that I have been squinting at the papers scattered on my desk in the enveloping dark. It is plausible to think that my wanting to continue reading guided my squinting behavior although there was no phenomenology of that occurrent state prior to my reflecting on my behavior. (How could there be?. I was not aware of this aspect of my behavior, and it came as a surprise to me to realize just what I was doing.) This suggests that there are at least two "levels" of intentional states for which there is no phenomenology, the purely dispositional and the merely occurrent.

Horgan and Tienson explicitly restrict their thesis of the phenomenology of intentionality to intentional states when they are conscious. But one wonders what this amounts to in light of the following passage:

The full-fledged phenomenal character of sensory experience is an extraordinarily rich synthetic unity that involves complex, richly intentional, total phenomenal characters of visual-mode phenomenology, tactile-mode phenomenology, kinesthetic body-control phenomenology, auditory and olfactory phenomenology, and so forth – each of which can be abstracted more or less from the total experience to be the focus of attention.[11]

On this conception, phenomenology outstrips attention. On one reading, one that equates attention with consciousness, there is phenomenology of which one is not conscious. (But how then do you tell what *its* content is?)

Alternatively, if Horgan and Tienson are equating consciousness with phenomenology, they are saying that we only attend to a portion of our conscious experience. But what is the status of the phenomenal content of that unattended portion of our conscious experience? Does it exist, and if so, how do we know its nature (since, by hypothesis, we do not attend to it)? If we do not know the phenomenal content, then it is plausible to think that such states have no more specific phenomenal content than do dispositional states.

The point here is that the phenomenology of intentionality begins to look more restricted in the range of states it applies to at any given time than one might initially think: The dispositional, the merely occurrent, and the unattended all seem to be precluded. If the inseparability thesis is true, then it seems that it is true of a much more restricted set of states than simply all intentional and all phenomenological states. In light of that, the thesis loses a lot of the punch that it packs vis-à-vis traditional views of the mind that operate on the assumption that there is no necessary or deep connection between the intentional and the phenomenal.

A different sort of problem in the scope of the thesis arises in Loar's discussion of phenomenal intentionality and intentional qualia. Loar builds up a case for phenomenal intentionality by considering perceptually based concepts, then generalizes to recognitional concepts, spatial concepts, and socially deferential concepts. But apart from the special case where we reflect on such concepts and their instances we do not, in our everyday experience, have any phenomenology of these concepts, any more than we have any phenomenology of the individual phonemes or distinctive features that make up the stream of speech we have auditory experience of. The stream of consciousness is not, without special prodding, segmented into constituents such as concepts. It seems primarily in the hands of philosophers that our experience can become segmented and particularized, in much the way that it was atomized in the hands of that master introspectionist, Wilhelm Wundt.[12]

This reference to Wundt may remind us that neither introspection nor phenomenology is simply a matter of turning one's mind inward and reporting what one finds. What one finds in one's own experience will depend in part on what one is looking for, the background perspective that one brings to this first-person task.

4 PHENOMENAL INTENTIONALITY

So far I have not argued that the inseparability thesis is false, but that there are three dimensions of strength – scope, modality, and determinateness – on which it rates lowly. The point here is to deflate (not refute) the inseparability thesis, for surely only a skeptic about the phenomenal would refuse to concede that versions of the thesis weakened on each of the forgoing dimensions are true. Along the way I have raised, in passing, some doubts about the suitability of the methodology on which Horgan, Tienson, and Loar rely in gesturing at what the phenomenology of intentionality is. Rather than develop these doubts here directly I shall turn to consider phenomenal intentionality itself and what its proponents claim about it. Again, the chief point will not be to show that such a property does not exist, but that the most plausible way of understanding it makes it unlikely that what its proponents claim about it is true.

Horgan and Tienson characterize phenomenal intentionality as a "kind of intentional content, pervasive in human life, such that any two possible phenomenal duplicates have exactly similar intentional states vis-à-vis such content." Although this sounds stipulative, it is not, since Horgan and Tienson continue by arguing for the thesis through imaginative evocation. As I noted in the previous section, this style of argument does not seem well suited to the modally strong conclusion they seek, except insofar as it shifts the burden of proof to those who deny phenomenal intentionality. Yet it remains open to skeptics here to concede that there can be a sort of phenomenal intentionality that is nonconceptual but balk at the claim that the same is true of the conceptual realm. For although *ex hypothesi* phenomenal duplicates share all their phenomenal states, we are to show, not assume, that they share cognitive structures that are genuinely intentional, such as concepts or beliefs.[13]

As part of their bridge from phenomenology to intentionality, Horgan and Tienson distinguish between "two ways of thinking about truth conditions: as determined wholly by phenomenology, and as determined in part by items in the experiencer's environment that satisfy the experiencer's phenomenology." The former of these, they argue, are narrow and more fundamental than the latter. In their discussion, through imaginative evocation, they invite each reader to compare him or herself to both a Twin Earth *doppelgänger* and a "Cartesian duplicate." The latter of these has thoughts purporting to refer to someone named "Bill Clinton," but these lack reference altogether since there is no thing at all that satisfies that putative reference for a Cartesian duplicate. Here it seems that Horgan and Tienson allow that some phenomenal duplicates (for example, those in no environment) may have mental states that have no wide truth conditions and so no wide intentionality at all. But if at least some phenomenal duplicates differ in that one has concepts with ordinary (wide) satisfaction conditions, while the other does not, then the defender of phenomenal intentionality must have available a way of articulating the intentionality that such duplicates share that is independent of their wide intentionality. Whether phenomenology alone suffices for intentionality given the complete severance to wide intentionality, as in the case of Cartesian duplicates or brains-in-a-vat, might reasonably be questioned. A more developed account of something like "phenomenal intentionality" could silence doubts here.[14]

Brian Loar provides an account that purports to do the trick. To bridge from phenomenal identity to intentional identity he appeals to (i) brains in vats that (ii) share perceptually based concepts and (iii) share all other concepts in virtue of their sharing their conceptual roles. As Loar notes, in effect, (i) is required to ensure that any shared intentionality does not hold in virtue of shared (or similar) environments, and so is narrow; (ii) provides a base case that Loar takes externalists to be committed to in virtue of the phenomenon of failed perceptual demonstratives (and the inadequacy of representationalist accounts of it); and the extension in (iii) appeals primarily to another resource, conceptual roles, that Loar takes externalists to view as shared across contexts, no matter how radically different those contexts may be. I want to put aside concerns about (i) for now and concentrate on (ii) and (iii).[15]

Loar's view here is programmatic and sketches a large-scale view of phenomenal intentionality, rather than presenting detailed analyses for any concept that putatively has phenomenal intentionality. My comments are correspondingly cast. The idea of starting with individual concepts, rather than phenomenal experience in its fullness, is a good one prima facie, for part of the problem with sensory experience as a whole is that it must be articulated ultimately in terms of a range of concepts, and many of these are conceded by nearly everyone as being externalist, as having wide content. This seems true even when the articulation we are interested in is done from the first-person perspective. Begin, then, with concepts whose phenomenal intentionality (and its nature) is not in serious dispute, and then constructively build a full account of experience as having phenomenal intentionality.

I have already flagged one problem with this approach, however: It seems to fragment the actual phenomenology we have, and so to be at best a philosophical analysis rather than a description of it. Suppose that we put this aside, and suppose that we simply grant, for now, that there are perceptually based concepts that are individualistic. How do we constructively build from here? After discussing perceptually based concepts, including "recognitional concepts" that "purport to pick out, perceptually, kinds and properties rather than individuals," Loar considers concepts that seem neither paradigmatically narrow nor wide, including the general concepts of physical objects and spatial relations. Here he argues that these contain a recognitional component or aspect, and so can be assimilated to recognitional concepts, which he has argued to have phenomenal intentionality. While Loar concedes that this is not true of socially deferential concepts – which have paradigmatically wide content – he argues that their phenomenal intentionality derives from the conceptual roles they play in the systematic internal economy of the individual.[16]

There are thus two distinct paths to phenomenal intentionality: via assimilation to perceptually based concepts (paradigms of the recognitional), and via assimilation to the logical connectives (paradigms of the Conceptual Roley). These paths are very different from one another. One concern is whether they in fact converge. That is, what grounds are there for thinking that they determine the very same kind of property, phenomenal intentionality? This question seems to me to need a nonstipulative answer for Loar's program to be successful. Unlike Horgan and Tienson, Loar explicitly rules out the possibility that this unifying feature is the availability of phenomenal intentionality to introspection, for he wants conceptual roles to determine phenomenal intentionality independent of its relationship to "an introspective glance." Indeed, in light of this concession, and independently, one wonders what makes intentionality determined solely by conceptual roles phenomenological. What is the *phenomenology* of the concept "and," one wonders?[17]

There is a parallel here with wide content that might be drawn. Why think that "environmentally-determined" content (for example, that of water, à la Putnam) and "socially-determined" content (for example, that of arthritis, à la Burge) are examples of a single kind of content, wide content? There are, however, two important dissimilarities between this pair of factors, the physical and social environments, and those determining phenomenal intentionality. First, one might plausibly (even if not definitively) make the case that environmentally determined content of these is really an instance of socially determined content, in that the key feature of Putnam externalism is the pattern of social deference that it identifies in the use of language. Second, and short of this, these "paths to externalism," while distinct, are not independent in that they share key elements, such as the idea of a social division of labor between regular folk and experts and that of individual knowledge as being incomplete or partial.

A final general question: If phenomenology determines a kind of intentionality (albeit via two distinct paths), what is phenomenology's relationship to wide content? Given the diaphanous character of much of our introspective experience, a point (as we have seen) upon which representationalists have seized, it seems implausible to think that our everyday phenomenology is never of, or never leads us to, intentionality that is wide. I suspect that Loar, Horgan, and Tienson would concur. If that is so, then there is nothing special about the path from phenomenology to individualism, for there is also a path from phenomenology to at least some kind of externalism. We should consider this issue more fully by turning explicitly to the focus on twin cases, and to Loar's appeal to the brain in a vat.

5 INDIVIDUALISM AND PHENOMENAL INTENTIONALITY

Loar's view, like that of Horgan and Tienson, moves from claims about the phenomenology of perceptual experience to a conclusion about its (phenomenal) intentionality. As we saw in the previous chapter, Fred Dretske has argued that even conscious perceptual experience is externalist, and a reminder of the dilemma that Dretske poses is useful in understanding a challenge facing those, such as Horgan, Tienson, and Loar, who deny this. What are perceptual experiences? If they are or involve conceptual or thoughtlike entities, then those experiences inherit their width from that of the concepts or thoughts they involve. But if perceptual experiences are completely divorced from conceptual or thoughtlike entities, then perceptual states are unknowable, even by their bearers.

In terms of this "wide or unknowable" dilemma posed by Dretske, defenders of phenomenal intentionality attempt to grasp the first horn by articulating concepts that are not externalist. This is a strategy that Colin McGinn has also pursued in demarcating the limits of externalism, arguing that perceptual content in general is not what he calls strongly externalist, that is, dependent on nonmental features of a subject's environment. Central to McGinn's argument is the imagined case of Percy and his *doppelgänger*. Percy is behaviorally disposed to respond to round things in round-thing appropriate ways, and so is said unproblematically to have an internal state that corresponds to the concept round; Percy's duplicate, by contrast, is behaviorally disposed to respond to square things in round-thing appropriate ways. The question is whether Percy's duplicate has the concept *round* or the concept *square*.

McGinn thinks that an externalist is committed to the latter, since square things are the distal causes of the relevant internal state, and externalists individuate mental states by such external causes rather than by internal features of an organism. Yet the intuition that Percy's duplicate is a creature that misperceives square things as round is strong, and so, according to McGinn, only an individualistic view of the case will do. Here is a case where perceptual content is individualistic. It seems to both Percy and his duplicate that they are both seeing a round thing, and this shared seeming is what explains their shared behavioral dispositions. Hence, there is a need for an account of the constituents of perceptual contents that is individualistic. The argument here is similar to Loar's chief reason for appealing to phenomenal intentionality: cases of failed perceptual demonstrative reference.[18]

Such pair-wise twin comparisons may be a first step in articulating a conception of phenomenal intentionality via the claim that there are at least some narrow concepts. Yet one needs to be able to generalize from them to reach a conclusion about any pair of physical twins, which is required to show that the corresponding concepts, and thus phenomenal intentionality, are individualistic. In the remainder of this section I shall argue for a contestability to what we can imagine in twin cases that poses a problem for this generalization, and so for this putative link between individualism and phenomenal intentionality.

In discussing a variant of McGinn's example, Martin Davies implicitly defends the view that a generalization from particular twin cases to arbitrary twins will fail, since it is relatively easy to find examples where it seems more plausible to characterize the twin's concepts in terms of our ordinary notion of (wide) content. More telling, in my view, is Davies' suggestion, following Fricker, that it may be most plausible to refrain from ascriptions of content at all. Suppose, for example, that Percy's twin is a brain in a vat, in an internal state identical to the one that Percy has when he sees round objects, but which is caused by square objects, and that Percy's twin (not ever having been embodied at all) has no behavior at all. In such a case, we have no basis to ascribe even behavioral dispositions to Percy's twin. It seems to me very hard to ascribe a specific content to Percy's twin in such a case without already supposing that only what's "in the head" determines content. Thus, what ascriptions one is prepared to make turn largely on one's prior theoretical commitments regarding the individualism-externalism debate.[19]

If this is true, then defenders of phenomenal intentionality have seriously underestimated the task before them, for they have been content to find relatively few cases in which perceptual content is (plausibly) shared by *doppelgängers*, and then simply supposed that the generalization from such cases is unproblematic. But consider the range of possible duplicates there are for any given individual. There is Rex and there is his *doppelgänger* T-Rex on Twin Earth. But there is also brain-in-a-vat Rex, entirely virtual Rex, Rex the happy android, Rex the purely immaterial substance, and multiperson Rex, whose phenomenal life is the fusion of half-lives of two other individuals. In each case, we can imagine Rex's phenomenal mental life being present in some radically (or not so radically) different circumstance.

Or can we? Can we really imagine a purely immaterial substance having exactly the same phenomenal life as regular embodied and embedded Rex here on Earth? As I intimated in section 3, we may conceptualize phenomenology not simply as the result of passive experience and active introspection but as the active exploration of one's environment through one's body. Given that, whatever it is we are imagining in these cases, it is not a scenario in which the phenomenology remains constant across the two scenarios. In fact, if one adopts such a view of at least the phenomenology of perceptual experience, as I think is plausible, then it is difficult to imagine disembodied minds having the corresponding phenomenology at all. In this respect, to draw on a Wittgensteinian example, comparing Rex and disembodied Rex is like comparing the time on Earth with that on the sun. Worse, comparing disembodied Rex and his differently situated but equally disembodied duplicate is like comparing the time it is on two places on the sun's surface.[20]

This is to flip around a response that physicalists have made to an objection based on the conceivability of zombies: That in fact they are not conceivable, or their conceivability (if it implies possibility) presupposes, rather than indicates, the falsity of physicalism. Here I am suggesting that the conceivability of phenomenal duplicates itself presupposes, rather than indicates, the narrowness of phenomenal intentionality, by assuming that embodiment does not go to the heart of phenomenal experience. Loar, Horgan, and Tienson seem to me to adopt a misleading view of what phenomenology is, of how it is merely contingently or extraneously both embodied and embedded, and so make the task of imagining the phenomenal experience of radically different individuals appear easier than it in fact is. But the more basic point is that phenomenology itself is a contestable phenomenon, and what one can and can't imagine about it inherits that contestability.

Thus, the claim that phenomenal duplicates share all phenomenal states is more problematic than it initially appears. We can come at this point in another way that brings us back to my initial discussion of the three dimensions to the inseparability thesis in sections 2 and 3. If we allow that at least some phenomenal experience is the active exploration by an embodied agent of its environment, then there are far fewer possible phenomenal duplicates of any given individual than one might initially suppose, since such duplicates are constrained by having to have suitably similar bodies and environments. Given that, the focus on brains in a vat is misplaced, and will tell us little about phenomenal intentionality.

But should the proponent of phenomenal intentionality adopt this concessive view of phenomenal experience, whereby at least some such experience is that of an essentially embodied agent? Neither Loar nor Horgan and Tienson are as explicit as one might like about this issue, and different things that they say suggest different answers here. Consider what Horgan and Tienson say about the thesis of phenomenal intentionality. In arguing for the narrowness of phenomenal intentionality through imaginative evocation, they seem to reject the concession, suggesting that all phenomenal intentionality is narrow. Yet the inseparability thesis itself posits a necessary connection (of some type) between phenomenology and ordinary wide intentional states, suggesting that they may be happy with the concession. This would mean, in turn, that the quantifier in the thesis of phenomenal intentionality ("there is a kind of intentionality...") should be understood at face value as an existential quantifier. In either case, there is a problem that can be expressed as a (somewhat complicated) dilemma.

Consider whether ordinary wide intentional states have a phenomenology. If they do not, then while the claim that phenomenal intentionality is narrow remains general in scope, the inseparability thesis loses its quantificational range, since Loar, Horgan, and Tienson all concede that wide intentional states exist. By contrast, if they do have a phenomenology (as the inseparability thesis prima facie suggests), then either that phenomenology is narrow or it is wide. Given that the first half of the inseparability thesis has been endorsed by proponents of representational accounts of phenomenology as part of their phenomenal externalism, as Horgan and Tienson recognize, and that both halves of the thesis are concerned with intentional states as they are ordinarily conceived, such phenomenology would seem to be wide. Certainly, when I describe my current visual phenomenology – in terms of the books, paper, walls, and computer screen I am currently looking at – it seems to me that my experience is not merely "as of" a world beyond my body but it is in fact so. The intuition that drives both Horgan and Tienson's "intentionality of phenomenology" thesis and phenomenal externalism – that at least much of our experience seems to be experience of the actual world – together with respect for the link between phenomenology and first-person reports of its content, suggests that there is such a thing as wide phenomenology. But then, clearly, the claim that phenomenal intentionality is individualistic implies that the thesis of phenomenal intentionality does not apply to all phenomenology. And so the inseparability thesis has a more restricted quantificational range and a weaker modality than one might initially think.

The remaining option, that ordinary wide intentional states have a narrow phenomenology might seem obviously the right thing to say dialectically, but it is also fraught with problems. Not only does it sever the relationship between the first-person perspective and phenomenology, but it makes little sense of those aspects of phenomenology that at least seem – from both first- and third-person perspectives – to be bodily in nature. For example, both haptic perception and proprioception are difficult to make sense of for the case of immaterial entities, and I think they are almost as problematic for material entities, such as brains in vats or Cartesian duplicates, that have no body at all. As I sit in my chair I feel a pressure exerted on my lower back by the back of the chair, and there is a certain feel to my body when I flex the muscles in my leg. But how would these feel to a being without a body, even a being that was molecule-for-molecule to me from the neck up? And, picking up on the grain of determinateness dimension to the inseparability thesis, what reason is there to think that these would have the same feeling that they have in me? Try as I may, I find these questions very hard to answer without simply assuming that they must feel just as they feel in my own case. I suspect that I am burdened by my externalist commitments, and proponents of the individualistic view of phenomenology may find it easier to make sense of these cases than do those hampered by externalism. But if so, this reinforces my more general concern that the method of imaginative evocation generates views of phenomenology that are subject to philosophical contamination.

To sum up this part of the argument: There may be aspects to phenomenology that are individualistic, but, more importantly, there are aspects that likely are externalist. Critically, some cases of perceptual experience, a general category that plays a central role in both the arguments of Loar and of Horgan and Tienson, appear to fall into the latter category. Again, this is not to say that there is no truth to the claim that phenomenal intentionality is narrow, but to suggest that it is a claim true in a significantly more restricted range of cases than its proponents have thought.

To have come this far, however, is to be in a position to comment on the further claim that phenomenal intentionality is more fundamental in at least certain respects than wide content. Minimally, this claim will have to be hedged in ways that correspond to the restricted domain in which phenomenal intentionality is narrow. But if externalism is true of the phenomenology of at least some perceptual content, as I have suggested, then given the centrality of this case to the conception of phenomenal intentionality more generally, we may wonder what content remains to this claim of "basicness."

6 HOW TO BE A GOOD PHENOMENOLOGIST

The fruitfulness of applying a divide and conquer strategy to the intentional and the phenomenal can be, and has been, reasonably questioned. In this chapter, I have been concerned primarily with recent views that go further than such questioning and make specific proposals about the relationship between the intentionality and phenomenology of particular mental states. In particular, I have focused on the inseparability thesis that Horgan and Tienson have articulated, and the general conception of phenomenal intentionality that they share with Loar, as well as Loar's own way of further articulating that conception. There are certainly other possible ways to articulate the relationship between intentionality and phenomenology. Because I have not tried to show that the views of Horgan, Tienson, and Loar suffer from some deep, general, flaw, I do not take the argument of this chapter to express any overarching skepticism about such work on these two pillars of the mental. However, there are several general problem areas that can be marked "fragile" for now for those making alternative proposals.

Past attempts to defend or chalk out an individualistic perspective on the mental have typically focused on intentionality. The main concern about such positions is with the notion of narrow content itself. Because the chief proposals for articulating that notion – the narrow function theory of Stephen White and Jerry Fodor and the narrow conceptual role semantics of Ned Block and Brian Loar – are generally acknowledged as failures, one hope has been that the renewed attention to the phenomenal would provide the basis for a reinvigorated expression of the narrow content program. But it seems that the notion of phenomenology itself is as contestable between individualists and externalists as is that of intentionality. If this is true, then the idea of reviving narrow intentionality via an appeal to phenomenology inherits that contestability.[21]

Perhaps the initial surprise is that phenomenology itself should remain somewhat mysterious. In the good old days, when the phenomenal was neglected for the intentional, intentionality was thought to be more theoretically loaded, subject more to the grinding of particular axes, than phenomenology, in part because of the immediacy, directness, and first-person intimacy of the phenomenal. Recall Horgan and Tienson's claim that "[i]f you pay attention to your own experience, we think you will come to appreciate" the inseparability of the intentional and the phenomenal. Yet such attention and what it reveals are more contestable than one might initially think.

In mentioning Wilhelm Wundt in passing at the end of section 3, and so alluding to the introspectionist tradition in early experimental psychology to which Wundt was central, I have implicitly suggested that some of the problems in thinking about phenomenology parallel those that Wundt faced in thinking about the nature of introspection. If the advice on how to be a good introspectionist early in the twentieth century was "Don't listen to the psychologists," then the corresponding advice on how to be a good phenomenologist early in the twenty-first century might be "Don't listen to the philosophers." And while it is not all the advice one could hope for, it might be advice enough for now.

PART FOUR

THE COGNITIVE METAPHOR IN THE BIOLOGICAL AND SOCIAL SCIENCES

11

Group Minds in Historical Perspective

1 GROUP MINDS AND THE COGNITIVE METAPHOR IN THE BIOLOGICAL AND SOCIAL SCIENCES

There are two strands to our thinking about where the mind begins and ends, and the role that the individual has in demarcating that boundary. First, individuals are paradigmatic subjects or bearers of mental states: They are the things, perhaps the only things, which literally have minds. Minds are where, perhaps only where, people are. Second, individuals serve as some sort of boundary for the mind. Minds are located inside individuals, and we do not need to consider the body or the extracranial world in theorizing about the nature of minds. The argument of Parts Two and Three has been that this second strand to our thinking about the relationship between individuals and minds is mistaken. For the most part, this rejection of the mind as bounded by the individual has gone hand-in-hand with the acceptance of minds as belonging to individuals.

In Part Four, this latter view is reconsidered. I shall take seriously the idea that individuals are not the only kind of thing that has a mind. In particular, I focus on the idea that groups can have minds. Doing so will take us beyond the cognitive to both the biological and social sciences. For it is in these other parts of the fragile sciences that the idea of a group mind has a rich history, a history appealed to in a number of contemporary discussions. In this chapter, I consider the history, and in the next the contemporary revivals of what I shall call the *group mind hypothesis*.

The idea that there are group minds will strike many as either an ontological extravagance or an outright absurdity with which neither biologists nor social scientists need concern themselves. Yet this group mind hypothesis was prominent both in social psychology and social theory in the late nineteenth century and in the study of social insects and community ecology in the early part of the twentieth century. It was defended by many of the most influential thinkers in these fields and gained support from the range of resemblances between at least some groups and individuals. Moreover, contemporary defenses of the group mind hypothesis can be (and have been) expressed as part of more systematic views that are reshaping the ways in which parts of biology and the social sciences are conceptualized and practiced.

For example, expressed as the idea that there may be group-level cognitive adaptations, the group mind hypothesis forges a link between evolutionary psychology, based on the claim that the mind is subject to natural selection in much the way that the body is, and recent work on the levels of selection, which has become pluralistic and endorsed the claim that natural selection operates at multiple levels, including that of the group. And when expressed in terms of the place of institutions and cultural practices in directing whole styles of cognition, the group mind hypothesis finds application both in anthropology and in the social studies of science. Taking such claims seriously will enable a discussion of some deep and interesting issues, such as the sense in which cognition is social, the relationship between evolution and cognition, and the interplay between culture, evolution, and sociality.

The group mind hypothesis itself can be understood in two different ways. First, it can be interpreted literally as saying that groups have minds in just the way that individuals have minds. On this view, groups are a sort of higher-order individual and function as such in biological or social theory. Groups may not have minds as rich as our own, much as we might think animals have just some subset of the full range of psychological states that we have. But however restricted be their minds in scope and richness, groups are mind-laden entities of some kind. Alternatively, the hypothesis can be viewed as an instance of what I shall call the *cognitive metaphor*, the idea that we metaphorically extend our conception of mind from our paradigmatic individuals to things – in this case groups – that are not individuals. On this view, groups are sufficiently like individuals in some respects that it proves useful or even inevitable that we treat them as if they had a psychology, as if they were mind-laden entities, when in fact they are not. They have minds by courtesy, in virtue of our application of a cognitive metaphor to them.

The literalist interpretation makes most direct sense of the appeal to certain groups of organisms as superorganisms in the biological sciences, and some human social groups (the crowd, an army, the nation) as manifesting a variety of individual-level properties. The metaphorical reading locates the group mind hypothesis among other instances of the cognitive metaphor within the biological and social sciences. These range from the view of genes as selfish or competitive and cells as recognizing, remembering, preferring, and seeking certain other cells or molecules within the biological sciences, to ascriptions of institutions and human social practices as rational or friendly and corporations as morally and legally responsible.

Each of these interpretations is helpful in making sense of the uses to which the group mind hypothesis is put in the fragile sciences, and the differences between them is not as sharp as one might initially suspect. Hence, I shall rely on both of them over the next pair of chapters and will not enter into any discussion of which is the correct way to understand the group mind hypothesis.

2 TWO TRADITIONS

In order to articulate a thesis that can be understood in these two ways, consider the following as a statement of the group mind hypothesis:

Groups of individual organisms can have or can be thought of as having minds in something like the way in which individual organisms themselves can have minds.

The literalist can be seen as removing the "can be thought of" caveat and sharpening the "something like" in this statement of the group mind hypothesis, while the proponent of the cognitive metaphor interpretation places emphasis on these softening aspects of the hypothesis.

The group mind hypothesis was held by many of the founders of social psychology and the social sciences more generally, including William McDougall in the former and Emile Durkheim in the latter. The view of human social groups acting in ways that were guided by their distinctive mental characteristics and activity was widespread at the turn of the twentieth century in thinking about society and its foundations. All such views, which I will refer to as forming part of the *collective psychology* tradition of thought, were developed in the broader context of reflecting on the relationship between individuals and the societies that they constituted. Those in the collective psychology tradition typically defended some sort of nonreductionist view of this relationship. Despite their otherwise diverging interests and orientations, these theorists were committed to a view of the communal or collective aspect of psychology as autonomous and separable from both physiology and the experimental psychology that derived from it. The psychology of collectives was emergent from and thus not reducible to the psychology of the individuals in those collectives and was to be studied as such.

The group mind hypothesis has an independent origin in foundational work in ecology and on the social insects in the first twenty years of the twentieth century. The idea at the core of what I shall call the *superorganism* tradition in evolutionary biology is that in certain groups of living things – in particular, in colonies of *Hymenoptera* – it is the group rather than the individual organism that lives in those groups, that functions as an integrated unit, having many of the properties that individual organisms possess in other species. Individual bees, ants, and wasps function more like organs or (parts of) bodily systems in those species. Colonies are independent, self-regulating groups organized to achieve specific biological goals – such as food collection and distribution, nest construction and maintenance, and reproduction – via dedicated behavioral strategies, where some of these strategies and the functions they perform can properly be thought of as psychological or mental. Since the members of these colonies often lack any or all of these goals and the accompanying strategies, these strategies and goals are emergent properties of the colony, in much the way that a group mind was thought to be emergent from individual minds in the collective psychology tradition in the social sciences. The Harvard entomologist William Morton Wheeler is arguably the key figure in the superorganism tradition.

The collective psychology and superorganism traditions, as I am construing them, each extend over some fifty or so years, and receive their most crisp and influential expressions, respectively, shortly before and shortly after the turn of the twentieth century. Authors working within each of these traditions are diverse both in how they introduce the idea of a group mind and what they are motivated by. The work within the two traditions that seems to me of most interest in discussing the contemporary revival of the group mind hypothesis shares with that revival a common explanationist commitment that might be characterized as follows. Certain sorts of groups themselves have the ability to behave or act as unified, integrated units, and these actions are properly or best explained by positing a group mind. It is the group that behaves or acts, and the cause of this behavior or action is, in some sense, the mind of the group. I shall use this commitment to explanationism to help bring out aspects of both traditions of relevance to contemporary claims about group minds or group-level cognitive adaptations.

Both of these traditions postulate minds as emergent properties of groups of organisms. Within these traditions, minds are *group-level* traits. Yet there is a difference between the traditions on this issue important enough for our later discussion to mark with some further terminology now. In the collective psychology tradition, a group mind is what I will call a *multilevel* trait, since the mind is claimed to exist at both the level of the group and at the level of the individuals comprising the group. By contrast, in the superorganism tradition, a group mind is a *group-only* trait, in that it is claimed that it is only groups of social insects, not individual members of those groups, which possess a mind.

Consider a nonpsychological illustration of the distinction between multilevel and group-only traits, adapting an example suggested to me by the biologist Michael Wade involving the fire ant (*Solenopsis geminata*). Individual fire ants have the ability to sting small predators when they are threatened, and this stinging action is often effective in deterring such predators. Fire ants can also coordinate their individual stinging behavior by emitting pheromonal signals, and this coordinated stinging repels larger predators that attack the ant colony. The ability to sting is a multilevel trait in that it is a trait that both individuals and groups can possess. The capacity to deter large predators (or defend the nest) by stinging, however, is a group-only trait, a trait that only groups of fire ants possess.

In the next two sections, I shall discuss each of these traditions in some detail before moving, in sections 5 and 6, to some critical analysis that I hope sheds light on both. These traditions represent both insight and confusion. Part of my aim will be to distinguish discovery from disease.

3 THE COLLECTIVE PSYCHOLOGY TRADITION

The collective psychology tradition has its roots in post-Commune Parisian and French social thought, lasting roughly from 1870 to 1920. Aspects of the tradition have been the subject of a number of detailed professional historical studies. The tradition was largely motivated by and focused on two sweeping, related social changes: a heightening in the visible actions of politically disenfranchised or marginalized individuals in groups, including industrial strikes and peasant uprisings, and the increased activity of socialist and anarchist political organizations. These two social changes were certainly related, and the collective psychology tradition conceptualized their relationship in a particular way: as two manifestations of the very same phenomena, the rise of the crowd.[1]

Although there are works that foreshadow the dark picture of the crowd painted within the collective psychology tradition, the tradition itself begins with the historical accounts of the French Revolution and the Paris Commune by Hippolyte Taine in the mid-1870s, and reaches an apotheosis in 1895 with the publication of the best-known work in the tradition, Gustav Le Bon's *La Psychologie des foules*, translated into English simply as *The Crowd*. Authors in the tradition include Scipio Sighele and Pasquale Rossi in Italy and Gabriel Tarde, Emile Zola, Henry Fournial, and Emile Durkheim in France. The collective psychology tradition significantly shaped theories of "mass psychology" and political groups in the early twentieth century.[2]

The above mixture of lawyers and criminologists, historians and anthropologists, novelists and social theorists here suggests that although authors in the collective psychology tradition often invoked the name of science, theirs was not simply a distant, academic perspective on new social phenomena. Rather, the collective psychology tradition actively influenced how the very phenomena it studied was perceived and acted on by those in political power. It was as much constituted by and appealed to the popular imagination as by and to scholarly research. A number of the works in the tradition became instant and long-term national bestsellers, and their authors were consulted by a range of political authorities about how best to deal with "the crowd" as a political entity.

The importance of the concept of a crowd, its unifying force for understanding contemporary social life in France in the last quarter of the nineteenth century, and why the *psychology* of crowds provides the key to understanding social upheaval, are brought out clearly in the early pages of Le Bon's *The Crowd*. Near the outset of his Preface, after linking individuals and their abilities to the "genius of the race," Le Bon says,

> When, however, a certain number of these individuals are gathered together in a crowd for purposes of action, observation proves that, from the mere fact of their being assembled, there result certain new psychological characteristics, which are added to the racial characteristics and differ from them at times to a very considerable degree.
>
> Organized crowds have always played an important part in the life of peoples, but this part has never been of such moment as at present. The substitution of the unconscious action of crowds for the conscious activity of individuals is one of the principle characteristics of the present age.

The concept of the crowd is important to making sense of the social and political climate, claims Le Bon, because the action of crowds has come to replace that of individuals in the public sphere. The "crowd" that Le Bon is interested in is not simply a gathering of individuals, however: It is an organized crowd or a psychological crowd that exercises this heightened influence over the direction of history. As such, crowds are not simply aggregations of individuals, but something more. When such a crowd exists,

> The sentiments and ideas of all the persons in the gathering take one and the same direction, and their conscious personality vanishes. A collective mind is formed, doubtless transitory, but presenting very clearly defined characteristics.... It [the organized crowd] forms a single being, and is subjected to the *law of the mental unity of crowds.*[3]

This putative law claims that crowds destroy the individuality of those that belong to them, psychologically unifying and homogenizing them so that they act as one. We shall return to consider this law shortly. The point to note for now is that the key to understanding the behavior of crowds, something that we can all observe, is their underlying psychology, something to be discovered. It is a putatively empirical or observational study of the mental life of the crowd that is provided by the collective psychology tradition.

Le Bon's focus on a special type of gathering of individuals, one in which an organized or psychological crowd is formed, however, is only apparent, as both the rest of his introduction and Part III of *The Crowd* make clear. Having stated that his time is the era of crowds, Le Bon elaborates by discussing the ways in which the "popular classes" have entered into political life, and the increased "power of the masses" via labor unions, parliamentary assemblies, and electorates. Understanding the psychology of crowds is crucial to ameliorating the effects of these political changes, and to controlling the masses within the existing political framework because such groups are themselves organized crowds. What we might call gatherings represent but one type of crowd, but their psychology is also the psychology of a diverse variety of collectivities, "crowds," many of which need not be aggregated in space and time as gatherings are.

Whether there is such a kind of thing as a crowd in this broad sense, a kind that could be the object of scientific study and generalization, particularly of psychological study and generalization, requires closer scrutiny. The term "crowd" within the collective psychology tradition encompassed small and large gatherings (for example, mobs), associations constituted chiefly by membership (for example, political parties), and institutional abstractions (for example, electorates). All of these are social collectivities, but very different sorts of collectivities. The idea that each has a psychology, a set of underlying mental dispositions and traits, let alone that all share a common psychology, seems doubtful.[4]

By contrast, the idea that there might be general rules about social control that can be formulated and exercised on a variety of levels, and that these can be uncovered by understanding the "psychology of crowds," does provide an arena in which a broad, unified concept of crowds may operate. Thus, we can use our crowd psychology, claims Le Bon, to see why an indirect tax on consumer goods will be more readily accepted by "the crowd" than will a direct tax on income; to select and influence a jury in a trial; or to counteract the illegal acts of an angry mob.[5]

As several authors have noted, although Le Bon borrowed heavily from earlier authors in what I am calling the collective psychology tradition, his emphasis on the practical political consequences of "understanding crowds" shows a greater debt to Machiavelli's *The Prince* than to the nascent social science to which he explicitly gestures. Indeed, *The Crowd* became extremely influential within French military circles prior to the outbreak of the First World War, and a number of Le Bon's subsequent books placed less emphasis on the danger posed by the crowd than on delivering advice on how to regiment and control groups of people in order to extract the best from them.[6]

As might be expected, the collective psychology tradition drew on developing psychological views of the individual, particularly on what have been called irrationalist theories of individual psychology invoked to explain individual deviance and breakdown. The real key to understanding the psychology of crowds was to recognize the ways in which such a psychology was diametrically opposed to the psychology of individuals in everyday life. The fundamental juxtaposition, which we have already seen in passing in the quotes I have given from Le Bon, was between the rational, conscious, and controlled individual, and the irrational, unconscious, and potentially uncontrollable crowd.[7]

The historian Susanna Barrows has pointed to the debt that the collective psychology tradition owed to Jean-Martin Charcot's work on hypnosis and suggestibility as a rising form of irrationalism about individual behavior. Hypnosis provides one mechanism by which an individual can be changed from a thinking, rational, conscious agent to an unthinking, irrational, unconscious agent. Forming a part of an organized crowd, claims Le Bon, constitutes another way of bringing about such a transformation in an individual. This contrast appeals to the emotionality of individuals in a crowd, as opposed to their rationality as individuals in everyday life. Le Bon cites the sentiment of invincibility, the contagiousness of sentiments, and suggestibility as the three features of individuals that cause crowds to have the psychology they do.[8]

The attribution of a heightened level of emotion in crowds, and the contrast between rational individuals and emotional crowds, allowed the crowd to be characterized in terms of dominant social dichotomies. The psychology of the crowd was described as both feminine (rather than masculine), and primitive (rather than civilized), ascribing to crowds the putatively inferior minds of women and savages. This identification of crowd psychology, manifesting and reinforcing a form of democraphobia in the late nineteenth century, was seized on in the fascistic appropriations of theories of mass psychology in the early twentieth century, including in the work and thought of both Benito Mussolini and Adolph Hitler.[9]

This brief summary suggests three components at the core of the collective psychology tradition: (i) a deep mistrust and fear of crowds; (ii) a way of understanding their psychology that drew on and reified existing gendered and racialist thinking; and (iii) a self-image of thinkers from diverse intellectual backgrounds forging a new science, that of the psychology of the crowd, of real practical import. As some of my comments about Le Bon suggest, however, alongside this overarching framework in which crowds were viewed negatively and pessimistically, there was also a more enthusiastic view of what only crowds, not individuals, could accomplish, a view that built on perhaps only (iii) above.

This optimistic strand to the collective psychology tradition extended from roughly 1895 until 1920 and was particularly prevalent amongst some of the most influential participants in the emergence of both psychology and the social sciences, particularly sociology and anthropology. Wilhelm Wundt's *Völkerpsychologie* (discussed briefly in Part One), Emile Durkheim's collective representations, and William McDougall's group mind each constituted the basis for conceptualizing groups of individuals in a more positive light. Collectively they represent attempts to theorize about the products of group actions as cultural achievements. The processes leading to these achievements remain conceptualized as the underlying psychology of the group, and as such were viewed as irreducible to changes wrought on an individual. But this group mind was no longer dark, destructive, and to be feared. Rather, its formation was crucial to diverse social accomplishments, ranging from military victories to the distinctive progressive political organizations associated with the established and emerging democracies within Europe.[10]

It was within this more optimistic strand within the collective psychology tradition that one finds a vision of group psychology as both a meliorative and celebratory project. In the work of McDougall and William Trotter written self-consciously in the shadow of the First World War, groups constitute a natural environment for the development of basic human instincts, the foremost of which is *gregariousness*. It is thus important that humans form the right sorts of group so that this positive instinct can be properly nourished and play a role in directing individuals into constructive rather than destructive behavior. This conception of the group mind gave its study within social psychology and the social sciences a sense of importance and urgency in making a difference to the directions in which society as a whole were to take.[11]

These two strands to the collective psychology tradition – its original, pessimistic view of the group mind and its later, more encouraging view – both need to be kept in mind as we consider other developments of the group mind hypothesis. For the group mind has often been viewed exclusively through either a negative or a positive lens.

4 THE SUPERORGANISM TRADITION

Unlike the collective psychology tradition, the superorganism tradition has not been well canvassed by professional historians. The chief discussions of this tradition are to be found in brief historical precursors or afterthoughts in broader reviews by contemporary biologists, such as that on social insects by Edward O. Wilson and on group selection by David Sloan Wilson. As such, these are very much internal histories, told by those who form part of the research lineage whose history is recounted.[12]

The apparent intricate social harmony and organization of the "social insects" – the *Hymenopteran* wasps, bees, and ants, as well as the termites – has been invoked as a model for moralizing about human society for centuries. Unsurprisingly, it is these social insects that have been taken as paradigms within the superorganism tradition, a tradition central to the biological sciences from roughly 1900 until 1950. Rather than viewing itself as intermeshed with and building on existing common-sense beliefs about group behavior, as did the collective psychology tradition, the superorganism tradition set itself apart from prescientific, anecdotal reports on and claims about the groups on which it focused, insect colonies. Their systematic study, however, remained linked to a wider understanding of prosocial life.

The earliest work in this tradition that utilized the concept (even if not the name) of the superorganism was in plant ecology, where there was an early recognition of the various levels at which living systems were organized. Ecology was founded as a science studying communities of organisms, and early in its history Frederic E. Clements introduced the idea that the key to understanding changes in communities of plants over time was to view those communities themselves as complex organisms. These complex organisms themselves formed plant-animal communities or *biomes*, and both complex organisms and biomes were viewed as more than the sum of their individual parts. Clements treated biomes as subject to developmental regularities, whereby the succession of biomes could be viewed as moving from immature to mature states, something that could be measured in terms of species numbers and diversity. This is the beginning of the treatment of populations of organisms as having a physiology. This physiological, organismic view of populations came to play a crucial role in the Chicago school of ecology headed by Warder Clyde Allee, especially in the experimental work of Thomas Park.[13]

Ecological work in this broader tradition was in part inspired by observation of the ways in which individual organisms, both within a species and between species, operated in harmony with one another, from mere coexistence to active cooperation. This observation, made initially in plant communities, also motivated similar developments in animal ecology. Although research in both plant and animal ecology covered a wide range of forms that sociality and community took across the living world, it was the special form of social organization found in the social insects that became the focus of the superorganism tradition itself.[14]

I have already said that perhaps the most central figure in the superorganism tradition as it developed with this focus on the social insects was William Morton Wheeler, professor of entomology at the Museum of Comparative Zoology at Harvard University from 1908 until his retirement in 1933. Wheeler's career extended over a fifty-year period, and his publications included many popular articles as well as a dozen books, including the influential *Ants* (1910), *Social Life Among the Insects* (1923), and *The Social Insects* (1928).[15]

Wheeler's 1911 essay "The Ant-Colony as an Organism" developed the idea of thinking of an insect colony as a higher-order organism, and his hilarious, insightful 1920 essay "The Termitodoxa, or Biology and Society" introduced the term "superorganism" to describe such a colony. In the earlier essay, Wheeler had defined an organism functionally as anything that has the capacities for nutrition, reproduction, and protection, and in the latter essay he correlates these with "castes" in termite societies: respectively, workers, "the royal couple," and soldiers.

Edward O. Wilson has implied that Wheeler's defense of his thesis that the ant colony is literally an organism follows readily from the breadth of Wheeler's functional definition of an organism. In fact, Wheeler actually places little weight on the definition and concentrates instead on showing that ant colonies share a wider range of features with paradigmatic organisms, such as individuality and regeneration. Wheeler's deepest motivation for defending the idea of colonies as superorganisms derives from and in turn further stimulates his sustained interest in the sociality of organisms in general and of certain species of organism in particular. Wheeler concludes "The Ant-Colony as an Organism" with some brief comments about what he calls "the problem of the correlation and cooperation of parts," saying

> If the cell is a colony of lower physiological units, or biophores, as some cytologists believe, we must face the fact that all organisms are colonial or social and that one of the fundamental tendencies of life is sociogenic. Every organism manifests a strong predilection for seeking out other organisms and either assimilating them or cooperating with them to form a more comprehensive and efficient individual.

Wheeler suggests that colonies provide a more experimentally tractable system for exploring this problem than do individual organisms. This, in turn, presupposes that there is one and the same problem about colonial and organismic life, and that it may have been solved in one and the same way.[16]

We can express this problem in two ways, one suggested by Wheeler's name for it, the other by his putative solution, "egoistic altruism." As a *problem of the parts*, we might see this as the problem of why and how an individual, whether it be a colony or an individual organism, manages to coordinate its parts harmoniously, given their own organismic character. Conversely, the idea of egoistic altruism – of cooperating in order to benefit one's own self – suggests not a problem of parts but of wholes, of why and how it is that individual units, each satisfying Wheeler's three-pronged functional characterization of organisms, come to form larger such units. So expressed, this approximates the *problem of altruism*, the problem of explaining how phenotypes that differentially benefit individuals other than their bearer could have evolved through a process of natural selection.

Although the superorganism tradition has maintained a focus on the forms of sociality within the social insects, it has also shared Wheeler's interest in sociality in general, often with the social insects serving as a paradigm for thought. Wheeler himself adapted his trichotomy between nutritive, reproductive, and protective organismic functions from their role in defining the organism and in characterizing castes within termite colonies, to serving as the basis for distinguishing three forms that "true societies" could take, depending on which of these functions predominate in the emergent social behavior. Corals and higher vascular plants, in which a colony is formed asexually by reiterated budding, exemplify a *nutritive* society, the social insects a *reproductive* society, and flocks and herds of birds, mammals, and fishes, in which unrelated individuals act together for protection, a *defensive* society. For Wheeler, such eusocial groups both benefit the individuals that participate in them and, since Wheeler thinks of them as types of superorganisms, they also benefit themselves. Since there are two levels of adaptedness, the individual organism and the superorganism, there are two recipients of evolutionary benefit.[17]

This concern with the form and levels of sociality within the superorganism tradition has been guided by a pluralistic view of the level at which natural selection operates, that is, the view that natural selection can operate on a range of biological entities, from genes and cells to organisms and groups. Group selection has been viewed as the most problematic part of such a pluralistic view. Although I do not want to enter into substantive discussion of the debate over the agents of selection here, note briefly two paths for exploring the interplay between group selection and the conception of sociality within the superorganism tradition.[18]

First, consider the path from the superorganism tradition to a conception of sociality via the appeal to group selection. The extension of Darwinian natural selection to at least groups of organisms that are themselves organisms, superorganisms, makes perfect sense, since the natural selection that takes place between such groups is a form of competition between individuals. Since group selection has been traditionally invoked principally to explain the origins and evolution of sociality, within the superorganism tradition sociality itself is a natural feature of individual organisms, one that arises from the process of group selection. Here group selection promotes sociality.

Second, consider a more direct connection between the superorganism tradition and the naturalness of individual sociality, and its implications for thinking about the "levels" at which natural selection operates. For superorganisms there is a clear sense in which the opposition between individual and group interest that has been taken as constitutive of the classic problem of altruism does not exist. The competing interests of individual organisms within a superorganism have played to a draw, much as one might think that they have in the case of individual cells within an organism. Thus, the very existence of sociality is deemed unproblematic or natural. Superorganisms are just those groups of organisms that have overcome Wheeler's "problem of the coordination and cooperation of parts," where the parts are individual organisms. Individual organisms that are part of superorganisms have, in effect, eliminated or minimized within-group selection. This suggests that in superorganisms the force of natural selection between groups should be greater than that within groups. Thus, if the sociality of individuals within superorganisms is to be expected, so too is the strength of group selection, which is just selection between groups. Here sociality promotes group selection.

The group mind hypothesis goes beyond a defense of the superorganism in that it postulates the emergence not simply of a novel individual but one with a mind. Yet so far nothing I have said about the superorganism tradition bears on the attribution of *mental* properties to superorganisms, and thus on the concept of a group mind. Indeed, explicit endorsement of the group mind hypothesis has been less pronounced in, and the hypothesis itself has been more peripheral to, the superorganism tradition than to the collective psychology tradition. This has been so for at least three related reasons.

In the first place, there has been a greater concentration on defending the integrity of the concept of the superorganism itself, introduced as a novel scientific concept for understanding ecological and social organization, particularly that found in eusocial insect colonies. The collective psychologists, by contrast, devoted little attention to defending "the crowd" as a legitimate, scientific construal of human groups – perhaps, as I have implied above, less attention than they should have. Moreover, for those like Clements who adapted the superorganism concept to make sense of plant communities and biomes, even if populations were understood as having something like an ontogeny and a physiology, there was little motivation to ascribe distinctly psychological functions to their objects of study. Even for someone like Wheeler, the superorganism concept primarily allowed one to treat insect colonies via a form of social physiology, not psychology. Thus, with attention focused on whether insect colonies were organisms at all, and an emphasis on adapting a physiological rather than a psychological methodology, the question of whether they were organisms with minds became secondary.

Second, the predication of mental or psychological characteristics to individual animals has generally been more circumspect than it has been to individual humans. This has constrained the readiness with which those characteristics could be extended to groups of animals. Particularly for those setting out to go beyond folklore about the social insects and engage in systematic scientific study of them, C. Lloyd Morgan's canon – never invoke a higher-level mechanism when a lower-level mechanism will do – has loomed large. Both the behaviorist tradition within psychology and the developing ethological tradition associated with Konrad Lorenz and Nikko Tinbergen offered accounts of animal behavior that, in their different ways, minimized appeals to the mind (as opposed simply to internal brain physiology). These traditions thus created a climate in which psychological attribution to animals was circumspect. When one's focus was animals whose individual members might be thought to have minds in only some trivial sense, these tendencies of cautious mental attribution were amplified. If insect societies were to have a group mind, then not only would they constitute a new entity, the superorganism, but this entity would itself have emergent properties of a kind that its constituents lacked. In terms that I introduced at the end of section 2, minds within the superorganism tradition are group-only, not multilevel traits, and this fact about them contributed to making endorsement of the group mind hypothesis circumspect within the superorganism tradition.

Third, as we have seen, collective psychologists readily helped themselves to the developing notion of unconscious mental processes in individuals to explain the behavior of crowds and how they produced something like group personality traits: They were fickle, foolhardy, emotional, irrational. For collective psychologists, the group mind was the seamy and steamy emotional underbelly of human nature, reducing complex and rich psychological processes and traits to simpler and more impoverished ones. But no such processes or traits could be invoked within the superorganism tradition without a charge of anthropomorphism being laid. This left those in the superorganism tradition with more limited resources for explaining the mechanisms through which a group mind could have evolved or operated, and truncated the list of human mental characteristics that could be literally or even metaphorically ascribed to animal societies.

Given the evolutionary and functional motivations for positing superorganisms, it is no surprise to find that the sorts of psychological properties that were postulated within the superorganism tradition were those that served some identifiable biological function in the relevant species. These include the perceptual and communicative abilities involved in gathering information about food sources and the motoric capacities to utilize resources and avoid predators and dangers in the world. Some of these abilities, such as the ability of the bee colony to locate distant sources of nectar and regulate the relative number of foragers and hive workers in accord with the richness of the source, or the ability of the termite colony to rapidly repair damage to its nest, manifest both some level of intentionality and a degree of concern over the integrity of the colony. In short, the behaviors of at least some groups is such that it seems directed at self-preservation, where the self here is a colony, and the means of achieving that goal is cognitive.

I have already intimated that both the collective psychology and superorganism traditions embody some confusions. It is to one of these that I turn next, the confusion between the group mind hypothesis and quite a distinct thesis about the nature of individual minds, the social manifestation thesis.

5 GROUP MINDS AND THE SOCIAL MANIFESTATION THESIS

As a way of introducing this confusion, let me begin with William McDougall's critical dissent from a reductionist claim made by Herbert Spencer. In defending a reductionist view of social groups, Spencer had introduced an analogy between how emergent properties of groups of physical entities, such as stable geometrical forms and crystallization, and those of social groups could be understood. To Spencer's claim that the "structure and properties of a society are determined by the properties of the units, the individual human beings, of which it is composed," McDougall says,

> the aggregate which is a society has, in virtue of its past history, positive qualities which it does not derive from the units which compose it at any one time; and in virtue of these qualities it acts upon its units in a manner very different from that in which the units as such interact with one another. Further, each unit, when it becomes a member of a group, displays properties or modes of reaction which it does not display, which remain latent or potential only, so long as it remains outside that group. It is possible, therefore, to discover the potentiality of the units only by studying them as elements in the life of the whole.

McDougall's continuation of this passage reexpresses his view in terms that make its location within the collective psychology tradition more explicit, as follows:

> That is to say, the aggregate which is a society has a certain individuality, is a true whole which in great measure determines the nature and the modes of activity of its parts; it is an organic whole. The society has a mental life which is not the mere sum of the mental lives of its units existing as independent units; and

a complete knowledge of the units, if and in so far as they could be known as isolated units, would not enable us to deduce the nature of the life of the whole, in the way that is implied by Spencer's analogies [to stable geometrical forms and crystallization].[19]

Melded together here, in both of the halves of this long passage, are two quite distinct ways in which one might develop the general nonreductionist view of the relationship between society and the individual that we have seen was characteristic of adherents to the collective psychology tradition.

First, there is the idea that groups have properties, including mental properties, which their individual members do not have, and which are not reducible to the properties of those members. This emergentist view of group properties, together with the further assumption that some of these properties are psychological, entails a version of the group mind hypothesis that postulates group psychological traits. These traits, while in a strict sense multilevel traits (such as being angry or irrational), are not actually possessed by the individuals in the corresponding social group prior to or simultaneous with their forming that group. In this sense, these group psychological properties are something other (or more) than the properties of those individuals. Thus, they are more like group-only than multilevel traits.

Second, there is the idea that individuals have properties, including psychological properties, that are manifest only when those individuals form part of a group of a certain type. This is what I shall call the *social manifestation thesis*. Precisely how this sits with the group mind hypothesis is far less clear, since it makes a claim about the character of individual minds, and it would seem that any group properties relevant to this claim could be and indeed would likely be nonpsychological in character.

For example, suppose that individual people become angry or aggressive in certain ways only when they form a certain type of group (for example, a crowd). Then, unless they do so only because the crowd itself has a specific psychological profile, there is no need to posit group psychological properties, and so no role for the group mind hypothesis. Moreover, insofar as the social manifestation thesis does lead to the group mind hypothesis, it does so only via the first view above, and so there is no independent path from the social manifestation thesis to the group mind hypothesis. It could still be true, of course, that there are important senses in which group behavior cannot be reduced to or be derivable from that of the individuals within it, even if what explains that behavior are the psychological states of individuals of whom the social manifestation thesis is true. These points together suggest that the nonreductionist motivation undergirding the collective psychology tradition, especially in its latter, more optimistic phase from 1895 to 1920, does not lead indelibly to the group mind hypothesis.

These two views, the group mind hypothesis and the social manifestation thesis, are logically independent. Clearly, the social manifestation thesis could be true without entailing the group mind hypothesis if group minds did not exist. Conversely, the group mind hypothesis could be true without entailing the social manifestation thesis if the relevant groups were comprised of individuals that did not have minds at all. We will return to further explore the relationship between the social manifestation thesis and the group mind hypothesis in the next chapter.

6 COLLECTIVE PSYCHOLOGY, SUPERORGANISMS, AND SOCIALLY MANIFESTED MINDS

Shifts between the group mind hypothesis and the social manifestation thesis are common in the collective psychology tradition. But I also want to make the perhaps initially puzzling proposal that those in this tradition in fact have a primary preoccupation with the social manifestation thesis. Although they often express their views in ways that suggest an explicit endorsement of the group mind hypothesis, the broader context in which their views are developed makes the social manifestation thesis a more plausible interpretation of what they mean. I shall provide support for this view by discussing a few representative quotations from the work of Wilhelm Wundt and Gustav Le Bon.

For example, consider the chapter headed "Mental communities" in Wundt's *Outlines of Psychology*. Here Wundt focuses on the importance of the environment, especially the social environment, both to the development of the child and to the sorts of properties that are significant for individual consciousness. This focus is appropriate if the social manifestation thesis is one's primary concern. But Wundt says, strikingly, that

> these social interconnections have just as much reality as the individual consciousness itself. In this sense we may speak of the interconnection of the ideas and feelings of a social community as a *collective consciousness*, and of the common volitional tendencies as a *collective will*.[20]

The first of Wundt's claims here, and the chapter in general, supports some version of the social manifestation thesis, even if his second claim expresses a version of the group mind hypothesis.

We see the same shift in Le Bon's *The Crowd*, which, as we have seen, attempted to delineate the psychological characteristics of crowds and their degrading effect on individual human cognitive performance. Le Bon says

> Whoever be the individuals that compose it, however like or unlike be their mode of life, their occupations, their character, or their intelligence, the fact that they have been transformed into a crowd puts them in possession of a sort of collective mind which makes them feel, think, and act in a manner quite different from that in which each individual of them would feel, think, and act were he in a state of isolation.[21]

This is a version of Le Bon's law of the mental unity of crowds, a view that the sociologist Clark McPhail has called the *transformation* thesis, since it is a thesis about how individuals are transformed when they belong to a crowd. As such, clearly this is a view about the abilities and proclivities of individuals, and should be subsumed under the social manifestation thesis. Le Bon's "law" implies the group mind hypothesis only if individuals are transformed because a group mind is formed, or a group mind is a consequence of this transformation. Each of these is a further claim, however, one for which no further argument is given. This is, in part, because Le Bon himself has not distinguished the two theses in the first place. If Le Bon's chief concern is with how the capacities and abilities of individuals are changed when they form a crowd, as I have been suggesting, then talk of group minds is a confusing (even if vivid) way of expressing something like the social manifestation thesis.[22]

Prima facie, there is less room in the superorganism tradition for this sort of melding of the social manifestation thesis with the group mind hypothesis, chiefly because of the more circumspect and limited appeal to psychological properties at all within it. For the social insects, recall that the group mind is posited as a group-only trait, and at least in this case there are no individual minds to have socially manifested, psychological properties. In the case of plant communities and plant-animal biomes, as I have already noted, there is a more circumscribed tendency to endorse either the group mind hypothesis or the social manifestation thesis at all.

There is in the superorganism tradition, however, a shift between two views that parallel our two theses, between a nonpsychological version of the social manifestation thesis, and the claim that certain communal groups are (literally) superorganisms. In addition, in at least places it is the former of these that captures the heart of work in the superorganism tradition, even when language that suggests the superorganism conception predominates. For example, although Clements and Shelford postulate the biome as a living entity in its own right, the primary concern of their book as a whole, and the guiding idea of early ecology more generally, is that animals and plants exert a ubiquitous, mutual influence on one another. By virtue of this influence that exists between individual organisms, each is able to act in ways that it could not otherwise act. This is a behavioral or physiological version of the social manifestation thesis, and is independent of the postulation of communities or biomes as superorganisms. Likewise, at the core of Allee et al.'s review of the social insects is a discussion of both "the division of labor between individuals composing the group and the integrative mechanisms that give unity to the group," both of which concern properties of and relations between individuals. Individuals take on these properties only in the "true social" insects. While Allee et al. take this to support the postulation of superorganismic status for the group, this is clearly a further claim, one that was particularly pronounced in the contributions of Alfred Emerson to the Chicago school's approach.[23]

I suspect that it is likely a fruitless endeavor to attempt to defend the idea that one or the other of these ideas is overall more fundamental within the superorganism tradition, for each of them was drawn on and developed for different purposes. My chief point here is that expressions that suggest something like the group mind hypothesis should not always be taken at face value, and that thinking about the ways in which groups serve as a crucial context for the manifestation of individual capacities, including cognitive capacities, may more adequately take one to issues at the heart of the both traditions.

7 FROM THE PAST TO THE PRESENT

In the previous section, I said that the superorganism tradition is motivated in large part by the observation of a variety of forms of social harmony in the living world. Peaceful coexistence and cooperation are viewed as constituting natural relationships between individual organisms, and there are mutual benefits gained by individuals through these forms of social organization. The social manifestation thesis and the group mind hypothesis represent the foundation of two different ways to develop a scientific account of this observation.

The social manifestation thesis places an emphasis on an individual's abilities and dispositions, albeit those manifest in particular social contexts. Here whether an individual belongs to a group of a certain kind becomes important to understanding that individual's psychology, but there is no need to posit psychological properties of groups themselves. The group mind hypothesis, by contrast, does look to levels of organization larger than the individual in identifying the subjects of psychology, claiming that groups themselves have minds. To many, this will seem to involve a more far-reaching revision to our ideas of what minds are, and what can possess them.

The argument in this chapter, particularly in the previous section, has been that many putative expressions of the group mind hypothesis, particularly those in the collective psychology tradition, are in fact better characterized as expressions of something like the social manifestation thesis. If this is correct, then, given the distinction between these two theses, there is a real issue as to whether the group mind hypothesis does form a central part of either the collective psychology or superorganism traditions. If the social manifestation thesis more accurately captures both traditions, then contemporary defenses of the group mind hypothesis are misconceived if they are viewed as revivals of lost traditions of thought.

I shall argue, in the next chapter, that there is an additional layer of misconception in such a construal of the group mind hypothesis. This is because its contemporary putative expression is also better construed as an expression of the social manifestation thesis.

12

The Group Mind Hypothesis in Contemporary Biology and Social Science

1 REVIVING THE GROUP MIND

In the previous chapter, I suggested that the group mind hypothesis could be understood either literally or metaphorically. Expressions of the group mind hypothesis, both in the collective psychology and superorganism traditions and in contemporary discussions, often lend themselves to the literalist interpretation, although I argued in the previous chapter that in fact many of these are better understood as somewhat confusing attempts to state a version of the social manifestation thesis. I shall begin by briefly looking at contemporary views that appear to express the group mind hypothesis, and then raise questions about how they are best understood, and what notion of mind they draw on.

David Sloan Wilson has been a key advocate of the group mind hypothesis in the biological sciences, casting his advocacy in terms of the notion of cognitive adaptations and locating it within his broader defense of group selection and group-level adaptations. Group-level adaptations are species-specific phenotypes, including behaviors, that evolved because they conferred a selective advantage on their bearers, that is, on the groups of organisms that have them. In extending the notion of a group-level adaptation to cognitive phenotypes, Wilson says,

> Group-level adaptations are usually studied in the context of physical activities such as resource utilization, predator defense, and so on. However, groups can also evolve into adaptive units with respect to cognitive activities such as decision making, memory, and learning. As one example, decision making is a process that involves identifying a problem, imagining a number of alternative solutions, evaluating the alternatives, and making the final decision on how to behave. Each of these activities can be performed by an individual as a self-contained cognitive

unit but might be performed even better by groups of individuals interacting in a coordinated fashion. At the extreme, groups might become so integrated and the contribution of any single member might become so partial that the group could literally be said to have a mind in a way that individuals do not, just as brains have a mind in a way that neurons do not.

Examples of group-level cognitive adaptations that Wilson cites here are foraging and resource allocation strategies in bee colonies, as discussed by David Seeley, human group decision making, and what Herbert Prins has called "voting behavior" in buffalo herds in deciding in which direction to move.[1]

As the passage above indicates, part of Wilson's argumentative strategy involves showing that cognitive or psychological processes are no exception to the general phenomenon of group-level adaptation. Wilson himself identifies his defense of the group mind hypothesis as a revival of an idea prominent in the foundations of the social sciences. What he finds problematic about "early views of the group mind in humans" is that they "were usually stated in a grandiose form and without attention to mechanisms, similar to naive group selectionism in biology during the same period." His revival of the group mind hypothesis within a sophisticated group selectionist framework is a remedy to at least the latter of these problems.[2]

The group mind hypothesis has received a more cautious and circumscribed defense within the social sciences, where intimations that groups can think, decide, remember, or cognize more generally are typically hedged in ways that invite the metaphorical interpretation of the hypothesis. For example, in her *How Institutions Think*, the anthropologist Mary Douglas urges a reconsideration of the contributions of Emile Durkheim and Ludwig Fleck to the foundations of sociology, particularly their treatment of institutions and social groups "as if they were individuals." Recognizing "a tendency to dismiss Durkheim and Fleck because they seem to be saying that institutions have minds of their own," Douglas attempts to "clarify the extent to which thinking depends upon institutions" through a discussion of Durkheim's priority of the social over the individual and Fleck's notion of a thought collective.[3]

Likewise, in attempting to demarcate cultural psychology as a distinct field, Richard Shweder introduces the idea of *intentional worlds* that individuals share, taking cultural psychology to be the study of such worlds. But Shweder is, like Douglas, wary of understanding such a study as implying that social structures in and of themselves have minds, apart from the intentionality that individual agents have. In both cases we need to move beyond the attribution of in-the-head representations to individuals in order to do full justice to the complexity of social and cultural phenomena, and this is done through making cognitive attributions to entities larger than the individual.[4]

David Sloan Wilson has recently extended his thinking about the group mind hypothesis to understand the nature of religious ideas and social organizations. In doing so, he bridges between the biological and social sciences, a trademark of some of his earlier work. Wilson's basic idea is to see society and particular social structures as organisms, with religion being one of the culturally evolved adaptations that maintain this social organism over time. Again locating his views against the background of his multilevel theory of selection, he says

> If the individual is no longer a privileged unit of selection, it is no longer a privileged unit of cognition. We are free to imagine individuals in a social group connected in a circuitry that gives the group the status of the brain and the individual the status of the neuron.[5]

While Wilson considers religion as a group-level adaptation, his expression here is interestingly neutral between a literal and a metaphorical reading of the group mind hypothesis.

Critical to determining the plausibility of the group mind hypothesis, on either the literal or metaphorical understanding, is some further discussion of what it means to have a mind at all.

2 ON HAVING A MIND

A common reaction to the group mind hypothesis, at least among many working biologists and social scientists, is that it has no real empirical content because mindedness, the property of having a mind, is so vague. This view of the group mind hypothesis is, no doubt, facilitated by the ubiquity of the cognitive metaphor in describing biological and social processes. But I suspect that this view also reflects an ignorance of and an insensitivity to the conceptual work necessary to articulate what it is to have a mind. Regarding this latter point, I think we can make some progress.

In order for some entity to have a mind it must possess at least some psychological properties (states, processes, dispositions). Rather than attempting to offer a definition or analysis of what a psychological property is, let the following incomplete list suffice to fix our ideas. There are classical faculties, such as perception, memory, and imagination; processes or abilities that are the focus of much contemporary work in the cognitive sciences, such as attention, motivation, consciousness, decision making, and problem solving; and common, folk psychological states, such as believing, desiring, intending, trying, willing, fearing, and hoping. But what is it to *possess* one or more of these?

Whatever else is involved in having a psychological property, it surely turns largely on how one is physically structured. In particular, one must have a physical structure that realizes or implements that property (state, process, disposition). If we put this together with the standard view of realization discussed in Part Two, to possess a psychological property is to contain what I called an entity-bounded system or systems that realize processes that generate or physically constitute that property (state, process, disposition). For at least our paradigms of cognitive agents – intact, fully functioning, normal human beings – I have argued that we need a more general concept of realization, what I called the context-sensitive view, which allows that the systems that realize cognition can extend beyond the boundary of the individual. Replacing the standard with the context-sensitive view of realization gives us the following account of what it is for a paradigm cognitive agent, A, to possess or have a psychological property:

A possesses psychological property (state, process, disposition) P just if A either physically contains an entity-bounded system or systems, or is part of a wide system or systems, that realize the processes that generate or physically constitute P.

So when Tom feels pain, he has that property by virtue of physically containing a nociceptive system that generates that feeling. When he believes that snow is white, it is by virtue of being part of a folk psychological system that extends beyond his own head.

This account does not tell us why it is A – rather than, say A plus or minus bits of the physical world (including bits of A) – that possesses a given psychological property. Rather, we start with paradigm cognitive agents and attempt to explain what it is *for them* to possess psychological properties. For the most part, both in common sense and in the cognitive sciences, individuals rather than their parts or larger units of which they are a part leap out at us as the bearers of psychological properties. It is Tom who feels the pain or has the belief, and not Tom's nociceptive system or the folk psychological system at large. In passing, I have suggested that this is because we physically contain at least the *core* realizations of our mental states (even if not their total realizations), and are the *locus of control* for the actions that result from those states. There are special cases, such as those of split-brain patients being tested under special conditions, or when people have specific neural deficits, where it may be compelling to attribute (say) folk psychological states to entity-bound systems themselves, and we might view the group mind hypothesis as implying the need to consider something like wide systems as the subjects of such states.

Clearly, in considering the idea that other kinds of entities, such as groups, might have minds, we need to attend more closely to the issue of just which entity it is that possesses psychological properties than in the case of individuals. In particular, I have been arguing that at least historically many of the claims about group minds and group psychology are best understood as making a claim about the role of groups in regulating, developing, or disabling individual minds. In moving from individuals to groups as putative bearers of psychological properties we need to ensure that the multilevel or group-only traits we are ascribing are not better understood simply as individual-level traits.

There is a second complication as we move from paradigm cognitive agents to groups as putative cognitive agents. We do not simply instantiate a few psychological properties but many, and these encompass a diverse range of states, processes, and dispositions. Following my discussion of folk psychology in Part Three, I shall say that in virtue of that we possess *full-blown* minds. But I know of no defense of the group mind hypothesis that has claimed that groups have full-blown minds. (Talk of "group consciousness," for example, is nearly always talk of an aspect of the consciousness of individuals.) That is clearly science fiction, not borderline science, in the league of Attack of the Killer Tomatoes rather than, say, the Gaia Hypothesis. Moreover, there would seem to be little explanationist motivation for adopting the full-blown group mind hypothesis.

As is the case in striving to make sense of the idea of animal minds or of artificial intelligence, we should probably start with something less than full-blown minds in trying to understand what group minds might be. Consider, then, in light of our account of what it is to have or possess a psychological property (state, process, disposition), the notion of *minimal mindedness*:

X has a minimal mind just in case X has at least one psychological property (state, process, disposition).

Given that we have full-blown minds, the group mind hypothesis and minimal mindedness together entail that groups either literally have at least one of the psychological properties that intact, functioning individual human beings have, or can usefully be treated as if they do. At least the literalist reading is quite a strong and striking thesis about groups, and seems the right way to understand the group mind hypothesis insofar as it has formed part of the collective psychology and superorganism traditions. Where discussion of "the group mind" concentrates on just one psychological process or ability, as it often does, I will refer to that as the *focal* process or ability.

Can a group literally have a minimal mind? I want to argue first that the most obvious reasons for saying "yes" or "no" won't do, and that resolving even this question is more difficult than it may initially appear.

Consider a problem with one "obvious" reason for answering "yes." It isn't sufficient for a group to have a minimal mind for that group to engage in some action that is caused by psychological processes. For if those processes take place entirely in the heads of the individuals in those groups, then the action can be explained adequately without positing a group minimal mind. This implies that socially distributed cognition, at least considered as a multilevel trait, does not entail even a group minimal mind. For example, consider the (by now familiar) action of navigating a navy vessel, which requires a complex social and mental division of labor, or decision making in medical diagnosis. In these cases, the social context is integral to the way in which individuals process information and cognize. If, despite this, it remains true that every psychological process or ability that causes and thus explains the acts that constitute the navigation of a navy vessel, or that leads to a medical diagnosis, belongs to an individual member of the group, then there is no group minimal mind in addition to these minds. The statement "the crew saw the oncoming ship and decided to change direction" might be made true simply by individual-level psychological facts, together with other, nonpsychological facts about social organization. Thus, those who grant the existence of group actions but think that psychological processes and activities are essentially individual-level (and so neither group-only nor multilevel) traits, could deny the existence of even group minimal minds. Such a position might be attractive to one who held the social manifestation thesis.[6]

As an aside, note how this points to an important disanalogy between spatially organized neurons in someone's head, and socially distributed agents in some larger system, an analogy that we have seen David Sloan Wilson make and that has been mentioned by others in conversation and correspondence. Individual neurons do not perceive (but fire in response to a stimulus), remember (but transmit information about the past), or plan. By contrast, individual agents do all of these things, and these are just the sorts of properties attributed in socially distributed cognition.

This suggests that the clearest cases that exemplify group minds will be those that involve group-only, rather than multilevel traits. For in identifying a cognitive process found only at the group level, we preempt the objection that the social manifestation thesis suffices to capture the phenomenon being explained. Group-only cognitive properties can be posited for human social groups, but as I implied in the previous chapter, they are clearest in the case of superorganismal colonies.

Consider the claim that beehives have the focal ability of planning to acquire honey from a given source. The action of the beehive is systematic, structured, and predictable, and involves the coordination of hundreds or thousands of bees. Putting together the group mind hypothesis, the notion of minimal mindedness, and the account of what it is to possess a psychological property, we have the following

A beehive possesses the psychological property of planning to acquire honey from a given source just if that beehive either physically contains an entity-bounded system or systems, or is part of a wide system or systems, that realize the processes that generate or physically constitute such planning.

In this case, the entity-bounded systems are the individual bees that participate in the process of planning to acquire honey from the given source. But since it is implausible to think that any single bee does this planning (so little brain, so little time...), such planning, if it exists, must be a group-only cognitive process.

Turn now to the objection that entities that never have full-blown minds – minds in much the sense that we have minds – cannot possess any psychological properties. Prima facie, some groups seem able to initiate actions or behaviors – such as collecting food, steering a steady course north, or building a school – that would be explained by positing underlying psychological processes and abilities were those actions performed solely by individual agents. This is true even if those individual agents lack full-blown minds, as is the case with current robotic and computational systems. We readily and typically speak, for example, of a chess program as knowing that you are about to take its bishop, of deciding to castle early, of trying to dominate the center of the board, of remembering where your Queen is. It may be that the best way to view these attributions is as metaphorical, but it seems to me that a literal understanding of them is not ruled out simply because we do not think of those programs as complete psychological agents, agents with full-blown minds.

Consider the case of a person who has lost a range of her psychological functioning, perhaps permanently. Suppose that significant parts of her visual system, her memory, and her language abilities are impaired in ways that limit what states we can plausibly attribute to her. Still, there are likely to be clear cases where she literally tries to do something (such as reach a glass or find the words to ask where your children are), or knows or decides to do something. In short, she might be literally, and not just metaphorically, minimally minded. And if individual agents can have minimal minds, then so too can groups.

3 MINIMAL MINDS, CONSCIOUSNESS, AND HOLISM

Construing the group mind hypothesis in terms of the notion of a minimal mind brings with it more specifically philosophical concerns. In implying that a minimal mind could lack consciousness, doesn't this construal of the group mind hypothesis stop short of giving us the real cognitive McCoy, leaving us only with *ersatz* minds? And by suggesting that there could be a group mind with a solitary psychological state, doesn't it likewise trade real for mere "as if" intentionality and cognition? While I think that both of these implications hold, I don't see them as problematic for this way of construing the group mind hypothesis.

In effect, we could see each of these objections as placing a putative constraint on the types of minimal minds that can exist without full-blown minds: Minimal minds must be conscious, and minimal minds must be holistic, respectively. Consider the latter of these first.

Mental processes and states in general are, I think, instantiated in clusters, and so are, in some sense, holistic. That seems to be true both of individual and group minds (compare the characterization of artificial minds above). Thus, I see no problem in group minimal minds satisfying some version of the holism constraint. Since my characterization of minimal mindedness allows for the possibility of a minded entity with just one psychological state, it is not compatible with one predominant view of the holism constraint in the philosophy of mind. On this view, holism is an *a priori* constraint on psychological attribution. But I reject this view of the constraint. Rather, the constraint is empirical in reflecting the way in which psychological states and processes cluster. This clustering should be viewed as underwritten by homeostatic mechanisms and constraints that determine the ways in which these properties are realized in the actual world.[7]

The consciousness constraint, however, would make the group mind hypothesis implausible, since the claim that groups have consciousness has been central to neither of the collective psychology and superorganism traditions, nor in their contemporary revivals. Indeed, at least those in the collective psychology tradition have often tried explicitly to distance themselves from such claims. For example, William McDougall devotes an early chapter of *The Group Mind* to criticizing those in the German idealist tradition who talk uncritically of "the consciousness of the group" as something more than the consciousness of each of the members of the group. Recall also the affinity between the collective psychology tradition and the postulation of unconscious processes in individuals.[8]

In Part Three, I distinguished between six mental phenomena that have been considered as exemplifying consciousness, classing these as processes of awareness (higher-order cognition, attention, and introspection and self-knowledge) and phenomenal states (pain, bodily sensation, and visual experience). While some groups might be thought to manifest something like processes of awareness – for example, consider quorum sensing in bacteria as a form of self-awareness – none plausibly instantiate phenomenal states.

This can be readily accounted for within the framework I have introduced in this chapter. If the chief motivation for positing a group mind is explanatory, and we start with the notion of a minimal mind and add specific focal abilities as they seem justified, then we would expect to find paradigmatic intentional states (including processes of awareness) among the constituents of the group mind, but not the "what it's like" of mental life. This reinforces two negative themes from Part Three: that "consciousness" is not the name of a unitary mental phenomenon, and the intentional and phenomenal aspects to the mind are not inseparably bound together.

Interesting in this regard is the status of emotions, which one might well think of as straddling the putative divide between intentional and qualitative states. Emotions are psychological states that are typically about something – they have intentionality or mental content – but there is also something phenomenally distinctive about them: there is something it is like to have them. Both McDougall, and before him, Le Bon, considered the ways in which emotions such as fear and panic can be properties of groups that degrade the abilities of the individuals in those groups via the sympathetic responses of individuals to one another's reactions. If this is to be consistent with the denial of group consciousness, then these group emotions must be nonqualitative, even though the emotions in individuals that give rise to them (the "sympathetic responses") need not be. The same would also have to be true of the "positive" emotional traits of a group mind, those which, claims McDougall, emerge from highly organized human societies, and which lead to heightened individual abilities.

4 THE CONTEMPORARY DEFENSE OF THE GROUP MIND HYPOTHESIS

I have noted that David Sloan Wilson's recent revival of the group mind hypothesis draws on his defense of group selection in evolutionary biology. I will eventually turn to some of the complexities to the relationship between these two views, but first I want to examine Wilson's views in light of our discussion so far in this and the previous chapter.

It is in his paper "Incorporating Group Selection into the Adaptationist Program: A Case Study Involving Human Decision Making" that Wilson develops his views of the group mind hypothesis most fully, and it is the "case study" part of that paper on which I will concentrate in what follows. Wilson's focus here is on the literature on human decision making, particularly on human decision making in groups, and this case study review is intended primarily to support the idea that human decision making has evolved both by individual selection and by group selection. He begins by distinguishing two ways

in which human decision making can evolve to maximize the fitness of whole groups. First, individuals might function as independent decision makers whose goal is to benefit the group. This is the way we usually think about altruism (Sober and Wilson [1998]). Second, individuals might cease to function as independent decision makers and become part of a group-level cognitive structure in which the tasks of generating, evaluating, and choosing among alternatives are distributed among the members of the group.... At the extreme, the role of any individual in the decision-making process might become so limited that the group truly becomes the decision-making unit, a group mind in every sense of the word.

Wilson illustrates the second of these alternatives with an example of decision making about food sources in honey beehives, going on to suggest that although "we should not expect group-level cognition in humans to resemble the social insects in every detail," human social groups can be said to constitute what he calls *adaptive decision-making units*. In the framework proposed in this chapter, we can see Wilson here as positing the existence of minimal minds in groups of animals, including humans, where the focal psychological ability is decision making.[9]

It should be clear that it is only the second of Wilson's alternatives that represents the sort of emergentist view of group psychological properties that I outlined in the previous chapter as part of the collective psychology and superorganism traditions, and only in "the extreme" would such a view support the group mind hypothesis. Insofar as the first of the alternatives Wilson presents states a view about psychology at all, it expresses a version of the social manifestation thesis. Individuals have a psychological character that confers benefits on the group as a whole, and do so only because of properties of that group, such as having a high proportion of altruists, or imposing severe social costs for nonaltruists.

Wilson continues by discussing the second of these alternatives, equating it with the idea that groups are adaptive decision-making units, and focusing on an assessment of the performance of group and individual decision making. This discussion is aimed largely at offering support for the idea that human decision making evolved in part by group selection. But before turning to that claim, and in light of the discussion in this chapter so far, it is worth asking whether the phenomenon to be explained by an appeal to group selection concerns the character of individual decision making or that of group decision making. Wilson seems to imply that it is both when he says if "human cognition is a product of group selection, we should expect individuals to be innately prepared to easily 'hook up' with other individuals to form an integrative cognitive network." There is the formation of an integrative cognitive network, and the innate preparedness of individuals to form such networks. But it is only the former and not the latter of these that is directly relevant to the group mind hypothesis, rather than to the social manifestation thesis.[10]

Much the same general point can be made about Wilson's application of the group mind perspective to understand religion. At the outset, Wilson says that he aims to "treat the organismic concept of religious groups as a serious scientific hypothesis" and that his "purpose is to see if human groups in general, and religious groups in particular, qualify as organismic in this sense." But he needs two things to be true if this project is to lead to a defense of the group mind hypothesis.[11]

First, it is not just religious groups but religious *ideas* or something appropriately psychological that must be adaptations (to get the *mind* in "group mind"). Second, this psychology must attach not simply to the individuals in religious groups but to the groups themselves (to get the *group* in "group mind"). Much of what Wilson says about the functionality and adaptedness of religious groups could be true without either of these assumptions being true. That is, religious groups could be viewed as organismic in nature, and be subject to natural (group) selection, even though the processes that drive selection are either individual-level psychological or group-only nonpsychological processes. Moreover, it may be plausible to think that some of these individual-level psychological processes, such as those that produce feelings of devotion, of a deep sense of commitment, or of Godly love, pretty much require social groups with the features that religious groups have. If that is so, then we have a view more accurately characterized by the social manifestation thesis than by the group mind hypothesis.

Let us return to the examples of integrated cognitive networks and decision making. We have already seen that both the collective psychology and superorganism traditions had an emergentist view of the nature of groups (and thus group minds): Groups are more than the sum of the individual parts, and having a group mind is more than having a group of individuals with minds. If Wilson is to keep with this aspect of these traditions, then it follows that the integrative cognitive networks that he postulates must be something more than individuals being innately prepared to hook up with one another. Likewise, for his multilevel property of decision making: Having a minimal mind with this focal ability must be more than having individual members with this focal ability.

In some trivial sense, a club makes a decision (say, by majority vote) about whom will be their next president simply by each of the members publicly expressing a decision on this matter. Even though the decision here is distinct from those of the individual voters – since individuals by themselves cannot elect a new leader – if there is a group mind here it is nothing over and above the minds of individuals. Given the independence of the group mind hypothesis and the social manifestation thesis, there must be something more to having a group mind than there being individuals with socially manifested psychological characteristics. There is a problem for Wilson's views here, at least construed as reviving the group mind hypothesis.

With respect to human decision making, he would seemingly need to show that this functions at the group level by individuals relinquishing their own decision-making activities. For it is only by doing so that he could point to a group-level psychological characteristic that is, in the relevant sense, emergent from individual-level activity. Now those in the collective psychology tradition, and especially those writing from 1870–1895, did think that this happened. Yet they claimed that it typically led to a degradation of individual abilities. Crowds, for example, had their own psychological character, one that involved the transformation of autonomous individuals into members of *madding crowds*. Wilson must distance himself from this aspect of the collective psychology tradition – and does so in his discussion of Irving Janis's more recent concept of *groupthink* – since he aims to defend the view that groups are *adaptive* decision-making units, that is, units that have properties that promote fitness. Thus, he must look for ways in which collectivities confer benefits. But if those benefits are nothing more than benefits to individuals, then collective behavior would seem to be explained by an individual, not a group, psychology. If individuals can simply enhance their own individual decision making by forming groups of a certain character, then we may have the beginnings of an interesting argument for the social manifestation thesis, but are no closer to the group mind hypothesis.[12]

Thus far, I have argued that the group mind hypothesis has been run together with the social manifestation thesis not only in the collective psychology and superorganism traditions, but also in David Sloan Wilson's revival of those traditions within contemporary biology. Finally and more briefly, I turn to the recent social sciences to explore whether such confusion exists there as well. As I have mentioned, Mary Douglas's *How Institutions Think* aims, in part, to call attention to Ludwig Fleck's neglected notion of a thought collective, and examining Fleck's own views and their (belated) reception is revealing.

Fleck introduced the terms *Denkstil* and *Denkkollektiv* in a 1935 book, finally translated and published as *Genesis and Development of a Scientific Fact*. Part of the excitement about that book within the social studies of science community, and a chief reason for its eventual translation, was its anticipation of much that was ushered in to the study of science following Thomas Kuhn's *The Structure of Scientific Revolutions*. For Fleck, science is always practiced within the confines of a thought collective, "a community of persons mutually exchanging ideas of maintaining intellectual interaction," with its own thought style. These provide the framework in which "facts" are determined, and alternative thought collectives may represent different and incompatible sets of facts. In short, Fleck is seen as anticipating some of Kuhn's most influential ideas – that of a paradigm, of the dependence of phenomena on paradigms, and the incommensurability between successive paradigms. More generally, in ascribing a central role to social dimensions to theory construction, hypothesis testing, and scientific change, Fleck made an early, decisive break from the positivist view of science that Kuhn's *Structure* called into question twenty-five years later.[13]

As condensed as this summary is, one can note immediately that thought collectives are not group minds but the social context in which individual minds function. In Kuhn's foreword to the English translation of

Fleck's book, he says that his own struggles with the book cluster around "the notion of a thought collective," saying that "a thought collective seems to function as an individual mind writ large." Although Kuhn does not come straight out and say that the basic problem with the notion is that collectives do not have thoughts, even if they are crucial to the having of thought, this seems clearly one of the problems that Kuhn himself has with the notion.[14]

If we turn to Douglas herself, what she sees as important in Fleck (and in Durkheim) is the idea that "true solidarity is only possible to the extent to which individuals share the categories of their thought." Both Fleck and Durkheim "were equally emphatic about the social basis of cognition." She concludes her monograph by saying that

> Durkheim and Fleck taught that each kind of community is a thought world, expressed in its own thought style, penetrating the minds of its members, defining their experience, and setting the poles of their moral understanding.... individuals really do share their thoughts and they do to some extent harmonize their preferences, and they have no other way to make the big decisions except within the scope of institutions they build.[15]

What each of these expressions points to, and contrary to the title of Douglas's book, is not "how institutions think," but how individuals think (only) within institutional frameworks. Again, this is in keeping with the social manifestation thesis, rather than the group mind hypothesis.

We have been led to the social manifestation thesis both by reflection on our two traditions and their contemporary revivals in the biological and social sciences. So I round out this chapter with some further thoughts on that thesis that reach back to earlier discussions in *Boundaries*.

5 THE SOCIAL MANIFESTATION THESIS

The social manifestation thesis says something significant about the nature of cognition and its relation to individuals, but it is important to be clear about what is significant here. It says that some psychological states of individuals are manifested only when those individuals form part of a social group of a certain type. Both the "social" and "manifestation" parts of the thesis require further explanation.

One way of understanding this pair of notions is in terms of the idea that individuals have their psychology transformed through social membership. We have seen that this view has been thought of as playing a central role within the collective psychology tradition, where the relevant social groups were "crowds." Yet there are different ways to think about individuals, cognition, and this process of transformation.

First, we may take an individual as a self-contained bundle of psychological dispositions, with membership in a crowd temporarily causing some of these – the irrational, the unconscious, the emotional – to be manifested, and those of the rational individual to remain merely dispositional. Here an individual's social circumstances play a triggering role in the expression of preexisting psychological dispositions. They do not themselves bring about any new dispositions in individuals, just the manifestation of dispositions already latent in the individual. What are transformed are the dispositions that are manifested.

Second and alternatively, we can think of individuals who are normally constituted by psychologically rational states as acquiring a new, distinct set of psychological dispositions when they become part of a crowd. On this view, the social circumstances change not simply what dispositions become manifested, but what dispositions an individual has. Thus, being part of a crowd brings about a more radical form of transformation of the individual than is suggested by the triggering view. What is transformed is the dispositions themselves.

These two kinds of transformation that social circumstances can bring about in an individual contrast in roughly the way in which memory and learning do in common sense thought. In the case of memory, our social circumstances may cause us to manifest something we already have inside of us, something that we already know; in the case of learning, those circumstances change the knowledge structures that we have. The role of the environment in memory is epitomized as the asking of a question, while in learning it is the giving of instruction.

Despite this sort of difference, both of these conceptions of the social manifestation thesis are continuous with the individualistic tradition of thinking about psychological states that has been the focus of earlier parts of *Boundaries*. This is because, on these views, an individual's psychology itself can (and should) still be understood in abstraction from that individual's social environment. In neither of these cases do the psychological dispositions themselves become social, that is, become constituted by the social circumstances in which they are manifested. Both views draw on a conception of the role of social circumstances in cognition akin to that held by strong nativists (Chapter 3), whereby cognition is internally richly structured antecedent to, and independent of, social circumstances. And they both accept a view of (psychological) dispositions as always being intrinsic to the bearer of the disposition. In Chapter 6, I argued that the very idea of intrinsic dispositions sat uneasily with the prima facie existence of wide realizations of an individual's properties. In effect, the argument there claimed that properties and their realizations are either both individualistic or both wide, and thus that the width of the realizations for at least some psychological properties implied that those properties (including dispositional properties) were not individualistic. In light of this, the idea of an individualistic but socially manifested property is an unstable hybrid.

This places some conceptual stress on individualistic understandings of the social manifestation thesis, pushing us instead toward a view of it as positing psychological states that are socially constituted and so externalist in nature. This would make the thesis suitable for offering a general characterization of the externalist view of psychology, which depicts cognitive processes as themselves intrinsically social in nature. In Chapter 8, I introduced such a view of memory, cognitive development, and folk psychology as a way of providing a more concrete idea of how integrative synthesis should proceed in the psychological case. There I was chiefly concerned to show how to generalize the wide computational approach that I defended as a view of computational cognitive science to noncomputational approaches within psychology, and to indicate the sorts of questions and issues that arise on the resulting, externalist view of psychological inquiry.

This interpretation of the social manifestation thesis thus provides a middle ground between an individualistic psychology and the group mind hypothesis. In contrast to individualism, the externalist psychology demarcated by the social manifestation thesis views psychological states as both taxonomically and locationally embedded in broader social systems. In contrast to the group mind hypothesis, it does not ascribe psychological states themselves to entities, such as the group, the community, or the nation, larger than the individual and to which the individual belongs. On this view, while the individual is not a boundary for psychological theorizing, psychology does posit individual-level, rather than group-level, traits. We can put this the other way around. Socially manifested psychological traits are properties of individuals, but since they occur only in certain group environments, they cannot be understood in purely individualistic terms.

I like to think that this is just what recent advocates of group mind thinking, such as Mary Douglas and David Sloan Wilson, have themselves had in mind. If so, then the externalism about the mind articulated in

Parts Two and Three provides a framework for further explorations of the phenomena to which they have called attention.

6 THE COGNITIVE AND THE SOCIAL

The evolutionary and cultural conditions that give rise to psychological states have been recognized in a number of approaches to cognition, such as evolutionary psychology and explorations of social intelligence in the framework of the Machiavellian hypothesis. The social manifestation thesis suggests, however, that cognition and sociality are more intimately connected than even these research programs allow. For example, research on the Machiavellian hypothesis has been focused on the role of social complexity in producing mental complexity in the individual. It is largely devoted to exploring the forms that both types of complexity take and the relations between them. But if cognition itself is social, as the social manifestation thesis implies, not simply a product of the social, there will be a deeper connection between forms of (say) group living and intelligence. Research programs that attempt to isolate and then explain individualistic modules for intelligent cognitive performance will not go far enough in articulating the social and cultural dimensions to cognition.[16]

The social manifestation thesis should also lead us to rethink some of our ways of thinking about the "levels" at which selection operates. For example, it has been common within debates over the agents of selection to contrast individual-benefiting traits that evolve by individual selection with group-benefiting traits that evolve by group selection. Furthermore, at least in the hands of those who think that the "selfish gene" is the unit of selection, the latter has been discounted altogether. But this putative dichotomy becomes less compellingly exhaustive once we consider traits, including psychological traits, which benefit individuals because those individuals are members of groups of a certain type. In this sort of case, individual-level and group-level traits are intrinsically woven together, and natural selection is not sufficiently fine-grained to distinguish between properties that, despite being distinct, are reliably coinstantiated and homeostatically reinforcing.[17]

Concentration on the social manifestation thesis in an evolutionary context thus may direct us to think about ways in which individual and group selection can be mutually reinforcing processes, rather than conceived of primarily as forces that are opposed in evolutionary change. An important species of case in which they work in the same direction is one in which socially manifested traits are selected at the level of the individual, while group-level traits, whether psychological or nonpsychological, are selected at the level of the group. This would be a sort of coevolutionary process in which there is a mutually reinforcing causal loop between socially manifested psychological traits and group-level traits. This possibility suggests that, although the social manifestation thesis and the group mind hypothesis are distinct views, they may be most interestingly defended together.

7 FROM GROUP MINDS TO GROUP SELECTION

I have mentioned that David Sloan Wilson's views of group minds are tied to his defense of group selection. I want to conclude the substantive part of this chapter by arguing that it is unlikely that there is any easy argumentative flow between claims about the kinds of minds there are and the types of selective processes that operate in nature.

There are two basic reasons for this. First, the level at which selection operates and the level at which its products, adaptations, are characterized are not as tightly connected as has typically been supposed. In particular, selection could take place at the group level but produce adaptations *at the individual level.* Thus, arguing from the existence of group selection on cognitive traits does not imply the group selection of group minds. Conversely, pointing to group-level cognitive adaptations does not entail that these arose through a process of group selection. Group selection even on groups of minimally minded individuals doesn't itself make minimally minded groups of individuals any more plausible than it does groups of minimally minded individuals with certain dispositions and abilities. Second, the failure to distinguish the group minds hypothesis from the social manifestation thesis, and the resulting missing discussion of the relationship between the two, represents a crucial hiatus in arguing from group selection to group minds.

This pair of points, and the complexities and possibilities they introduce, can be seen in Wilson's discussion of group minds. Wilson recognizes both individual- and group-level traits as possible outcomes of a process of group selection, a view implicit in the claim that altruism, an individual-level trait, could evolve by group selection. He argues that if group selection were the sole selective force shaping the psychological features of individuals and groups, then we would expect to find individuals whose decisions benefit the group (for example, altruism), and/or adaptive group decision making (for example, honey bee foraging). To put it slightly differently, group selection might lead either to individuals whose nature would support the social manifestation thesis or to group minds (or even both). But since Wilson doesn't explicitly distinguish the social manifestation thesis from the group mind hypothesis, he fails to consider the idea that traits of individuals that would support the former, at least when produced by group selection, sometimes or always lead to traits of groups that exemplify the latter. That is, missing here is any discussion of the relationship between socially manifested individual traits and the group-level cognitive adaptations. When Wilson does turn to discuss the factors that regulate group decision processes, these turn out to be nonpsychological characteristics of groups (for example, social organization, leader control) that could simply be construed as effecting whether and how certain individual-level characteristics are manifested.[18]

A corresponding problem arises in Wilson's sketch of what one would expect to find if individual selection were the sole process acting on individuals. He says,

> If group selection can truly be ignored as a factor in human evolution, we should expect individuals to be highly adaptive as autonomous decision-making units, capable of performing the full range of activities from framing the problem, to generating alternatives, to evaluating the alternatives, to making final decisions. When individuals exist as members of groups, we should expect them to use others as sources of information, but only in ways that increase the individual's relative fitness within the group.[19]

Here Wilson rests on the assumption that selection at a given level – in this case, that of the individual – will produce certain traits at that level. But suppose that the social manifestation thesis is true: Certain psychological properties are manifest only when individuals form a group of a certain type. And suppose that these evolve through a process of individual selection. Must such traits promote autonomous decision making? One reason to think not is that what kinds of trait evolve in such a case will depend largely on the structure and internal dynamics of the group itself. For example, if the group is one in which autonomous decision making is punished, and shared, partial, and contributory decision making promoted, then it is more likely that these will be the socially manifested psychological traits that evolve, even if there is only individual selection. Thus, Wilson's claim above linking selection at a level with a certain kind of agency is mistaken.

Moreover, one could envision these social structures as diminishing the completeness and autonomy of individual decision making so radically that full decision making appears as a group-only trait. If this happened, then we would have a group mind (with respect to decision making) evolving through a process of individual selection. Wilson and Sober's own discussion of the Hutterites, a religious sect that has lived in relatively small communities in North America for almost two hundred years, prominent in Wilson and Sober's influential *Behavioral and Brain Sciences* paper, suggests to me something like this interpretation, whereby a process of within-group selection shifts the balance of decision making from individual to group. The more general point here is that taking the social manifestation thesis seriously complicates any argument from the character of natural selection to the character of the minds it produces.[20]

Two final complications concerning the relationship between group selection and group minds seem worth mentioning. First, there are very few, if any, established hypotheses about which individual-level psychological processes and abilities are the product of evolution by natural selection. Part of the problem here is the lack of consensus about how to characterize human and animal minds in the first place. In addition, there are doubts about each of the following: whether any of the processes and abilities so characterized are plausibly viewed as cognitive adaptations (versus by-products of other adaptations) at all; supposing that some are, about which these are; and supposing that some particular process or ability evolved, how it is to be characterized.

To take a classic example, there is a lively debate over whether social reasoning constitutes a well-defined psychological unit – something like a module – or whether the appropriate unit is better construed as a general-purpose reasoning module that can be applied to the social domain, or as a pragmatic reasoning module, or in some other way altogether. The same seems to me true of most postulated psychological units – including memory, language, and emotion. This is not to express skepticism about whether there are cognitive adaptations, but to report a view of the state of our knowledge here, despite (or perhaps because of) recent work that makes some progress here.[21]

Second, there is the general issue of the role of cultural selection in producing minds like ours and those of our recent ancestors. I have elsewhere followed Elliott Sober in viewing cultural selection as a family of views, each of which extends or generalizes one or more of the three conditions typically considered necessary for natural selection to occur: phenotypic variation, related fitness variation, and heritability. We may simply extend the concept of a phenotype to include cultural traits; in addition, we may consider fitness to have something more than a strictly reproductive dimension; and finally, we may conceptualize heritability as cultural in nature. The members of this family of views constitute views of cultural selection that increasingly depart from traditional natural selection.[22]

My hunch is that these forms of cultural selection, particularly those that involve more rather than less departure from traditional natural selection, are likely to have been instrumental in shaping up the kinds of minds that we beasts have. Given the demonstrated effectiveness of cultural group selection in shaping at least some behaviors, cultural selection needs to be factored in at both the individual and the group level. If that is right, then our exploration of the relationship between group selection and group minds must include these cultural varieties of natural selection.[23]

8 GROUPS, MINDS, AND INDIVIDUALS

I began the previous chapter by saying that I would seriously consider the idea that groups can have minds. While I have not argued that idea is mistaken, I have suggested that much of what proponents of the group mind hypothesis want to say about the mind can be expressed within the parameters of the externalist view of the mind that I have developed in Parts Two and Three. This involves reconceptualizing where the mind begins and ends, in ways that both build on but also depart from our sciences of the mind. But it does not require that we replace the individual as the locus of cognition, as the subject of mental states, with some larger unit, such as the group.

This deflationary view of appeals to group minds in the biological and social sciences has been developed within a framework that both provides some constructive tools for thinking about group minds, as well as a few conjectures, guesses, and hunches about where some of the complexities to thinking about the relationships between groups, minds, and individuals lie. I have argued that we should approach the question of whether cognition is a group-level trait in any particular case by identifying the focal processes or abilities, and so to some extent forego the more exciting-sounding question of whether "group minds" or "group consciousness" really exist. In the last few sections I have voiced a suspicion about general appeals – to the level at which selection occurs, or to the kinds of characteristics that we find in individuals – as the basis for defending the group mind hypothesis. But this, together with the overall deflationary message of Part Four, should not be taken to imply my skepticism about whether group minds exist.

There can be no group-level focal cognitive processes and abilities without the activities of individuals, and in at least some cases those individuals are cognitive agents, agents with minds. In articulating a view of the mind, however, in which the social embeddedness of the individual makes a crucial difference to the kind of mind that that individual has, I hope to have arrested the thought that the dependence relations here flow simply from "higher levels" (the group, the social) to "lower levels" (the individual, the cognitive). The minds that individuals have are already the minds of individuals in groups.

Notes

1: The Individual and the Mind

1. Richard Dawkins, *The Selfish Gene* (Oxford: Oxford University Press, 2nd edition, 1989), originally published in 1976; and *The Extended Phenotype* (Oxford: Oxford University Press, 1982). See also Elisabeth Lloyd, "Units and Levels of Selection: An Anatomy of the Units of Selection Debates" in R.S. Singh, C.B. Krimbos, D.B. Paul, and J. Beatty (editors), *Thinking About Evolution: Historical, Philosophical & Political Perspectives.* (New York: Cambridge University Press, 2001), for general discussion.
2. On higher-level selection, see Stephen Jay Gould, *The Structure of Evolutionary Theory* (Cambridge, MA: Harvard University Press, 2002), especially pages 714–744; and Elliott Sober and David Sloan Wilson, *Unto Others: The Evolution and Psychology of Unselfish Behavior* (Cambridge, MA: Harvard University Press, 1998).
3. See my *Cartesian Psychology and Physical Minds: Individualism and the Sciences of the Mind* (New York: Cambridge University Press, 1995), especially Chapter 10.
4. For Fodor's original views, see his "Methodological Solipsism Considered as a Research Strategy in Cognitive Psychology," *Behavioral and Brain Sciences*, 3 (1980), pages 63–73. Reprinted in his *Representations* (Sussex: Harvester Press, 1981).
5. See Jerry A. Fodor, *Psychosemantics: The Problem of Meaning in the Philosophy of Mind* (Cambridge, MA: MIT Press, 1987), Chapter 2; Tim Crane, "All the Difference in the World," *Philosophical Quarterly*, 41 (1991), pages 1–25; Michael Devitt, "A Narrow Representational Theory of Mind," in William G. Lycan, (editor), *Mind and Cognition: A Reader* (Cambridge, MA: Basil Blackwell, 1990); and Frances Egan, "Individualism, Computation, and Perceptual Content," *Mind*, 101 (1992), pages 443–459.
6. See the works cited earlier in notes 1 and 2, as well as George C. Williams, *Adaptation and Natural Selection* (Princeton, NJ: Princeton University Press, 1966); and David Sloan Wilson, "A Theory of Group Selection," *Proceedings*

of the National Academy of Sciences USA, 72 (1975), pages 143–146; and "The Group Selection Controversy: History and Current Status," *Annual Review of Ecology and Systematics*, 14 (1983), pages 159–187.

7. For a formulation of methodological individualism in terms of supervenience, see Daniel Little, *Varieties of Social Explanation* (Boulder, CO: Westview Press, 1991), Chapter 9.

8. For classic defenses of methodological individualism, see Karl Popper, *The Open Society and Its Enemies* (Princeton, NJ: Princeton University Press, 2nd edition, 1952); and J.W.N. Watkins, "Historical Explanation in the Social Sciences," *British Journal for the Philosophy of Science*, 8 (1957), pages 104–117.

9. Emile Durkheim, *Suicide: A Study in Sociology* (New York: Free Press, 1951), originally published in 1897; and Karl Marx, *Capital* (New York: International Publishers, 1967), originally published in 1867.

10. Richard Lewontin, *Biology as Ideology* (New York: Harper Collins, 1993 edition); and his *It Ain't Necessarily So: The Dream of the Human Genome and Other Illusions* (New York: New York Review of Books, 2000).

11. I discuss genetics and development more generally, particularly the sense in which genetics is individualistic, in Part Three of my *Genes and the Agents of Life* (New York: Cambridge University Press, 2004).

12. Adam Kuper, *Culture: The Anthropologists' Account* (Cambridge, MA: Harvard University Press, 1999).

13. John Tooby and Leda Cosmides, "The Psychological Foundations of Culture," in Jerome Barkow, Leda Cosmides, and John Tooby (editors), *The Adapted Mind: Evolutionary Psychology and the Generation of Culture* (New York: Oxford University Press, 1992); Edward O. Wilson, *Sociobiology: The New Synthesis* (Cambridge, MA: Harvard University Press, 1975); Steven Pinker, *The Language Instinct* (New York: Morrow, 1994); *How the Mind Works* (New York: Norton, 1997); and *The Blank Slate: The Modern Denial of Human Nature* (New York: Viking, 2002).

14. Clifford Geertz, *The Interpretation of Cultures* (New York: Basic Books, 1973), pages 12–13.

15. Since a number of people have asked me about "smallism," a brief note. I first coined "smallism" in the mid-1990s in unpublished work and talks characterizing the views of Jaegwon Kim and David Lewis in the philosophy of the mind. The term makes a brief appearance in my "The Individual in Biology and Psychology," in V. Hardcastle (editor), *Biology Meets Psychology: Philosophical Essays* (Cambridge, MA: MIT Press). See also Owen Flanagan, *The Problem of the Soul: Two Visions of Mind and How to Reconcile Them* (New York: Basic Books, 2002).

16. For atomism and corpuscularianism in general, see Daniel Garber, John Henry, Lynn Joy, and Alan Gabbey, "New Doctrines of Body and its Powers, Place, and Space," in Daniel Garber and Michael Ayers (editors), *The Cambridge History of Seventeenth-Century Philosophy*, (New York: Cambridge University Press, 1998), Volume 1. For an attempt to make sense of Locke's view of primary qualities, see my "Locke's Primary Qualities," *Journal of the History of Philosophy*, 40 (2002), pages 201–228.

17. For earlier efforts here, see Ron McClamrock, *Existential Cognition: Computational Minds in the World* (Chicago: University of Chicago Press, 1995); Mark Rowlands, *The Body in Mind* (New York: Cambridge University Press, 1999); and my *Cartesian Psychology*, Chapters 3 and 4.
18. On group minds, see David Sloan Wilson, "Incorporating Group Selection into the Adaptationist Program: A Case Study Involving Human Decision Making," in Jeffrey A. Simpson and Douglas T. Kendrick (editors), *Evolutionary Social Psychology* (Hillsdale, NJ: Erlbaum, 1997) and *Darwin's Cathedral: Evolution, Religion, and the Nature of Society* (Chicago: University of Chicago Press, 2002); Mary Douglas, *How Institutions Think* (Syracuse, NY: Syracuse University Press, 1986); and Ludwig Fleck, *Genesis and Development of a Scientific Fact* (Chicago: University of Chicago Press, 1979), originally published in German in 1935.

2: Individuals, Psychology, and the Mind

1. Michel Foucault, *The Order of Things: An Archaeology of the Human Sciences* (New York: Vintage Books, 1994 edition); *Discipline and Punish: The Birth of the Prison* (New York: Vintage Books, 1995 edition); and *The History of Sexuality: Volume 1, An Introduction* (New York: Vintage Books, 1990 edition). These were first published in English in 1970, 1975, and 1978, respectively.
2. For some recent discussion of the demarcation of psychology, see Gary Hatfield, "Psychology, Philosophy, and Cognitive Science: Reflections on the History and Philosophy of Experimental Psychology," *Mind and Language*, 17 (2002), pages 207–232; and Edward S. Reed, *From Soul to Mind: The Emergence of Psychology from Erasmus Darwin to William James* (New Haven, CT: Yale University Press, 1997).
3. Gustav Fechner, *Elemente der Psychophysik* (Leipzig: Breitkopf and Hartel, 1860); Wilhelm Wundt, *Grundzüge der physiologischen Psychologie* (Leipzig: W. Engelmann, 3rd edition, 1887), originally published in 1874; and William James, *The Principles of Psychology* (New York: Dover, 1950), originally published in 1890. The quote from James comes from the first sentence of *The Principles*.
4. On physiology's relation to anatomy, see J. Schiller, "Physiology's Struggle for Independence in the First Half of the Nineteenth Century," *History of Science*, 7 (1968), pages 64–89, especially pages 73–74.
5. My views of Wundt here are influenced by the writings of Arthur Blumenthal. See his "A Reappraisal of Wilhelm Wundt," *American Psychologist*, 30 (1975), pages 1081–1088; "The Founding Father We Never Knew," *Contemporary Psychology*, 24 (1979), pages 547–550; and "Leipzig, Wilhelm Wundt, and Psychology's Gilded Age," in G.A. Kimble and M. Wertheimer, *Portraits of Pioneers in Psychology*, Volume 3 (Mahwah, NJ: Erlbaum, 1998), pages 31–50.
6. *Beiträge zur Theorie der Sinneswahrnehmung* (Leipzig/Heidelberg: C.F. Winter, 1862); *Vorlesungen über die Menschen- und Tierseele* (Leipzig: Voss, 1863); and *Völkerpsychologie: Eine Untersuchung der Entwicklungsgesetzen, Mythus und Sitte* (Leipzig: Engelmann, ten volumes, 1900–1920).

7. On social cognition, see Ziva Kunda, *Social Cognition* (Cambridge, MA: MIT Press, 1999). For the revival of the work of Mead and Goffman, see for example: Rom Harré, *Social Being* (Oxford: Blackwell, 1979); J.D. Baldwin, *George Herbert Mead: A Unifying Theory for Sociology* (Beverly Hills, CA: Sage Publications, 1986); and Sheldon Stryker, "The Vitalization of Social Interaction," *Social Psychology Quarterly*, 50 (1987), pages 83–94. See also G. Collier, H.L. Minton, and G. Reynolds, *Currents of Thought in American Social Psychology* (New York: Oxford University Press, 1991).
8. See Roger Smith, *The Norton History of the Human Sciences* (New York: Norton, 1997), Chapter 12, for discussion of Comte. The quotation appears on page 427 and is taken from A. Comte, *The Essential Comte: Selected from the Course de philosophie positive* edited by S. Andreski (New York: Barnes and Noble, 1974), page 32.
9. Apart from Smith, *The Norton History of the Human Sciences*, see also Gordon Allport, "The Historical Background of Modern Social Psychology," in G. Lindzey and E. Aronson (editors), *The Handbook of Social Psychology* (Reading, MA: Addison-Wesley, 2nd edition, 1968). The quotation is taken from August Comte, *System of Positive Polity, Second Volume* (New York: Franklin, 1875), page 357.
10. Edward A. Ross, *Social Psychology: An Outline and Source Book* (New York: Macmillan, 1908). The quotations are taken from pages 1 and 2, with "suggestions" appearing on page 12.
11. Emile Durkheim, *Rules of Sociological Method* (New York: Free Press, 1938), at page 145, originally published in 1895. On collective representations, see his "Individual and Collective Representations," reprinted in his *Sociology and Philosophy* (Glencoe, IL: The Free Press, 1953), originally published in 1898. For a recent example of work that picks up this thread, see Bradd Shore, *Culture in Mind* (New York: Oxford University Press).
12. Kurt Danziger, *Constructing the Subject: Historical Origins of Psychological Research* (New York: Cambridge University Press, 1990).
13. The quotations are from *Constructing the Subject*, pages 48 and 58 respectively.
14. Francis Galton, *Hereditary Genius: An Inquiry into its Laws and Consequences* (London: Macmillan. 2nd edition, 1892), originally published in 1869; and *Inquiries into Human Faculty and its Development* (London: Macmillan, 1883).
15. Danziger, *Constructing the Subject*, page 77. Chapter 5 provides a more general discussion of the notion of a collective subject.
16. *Constructing the Subject*, page 56.
17. Influential early expressions of the continuity thesis include Noam Chomsky, *Cartesian Linguistics: A Chapter in the History of Rationalist Thought* (New York: Harper and Row, 1966) and Jerry Fodor, "The Present Status of the Innateness Controversy," in his *Representations* (Sussex: Harvester Press, 1981). For more recent expressions of it, see Fiona Cowie, *What's Within? Nativism Reconsidered* (New York: Oxford University Press, 1999); and Frank Keil, "Cognitive Science and the Origins of Thought and Knowledge," and Elizabeth Spelke and Elissa Newport, "Nativism, Empiricism, and the Development of Knowledge," both in W. Damon and R.M. Lerner (editors), *Handbook of Child Psychology, Volume 1: Theoretical Models of Human Development* (New York:

Wiley, 5th edition, 1998). For some discussion of this general issue in the context of theories of spatial perception, see Gary Hatfield, *The Natural and the Normative: Theories of Spatial Perception from Kant to Helmholtz* (Cambridge, MA: MIT Press, 1990).
18. For Locke's focus on ideas, see *An Essay Concerning Human Understanding* (Oxford: Clarendon Press, 1979), Book II, Chapters ii–xi. Locke discusses combination, association, and abstraction in the following chapter.
19. Jerome Bruner, *Acts of Meaning* (Cambridge, MA: MIT Press, 1990), page 20.

3: Nativism on My Mind

1. For recent attempts that use just one dimension, see Muhammad Ali Khalidi, "Innateness and Domain-Specificity," *Philosophical Studies*, 105 (2001), pages 191–210, and "Nature and Nurture in Cognition," *British Journal for the Philosophy of Science*, 53 (2002), pages 251–272; and Richard Samuels, "Nativism in Cognitive Science," *Mind and Language*, 17 (2002), pages 233–265. For an account that draws on more than two dimensions, see Fiona Cowie, *What's Within: Nativism Reconsidered* (New York: Oxford University Press, 1999).
2. Noam Chomsky, *Reflections on Language* (New York: Pantheon, 1975), pages 10–11.
3. See also Noam Chomsky, "Linguistics and Adjacent Fields: A Personal View," in Asa Kasher (editor), *The Chomskyan Turn* (Oxford: Basil Blackwell, 1991), one place where Chomsky draws this comparison directly.
4. For the classic articulation and defense of first of these views, see Jerry Fodor, *The Language of Thought* (Cambridge, MA: Harvard University Press, 1975). For the second, see his *The Modularity of Mind* (Cambridge, MA: MIT Press, 1983).
5. Fodor, *Language of Thought*, page 97.
6. Most recently in his *Concepts: Where Cognitive Science Went Wrong* (New York: Oxford University Press, 1998). See also Stephen Laurence and Eric Margolis, "Radical Concept Nativism," *Cognition*, 86 (2002), pages 25–55.
7. For the examples of the reach of the idea of modularity, see Jay Garfield (editor), *Modularity in Knowledge Representation and Natural-Language Understanding* (Cambridge, MA: MIT Press, 1987); Lawrence Hirschfeld and Susan Gelman (editors), *Mapping the Mind: Domain-Specificity in Cognition and Culture* (New York: Cambridge University Press, 1994); and Simon Baron-Cohen, *Mindblindness: An Essay on Autism and Theory of Mind* (Cambridge, MA: MIT Press, 1995). For Fodor's criticisms of such extensions, particularly within evolutionary psychology, see his *The Mind Doesn't Work That Way* (Cambridge, MA: MIT Press, 2000). I take a somewhat whimsical look at Fodor's views here in my "What Computations (Still, Still) Can't Do: Jerry Fodor on Computation and Modularity," *Canadian Journal of Philosophy*, in press.
8. Leda Cosmides, "The Logic of Social Exchange: Has Natural Selection Shaped How Humans Reason? Studies with the Wason Selection Task," *Cognition*, 31 (1989), pages 187–276; and Leda Cosmides and John Tooby, "Cognitive Adaptations for Social Exchange," in Jerome Barkow, Leda Cosmides,

and John Tooby (editors), *The Adapted Mind: Evolutionary Psychology and the Generation of Culture* (New York: Oxford University Press, 1992).

9. On the many modules, see Leda Cosmides and John Tooby, "Evolutionary Psychology," in R.A. Wilson and F.C. Keil (editors), *The MIT Encyclopedia of the Cognitive Sciences* (Cambridge, MA: MIT Press, 1999). On the adapted mind more generally, see their "Origins of Domain Specificity: The Evolution of Functional Organization," in L. Hirschfeld and S. Gelman (editors), *Mapping the Mind*. For a popular, enthusiastic treatment, see Steven Pinker, *How the Mind Works* (New York: Norton, 1997), and *The Blank Slate: The Modern Denial of Human Nature* (New York: Viking, 2002).

10. For Chomsky's critique, see his, "A Review of B. F. Skinner's 'Verbal behavior,'" *Language*, 24 (1959), pages 163–186.

11. The original Rumelhart and McClelland model is in their "On Learning the Past Tenses of English Verbs: Implicit Rules or Parallel Distributed Processing?," in David Rumelhart, James McClelland, and the PDP Research Group, *Parallel Distributed Processing: Explorations in the Microstructure of Cognition, Vol. 1: Foundations* (Cambridge, MA: MIT Press, 1986). The quotation from Pinker comes from his "Four Decades of Rules and Associations, or Whatever Happened to the Past Tense Debate?," in Emmanuel Dupoux (editor), *Language, Brain, and Cognitive Development: Essays in Honor of Jacques Mehler* (Cambridge, MA: MIT Press, 2001), page 159. For hybrid models, see Gary Marcus, *The Algebraic Mind: Integrating Connectionism and Cognitive Science* (Cambridge, MA: MIT Press, 2001); and Steve Pinker, *Words and Rules: The Ingredients of Language* (New York: Harper Collins, 1999).

12. Stephen Laurence and Eric Margolis, "The Poverty of the Stimulus Argument," *British Journal for the Philosophy of Science*, 52 (2001), pages 217–276, at page 219.

13. See Annette Karmiloff-Smith, *Beyond Modularity: A Developmental Perspective on Cognitive Science* (Cambridge, MA: MIT Press, 1992); and Edwin Hutchins, *Cognition in the Wild* (Cambridge, MA: MIT Press, 1995).

14. Steven Quartz and Terence Sejnowski, "The Neural Basis of Cognitive Development: A Constructivist Manifesto," *Behavioral and Brain Sciences*, 20 (1997), pages 537–596. For neural selection theory, see Jean-Pierre Changeux, *Neuronal Man* (New York: Pantheon Books, 1985); and Gerald Edelman, *Neural Darwinism: The Theory of Neuronal Group Selection* (New York: Basic Books, 1987). For Quartz's own recent analysis of innateness, see his "Innateness and the Brain," *Biology and Philosophy*, 18 (2003), pages 13–40.

15. On concepts, Greg Murphy, *The Big Book of Concepts* (Cambridge, MA: MIT Press, 2002). On word meanings, Paul Bloom, *How Children Learn the Meanings of Words* (Cambridge, MA: MIT Press, 2000). On syntax and other aspects of language, Ray Jackendoff, *Foundations of Language: Brain, Meaning, Grammar, Evolution* (New York: Oxford University Press, 2002).

16. Alan Leslie, "Pretense and Representation: The Origins of 'Theory of Mind,'" *Psychological Review*, 94 (1987), pages 412–426; Simon Baron-Cohen, *Mindblindness*; Peter Hobson, "The Grounding of Symbols: A Social-Developmental Account," in P. Mitchell and K.J. Riggs (editors), *Children's Reasoning and the Mind* (New York: Psychology Press, 1999); Michael

Tomasello, *The Cultural Origins of Human Cognition* (Cambridge, MA: Harvard University Press, 1999). See also Michael Tomasello and Hannes Rakoczy, "What Makes Human Cognition Unique: From Individual to Shared to Collective Intentionality," *Mind and Language*, 18 (2003), pages 121–147; and J.I.M. Carpendale and C. Lewis, "Constructing an Understanding of Mind: The Development of Children's Social Understanding within Social Interaction," *Behavioral and Brain Sciences*, in press.

17. See Alison Gopnik, "How We Know Our Minds: The Illusion of First-Person Knowledge of Intentionality," *Behavioral and Brain Sciences*, 16 (1993), pages 1–14; Josef Perner, *Understanding the Representational Mind* (Cambridge, MA: MIT Press, 1991); Jay Garfield, Candida C. Peterson, and T. Perry, "Social Cognition, Language Acquisition and the Development of the Theory of Mind," *Mind and Language*, 16 (2001), pages 494–541.

18. See the references in note 1.

19. The quotation is from his "Innateness and Domain-Specificity," page 193. Steve Stich articulates the triggering view in his introduction to his edited collection *Innateness* (Berkeley, CA: University of California Press, 1975).

20. The quotation is on page 247 of "Nativism in Cognitive Science." What I am calling "step two" can be found on page 246.

21. On intuitive mechanics, see Elizabeth Spelke, "Principles of Object Perception," *Cognitive Science*, 14 (1990), pages 29–56; and "Initial Knowledge: Six Suggestions," *Cognition*, 50 (1995) pages 433–447. On intuitive mathematics, see Karen Wynn, "Children's Understanding of Counting," *Cognition*, 36 (1990), pages 155–193; and "Origins of Numerical Knowledge," *Mathematical Cognition*, 1 (1995), pages 35–60.

22. Samuels, "Nativism in Cognitive Science," page 234.

23. See Frank C. Keil, "Nativism," in R.A. Wilson and F.C. Keil (editors), *The MIT Encyclopedia of the Cognitive Sciences* (Cambridge, MA: MIT Press, 1999), pages 583–586; and "Nurturing Nativism," *A Field Guide to the Philosophy of Mind*. Fall 1999–2000. E-journal at http://host.uniroma3.it/progetti/kant/field/cowiesymp.htm. Fiona Cowie, *What's Within*, Chapter 7. See especially her Figure 7.1 on page 159.

24. See also Peter Godfrey-Smith, *Complexity and the Function of Mind in Nature* (New York: Cambridge University Press, 1996) for a broad treatment of externalist views in epistemology and the philosophy of biology and mind.

25. Patrick Bateson, "Are There Principles of Behavioural Development?," in P. Bateson (editor), *The Development and Integration of Behaviour* (Cambridge: Cambridge University Press, 1991), pages 19–39, at page 21; Paul Griffiths, "What is Innateness?," *Monist*, 85 (2002), pages 70–85. The quotation is on page 70.

26. On canalization, see Andre Ariew, "Innateness and Canalization," *Philosophy of Science*, 63 (1996), pages S19–S27; and "Innateness is Canalization: In Defense of a Developmental Account of Innateness," in V.G. Hardcastle (editor), *Where Biology Meets Psychology: Philosophical Essays* (Cambridge, MA: MIT Press, 1999), pages 117–138. On generative entrenchment, see William Wimsatt, "Generativity, Entrenchment, Evolution, and Innateness:

Philosophy, Evolutionary Biology, and Conceptual Foundations of Science," also in the Hardcastle collection.

27. For the challenges, see Evelyn Fox Keller, *The Century of the Gene* (Cambridge, MA: Harvard University Press, 2000); and Lenny Moss, *What Genes Can't Do* (Cambridge, MA: MIT Press, 2003).

28. Brian Goodwin, *How the Leopard Got Its Spots* (New York: Simon and Schuster, 1993), and Gerry Webster and Brian Goodwin, *Form and Transformation: Generative and Relational Principles in Biology* (New York: Cambridge University Press, 1996).

29. For an introduction to developmental systems theory, see Susan Oyama, Paul Griffiths, and Russell Gray (editors), *Cycles of Contingency: Developmental Systems and Evolution* (Cambridge, MA: MIT Press, 2001). See also Eva Jablonka and Marion Lamb, *Epigenetic Inheritance and Evolution: The Lamarckian Dimension* (Oxford: Oxford University Press, 1995); Susan Oyama, *The Ontogeny of Information* (Durham, NC: Duke University Press, 2nd edition, 2000), originally published in 1985, and her *Evolution's Eye: A Systems View of the Biology-Culture Divide* (Durham, NC: Duke University Press, 2000); and Paul Griffiths and Russell Gray, "Developmental Systems and Evolutionary Explanation," *Journal of Philosophy*, 91 (1994), pages 277–304, and their "Darwinism and Developmental Systems," in *Cycles of Contingency*.

4: Individualism: Philosophical Foundations

1. Tyler Burge, "Individualism and the Mental," in Peter French, Thomas Uehling Jr., and Howard Wettstein (editors), *Midwest Studies in Philosophy, Volume 4, Metaphysics* (Minneapolis: University of Minnesota Press, 1979), at page 117.

2. For an insightful review see Tyler Burge, "Philosophy of Mind and Language: 1950–1990," *Philosophical Review*, 101 (1992), pages 3–51. For some thoughts on the influence of ordinary language philosophy and logical positivism on philosophy's contribution to the cognitive sciences, see section 3 of my "Philosophy" introduction in R.A. Wilson and F.C. Keil, *The MIT Encyclopedia of the Cognitive Sciences* (Cambridge, MA: MIT Press, 1999), pages xv–xxxvii.

3. Hilary Putnam, "The Meaning of 'Meaning,'" in Keith Gunderson (editor), *Language, Mind and Knowledge* (Minneapolis: University of Minnesota Press, 1975). Reprinted in Hilary Putnam, *Mind, Language, and Reality: Philosophical Papers, Volume 2* (New York: Cambridge University Press, 1975). See also Gary Ebbs, *Rule-Following and Realism* (Cambridge, MA: Harvard University Press, 1998), Chapters 7 and 8.

4. For Frege, see "On Sense and Reference," reprinted in Peter Geach and Max Black (editors), *Translations from the Philosophical Writings of Gottlob Frege* (Oxford: Blackwell, 2nd edition, 1960), originally published in 1892. For Carnap, see *Meaning and Necessity: A Study in Semantics and Modal Logic* (Chicago: University of Chicago Press, 2nd edition, 1956), originally published in 1947.

5. Carnap's *Logical Construction of the World* (Berkeley: University of California Press, 1967), originally published in German in 1928, was a key part of

that tradition, and is likely the source for Putnam's use of "methodological solipsism." For continuing controversies, see for example Nathan Salmon, *Reference and Essence* (Princeton, NJ: Princeton University Press, 1981) and Alan Sidelle, *Necessity, Essence, and Individuation: A Defense of Conventionalism* (Ithaca, NY: Cornell University Press, 1989).

6. This sort of point is made by Tyler Burge, both in passing in "Individualism and the Mental" (see his footnote 2) and in a more sustained way in "Other Bodies" in Andrew Woodfield (editor), *Thought and Object: Essays on Intentionality* (Oxford: Oxford University Press, 1982). The Putnam quotation is from "The Meaning of 'Meaning,'" page 227.

7. For "Individualism and the Mental," see the full reference given in note 1.

8. This contrast between physical and social externalism is drawn, for example, by Gabriel Segal, *A Slim Book About Narrow Content* (Cambridge, MA: MIT Press, 2000), in Chapters 2 and 3. See also Chapter 7 of Gary Ebbs, *Rule-Following and Realism*, for affinities between Putnam and Burge on the social dimension to having a mind.

9. H.P. Grice, "Meaning," *Philosophical Review*, 66 (1957), pages 377–388; David K. Lewis, *Convention: A Philosophical Study* (Oxford: Blackwell, 1968); Noam Chomsky, *Cartesian Linguistics: A Chapter in the History of Rationalist Thought* (New York: Harper and Row, 1966) and *Knowledge of Language* (New York: Praeger, 1986); and Roger Schank and Peter Abelson, *Scripts, Plans, Goals, and Understanding* (Hillsdale, NJ: Erlbaum, 1977).

10. For an anthology of significant papers on first-person knowledge and externalism, see Peter Ludlow and Nora Martin (editors), *Externalism and Self-Knowledge* (Palo Alto, CA: CSLI Publications, 1998). I discuss the putative conflict between externalism and self-knowledge in the final section of my "Individualism," in Stephen P. Stich and Ted A. Warfield, *The Blackwell Companion to Philosophy of Mind* (New York: Blackwell, 2003), pages 256–287.

11. Andy Clark's work over the past ten years reflects much in the embedded cognition movement. See in particular his *Being There: Putting Brain, Body, and World Together Again* (Cambridge, MA: MIT Press, 1997) and his recent *Natural-Born Cyborgs: Minds, Technologies, and the Future of Human Intelligence* (New York: Oxford University Press, 2003).

12. Donald Davidson, "Belief and the Basis of Meaning" and "Thought and Talk," both reprinted in his *Inquiries into Truth and Interpretation* (New York: Oxford University Press, 1984); Daniel C. Dennett, "Intentional Systems," reprinted in his *Brainstorms* (Cambridge, MA: MIT Press, 1978) and "True Believers: The Intentional Strategy and Why it Works," reprinted in his *The Intentional Stance* (Cambridge, MA: MIT Press, 1987).

13. For Burge, see "Individualism and Psychology," *Philosophical Review*, 95 (1986) pages 3–45; "Intellectual Norms and Foundations of Mind," *Journal of Philosophy*, 83 (1986), pages 697–720; and "Cartesian Error and the Objectivity of Perception," in R. Grimm and D. Merrill (editors), *Contents of Thought* (Tucson: University of Arizona Press, 1988). For Pettit, see his *The Common Mind: An Essay on Psychology, Society, and Politics* (New York: Oxford University Press, 1993), especially Chapter 2.

14. Wilfrid Sellars, "Empiricism and the Philosophy of Mind," *Minnesota Studies in the Philosophy of Science, Volume 1* (Minneapolis: University of Minnesota Press, 1956), pages 253–329; John McDowell, *Mind and World* (Cambridge, MA: Harvard University Press, 1994); Robert Brandom, *Making It Explicit* (Cambridge, MA: Harvard University Press, 1994); and John Haugeland, *Having Thought: Essays in the Metaphysics of Mind* (Cambridge, MA: MIT Press, 1998).
15. I have discussed these proposals previously in Chapter 9 of *Cartesian Psychology and Physical Minds*. In addition, some recent attempts to develop a phenomenological conception of narrow content are the subject of the final chapter in Part Three.
16. Fodor's original views on individualism and folk psychology are expressed in his "Cognitive Science and the Twin-Earth Problem," *Notre Dame Journal of Formal Logic*, 23 (1982), pages 98–118, and in *Psychosemantics*; his change on this front can be found in *The Elm and the Expert* (Cambridge, MA: MIT Press, 1994). Stich's eliminativism is developed in his *From Folk Psychology to Cognitive Science: The Case Against Belief* (Cambridge, MA: MIT Press, 1983).
17. See Fodor, *Psychosemantics*, Chapter 2. For critiques, see Robert van Gulick, "Metaphysical Arguments for Internalism and Why They Don't Work," in Stuart Silvers (editor), *Rerepresentation* (Dordrecht, The Netherlands: Kluwer, 1989); Frances Egan, "Must Psychology be Individualistic?," *Philosophical Review*, 100 (1991), pages 179–203; and my *Cartesian Psychology and Physical Minds*, Chapter 2. For views that draw on the underlying intuitions, see Colin McGinn, "Conceptual Causation: Some Elementary Reflections," *Mind*, 100 (1991), pages 573–586; Tim Crane, "All the Difference in the World," *Philosophical Quarterly*, 41 (1991), pages 1–25; Joseph Owens, "Content, Causation, and Psychophysical Supervenience," *Philosophy of Science*, 60 (1993), pages 242–261; and Denis Walsh, "Wide Content Individualism," *Mind*, 107 (1998), pages 625–651.
18. For criticism of the arguments of McGinn and Owens (cited in the previous note), see *Cartesian Psychology and Physical Minds*, Chapter 5. For a critique of Walsh's views, see my "Some Problems for 'Alternative Individualism,'" *Philosophy of Science*, 67 (2000), pages 671–679, and Part Three of my *Genes and the Agents of Life* (New York: Cambridge University Press, 2004).
19. See in particular the first half of "Individualism and Psychology"; much of the last third of "Philosophy of Mind and Language: 1950–1990" (from which the quotation is taken, page 39); "Individuation and Causation in Psychology," *Pacific Philosophical Quarterly*, 70 (1989), pages 303–322; and "Mind-Body Causation and Explanatory Practice," in John Heil and Alfred Mele (editors), *Mental Causation* (New York: Oxford University Press, 1993).

5: Metaphysics, Mind, and Science: Two Views of Realization

1. For Putnam's original discussion, see his "Minds and Machines," originally published in 1960. For Putnam's arguments invoking multiple realization,

see his "The Mental Life of Some Machines" and "The Nature of Mental States," both from 1967. All of these papers are reprinted in his *Mind, Language, and Reality: Philosophical Papers, Volume 2* (New York: Cambridge University Press, 1975).

2. Classic defenses of nonreductive materialism include Jerry Fodor, "Special Sciences," reprinted in his *Representations* (Brighton, Sussex: Harvester Press, 1981); and Richard Boyd, "Materialism without Reductionism: What Physicalism Does Not Entail," in Ned Block (editor), *Readings in the Philosophy of Psychology* (Cambridge, MA: Harvard University Press, 1980). For Kim's views, see his "The Myth of Nonreductive Materialism" and "Multiple Realization and the Metaphysics of Reduction," both reprinted in his *Supervenience and Mind* (New York: Cambridge University Press, 1993). Horgan's observation is made in his "From Supervenience to Superdupervenience: Meeting the Demands of a Material World," *Mind* 102 (1993), pages 555–586, in a footnote on page 573.

3. For recent work on realization, see Sydney Shoemaker, "Realization and Mental Causation," reprinted in Carl Gillett and Barry Loewer (editors), *Physicalism and it Discontents* (New York: Cambridge University Press, 2000); Carl Craver, "Saving Your Self: Dissociative Realization and Kind Splitting," *Philosophy of Science*, in press; Thomas Polger, "Neural Machinery and Realization," *Philosophy of Science*, in press; Larry Shapiro, "Multiple Realizations," *Journal of Philosophy*, 97 (2000), pages 635–654; and *The Mind Incarnate* (Cambridge, MA: MIT Press, 2004). See also my "Realization: Metaphysics, Mind, and Science," *Philosophy of Science*, in press.

4. A commitment to something like the sufficiency thesis can be found in Kim's, "Multiple Realization and the Metaphysics of Reduction" and his "The Nonreductivist's Troubles with Mental Causation," reprinted in his *Supervenience in Mind*; in Jeffrey Poland, *Physicalism: The Philosophical Foundations* (New York: Oxford University Press, 1994), Chapter 4; and in Stephen Yablo, "Mental Causation," *Philosophical Review* 101 (1992), pages 245–280.

5. For commitments to something like the constitutivity thesis, see Richard Boyd, "Materialism without Reductionism: What Physicalism Does Not Entail," page 100; David Lewis, "David Lewis: Reduction of Mind," in Samuel Guttenplan (editor), *A Companion to the Philosophy of Mind* (Oxford: Blackwell, 1994), pages 412–418; and Sydney Shoemaker, "Some Varieties of Functionalism," reprinted in his *Identity, Cause, and Mind* (New York: Cambridge University Press, 1984), page 265.

6. The long quotation and the short one that follows it are from "Multiple Realization and the Metaphysics of Reduction," pages 322 and 328, respectively. The final quote is from Kim's "Postscripts on Supervenience" in his *Supervenience and Mind*.

7. See Shoemaker, "Some Varieties of Functionalism."

8. On homuncular decomposition, see Robert Cummins, *The Nature of Psychological Explanation* (Cambridge, MA: MIT Press, 1983), Chapters 2 and 3; Daniel C. Dennett, "Artificial Intelligence as Philosophy and as Psychology," reprinted in his *Brainstorms* (Cambridge, MA: MIT Press, 1978); and William G. Lycan *Consciousness* (Cambridge, MA: MIT Press, 1987), Chapter 4.

9. See my *Cartesian Psychology and Physical Minds*, Chapters 3 and 4, and "The Mind Beyond Itself," in Dan Sperber (editor), *Metarepresentations: A Multidisciplinary Perspective* (New York: Oxford University Press, 2000).
10. For Clark's 007 Principle, see his *Microcognition: Philosophy, Cognitive Science, and Parallel Distributed Processing* (Cambridge, MA: MIT Press, 1989), at page 64. See also his "Reasons, Robots and the Extended Mind," *Mind and Language*, 16 (2001), pages 121–145; *Natural-Born Cyborgs*; and Andy Clark and David Chalmers, "The Extended Mind," *Analysis*, 58 (1998), pages 10–23. For Hutchins's views here, see his *Cognition in the Wild* (Cambridge, MA: MIT Press, 1995), Chapter 9.

6: Context-Sensitive Realizations

1. See Paul Teller, "Relational Holism and Quantum Mechanics," *British Journal for the Philosophy of Science*, 37 (1986), pages 71–81; and "Relativity, Relational Holism, and the Bell Inequalities," in James T. Cushing and Ernan McMullin (editors), *Philosophical Consequences of Quantum Theory* (Notre Dame, IN: University of Notre Dame Press, 1989).
2. David K. Lewis, *Philosophical Papers, Volume II* (New York: Oxford University Press, 1986), page xi.
3. The intrinsic view of dispositions is maintained by, amongst others: David Armstrong, *Belief, Truth, and Knowledge* (New York: Cambridge University Press, 1973); Elizabeth Prior, *Dispositions* (Aberdeen: University of Aberdeen Press, 1985); and David K. Lewis, "Finkish Dispositions," *Philosophical Quarterly*, 47 (1997), pages 143–158.
4. For these characterizations of acids, see W.H. Nebergall, H. Holtzclaw, and W. Robinson, *General Chemistry* (Lexington, MA: D.C. Heath and Co., 6th ed., 1980), chapter 14.
5. Jennifer McKitrick, "A Case for Extrinsic Dispositions," *Australasian Journal of Philosophy*, 81 (2003), pages 155–174.
6. For token identity, see Donald Davidson, "Mental Events" and "Philosophy as Psychology," both reprinted in his *Essays on Actions and Events* (New York: Oxford University Press, 1980). For compositional materialism, see Richard Boyd, "Materialism without Reductionism: What Physicalism Does Not Entail," in Ned Block (editor), *Readings in the Philosophy of Psychology* (Cambridge, MA: Harvard University Press, 1980).
7. See in particular his "The Myth of Nonreductive Materialism" and "Multiple Realization and the Metaphysics of Reduction," both reprinted in his *Supervenience and Mind* (New York: Cambridge University Press, 1993).
8. This seems to me also the situation that John Stuart Mill faced in his landmark discussion of causation in *A System of Logic, Ratiocinative and Inductive, Volume 1* (London: John W. Parker, 2nd edition, 1846). See especially Book III, Chapter 5.
9. See Terence Horgan, "Supervenience and Microphysics," *Pacific Philosophical Quarterly*, 63 (1982), pages 29–43, for the introduction; and "From Supervenience to Superdupervenience: Meeting the Demands of a Material World," *Mind*, 102 (1993), pages 555–586, at page 571, for the christening.

The examples I use in the text can be found on page 33 of the earlier paper.
10. See Denis Walsh, "Wide Content Individualism," *Mind*, 107 (1998), pages 625–651. The quotation is taken from page 626 and the numbered argument can be found on page 640.
11. "Wide Content Individualism," page 627, footnote.
12. See Denis Walsh, "Alternative Individualism," *Philosophy of Science*, 66 (1999), pages 628–648.
13. Thanks to Gary Ebbs, Paul Teller, and Andy Clark, respectively, for the discussions and exchanges that led to the following three sections.
14. On fear, see Joseph LeDoux, *The Emotional Brain* (New York: Simon and Schuster, 1996); and Joseph LeDoux and Michael Rogan, "Emotion and the Animal Brain," in R.A. Wilson and F.C. Keil (editors), *The MIT Encyclopedia of the Cognitive Sciences* (Cambridge, MA: MIT Press, 1999), pages 268–270. On motor imagery, see Marc Jeannerod, "The Representing Brain: Neural Correlates of Motor Intention and Imagery," *Behavioral and Brain Sciences*, 17 (1994), pages 187–245. On haptic perception, see R. Cholewiak and A. Collins, "Sensory and Physiological Bases of Touch," in M.A. Heller and W. Schiff (editors), *The Psychology of Touch* (Mahwah, NJ: Erlbaum, 1991); and Roberta Klatzky, "Haptic Perception," in *The MIT Encyclopedia of the Cognitive Sciences*, pages 359–360.
15. For standard presentations of the Ramsey-Lewis view, see Ned Block, Introduction to his *Readings in the Philosophy of Psychology* (Cambridge, MA: Harvard University Press, 1980); and David K. Lewis, "Psychophysical and Theoretical Identifications," *Australasian Journal of Philosophy*, 50 (1972), pages 291–315.
16. See Andy Clark and David Chalmers, "The Extended Mind," *Analysis*, 58 (1998), pages 10–23; Andy Clark, "Reasons, Robots and the Extended Mind," *Mind and Language*, 16 (2001), pages 121–145; and *Natural-Born Cyborgs: Minds, Technologies, and the Future of Human Intelligence* (New York: Oxford University Press, 2003).

7: Representation, Computation, and Cognitive Science

1. Ray Jackendoff, "The Problem of Reality," reprinted in his *Languages of the Mind* (Cambridge, MA: MIT Press, 1992), at pages 159–161. Jackendoff offers an extended approach to language within this framework in his recent *Foundations of Language: Brain, Meaning, Grammar, Evolution* (New York: Oxford University Press, 2002); see especially Part III.
2. Noam Chomsky, "Linguistics and Adjacent Fields: A Personal View" and "Linguistics and Cognitive Science: Problems and Mysteries," both in Asa Kasher (editor), *The Chomskyan Turn* (Cambridge, MA: Blackwell, 1991); "Language and Nature," *Mind*, 104 (1995), pages 1–61; and *New Horizons in the Study of Language and Mind* (New York: Cambridge University Press, 2000).
3. Leda Cosmides and John Tooby, Foreword to Simon Baron-Cohen, *Mindblindness: An Essay on Autism and Theory of Mind* (Cambridge, MA: MIT Press, 1995), pages xi–xii.

4. On distributed representation, Geoff Hinton, Jay McClelland, and David Rumelhart, "Distributed Representations," in David Rumelhart, Jay McClelland, and the PDP Research Group (editors), *Parallel Distributed Processing: Explorations in the Microstructure of Cognition, Volume 1: Foundations* (Cambridge, MA: MIT Press, 1986). On subsymbolic processing, Paul Smolensky, "On the Proper Treatment of Connectionism," *Behavioral and Brain Sciences*, 11 (1988), pages 1–74. And on dynamic approaches to cognition, Tim van Gelder and Robert Port (editors), *Mind as Motion* (Cambridge, MA: MIT Press, 1995), and Tim van Gelder, "Dynamic Approaches to Cognition," in R.A. Wilson and F.C. Keil (editors), *The MIT Encyclopedia of the Cognitive Sciences* (Cambridge, MA: MIT Press, 1999).

5. For a recent example, fashioned explicitly as a counter to the externalism that John Haugeland has articulated, see Rick Grush, "In Defense of Some Cartesian Assumptions Concerning the Brain and its Operation," *Biology and Philosophy*, 18 (2003), pages 53–93.

6. On computational symbols as conventional, see John Searle, *The Rediscovery of the Mind* (Cambridge, MA: MIT Press, 1992); and Steven Horst, *Symbols, Computation, and Intentionality: A Critique of the Computational Theory of Mind* (Berkeley: University of California Press, 1996).

7. For thermostats and fuel gauges, see Fred Dretske, *Knowledge and the Flow of Information* (Cambridge, MA: MIT Press, 1981), and *Explaining Behavior: Reasons in a World of Causes* (Cambridge, MA: MIT Press, 1988). For hearts and bee-dances, see Ruth Garrett Millikan, *Language, Thought, and Other Biological Categories: New Foundations for Realism* (Cambridge, MA: MIT Press, 1984), and *White Queen Psychology and Other Essays for Alice* (Cambridge, MA: MIT Press, 1993).

8. For a biographical sketch of Marr, see Whitman Richards, "Marr, David," in *The MIT Encyclopedia of the Cognitive Sciences*, pages 511–512. On Marr's continuing significance within vision research, see Stephen Palmer, *Vision Science: Photons to Phenomenology* (Cambridge, MA: MIT Press, 1999), Chapter 4. For a philosophically sensitive, introductory treatment, see Kim Sterelny, *The Representational Theory of Mind: An Introduction* (New York: Blackwell, 1990), Chapter 4.

9. Tyler Burge, "Individualism and the Mental," in Peter French, Thomas Uehling Jr., and Howard Wettstein (editors), *Midwest Studies in Philosophy, Volume 4, Metaphysics* (Minneapolis: University of Minnesota Press, 1979); and "Individualism and Psychology," *Philosophical Review*, 95 (1986), pages 3–45.

10. Marr, *Vision*, page 19.

11. For Marr's introduction of the three levels, see *Vision*, pages 24–25.

12. The quotation is from *Vision*, page 99.

13. "Individualism and Psychology," page 34.

14. The long quotation is from *Vision*, page 43. For other places where Marr talks like this, see pages 68, 103–105, and 265–266. Marr's characterization of blobs, lines, and so on is given on page 44.

15. Gabriel Segal, "Seeing What is Not There," *Philosophical Review*, 98 (1989), pages 189–214; Robert Matthews, "Comments on Burge," in Robert Grimm and David Merrill (editors), *Contents of Thought* (Tucson: University of

Arizona Press, 1988); and Frances Egan, "Must Psychology be Individualistic?," *Philosophical Review*, 100 (1991), pages 179–203; "Individualism, Computation, and Perceptual Content," *Mind*, 101 (1992), pages 443–459; "Computation and Content," *Philosophical Review*, 104 (1995), pages 181–203; and "In Defense of Narrow Mindedness," *Mind and Language*, 14 (1999), pages 177–194.

16. "Seeing What is Not There," page 207, my emphasis both times.
17. "Seeing What is Not There," page 197. For Segal's three general points, see pages 194–197, and his comments on zero-crossings, page 199.
18. "Seeing What is Not There," page 206.
19. For this sort of point and broader discussion, see Lawrence Shapiro, "Content, Kinds, and Individualism in Marr's Theory of Vision", *Philosophical Review*, 102 (1993), pages 489–513, especially pages 489–503.
20. Noam Chomsky, *New Horizons for the Study of Language*, page 159; the quotation that follows is from the same page. For his endorsement of Egan's interpretation, see "Language and Nature," page 55, footnote 25, and *New Horizons*, page 203, footnote 9.
21. See also Thomas Polger, "Neural Machinery and Realization," *Philosophy of Science*, in press.
22. For the knowledge level, see Alan Newell, "The Knowledge Level," *Artificial Intelligence*, 18 (1982), pages 87–127. For the semantic level, see Zenon Pylyshyn, *Computation and Cognition* (Cambridge, MA: MIT Press, 1984).
23. Egan, "Computation and Content," page 191.
24. Included here are Burge, "Individualism and Psychology," page 28; and Shapiro, "Content, Kinds, and Individualism in Marr's Theory of Vision," pages 499–500, and "A Clearer Vision," *Philosophy of Science*, 64 (1997), pages 131–153, at page 134. The table I draw the quotation from is Figure 1.4, *Vision*, page 25.
25. For Shapiro's point, see "A Clearer Vision," page 149; and for Egan's concession, "In Defense of Narrow Mindedness."
26. On the rigidity assumption, see Shimon Ullman, *The Interpretation of Visual Motion* (Cambridge, MA: MIT Press, 1979), page 146.
27. For brief discussions of exploitation and representation, see Larry Shapiro, "A Clearer Vision," pages 135 and 143. See also Mark Rowlands, *The Body in Mind* (New York: Cambridge University Press, 1999).
28. On the polar planimeter, see Sverker Runeson, "On the Possibility of 'Smart' Perceptual Mechanisms," *Scandinavian Journal of Psychology*, 18 (1977), pages 172–179, a paper that Frank Keil shoved in my hands nearly fifteen years ago (much to my puzzlement at the time).
29. See Robert Cummins, *The Nature of Psychological Explanation* (Cambridge, MA: MIT Press, 1983), and *Meaning and Mental Representation* (Cambridge, MA: MIT Press, 1989).
30. The three quotations in this paragraph are from *Vision*, pages 31, 68, and 104, respectively. On how representative these quotations are, see pages 43, 68, 99, 103–115, and 265–266.
31. For my views here, see *Cartesian Psychology and Physical Minds*, Chapter 3. For Marr's relation to the multiple spatial channels theory, see *Vision*, pages 61–64; and Palmer, *Vision Science*, pages 158–172.

32. On spatial coincidence, see *Vision*, pages 68–70; on stereopsis, pages 112–114.
33. See Rowlands, *The Body in Mind*, Chapter 5 in general and page 104 in particular for the continuum view. See also Edward Reed, *Encountering the World: Toward an Ecological Psychology* (New York: Oxford University Press, 1996).
34. For Segal's claim, see Gabriel Segal, "Defence of a Reasonable Individualism," *Mind*, 100 (1991), pages 485–494, at page 490.
35. On sneaking up on narrow content, see Jerry Fodor, *Psychosemantics* (Cambridge, MA: MIT Press, 1987), page 52; on anchoring, see pages 50–53 of the same. On realization conditions, see Brian Loar, "Social Content and Psychological Content," in Robert Grimm and David Merrill (editors), *Contents of Thought* (Tucson: University of Arizona Press, 1988).
36. *Cognition in the Wild*, page xiv.
37. See especially D. Ballard, M. Hayhoe, P.K. Pook, and R.P.N. Rao, "Deictic Codes for the Embodiment of Cognition," *Behavioral and Brain Sciences*, 20 (1997), pages 723–767. See also Dana Ballard, "Animate Vision," *Artificial Intelligence*, 48 (1991), pages 57–86; and "On the Function of Visual Representation," in Kathleen Akins (editor), *Perception: Vancouver Studies in Cognitive Science, Vol. 5* (New York: Oxford University Press, 1996), pages 111–131. Reprinted in Alva Noë and Evan Thompson (editors), *Vision and Mind: Selected Readings in the Philosophy of Perception* (Cambridge, MA: MIT Press, 2003), pages 459–479.
38. Rodney Brooks, "Intelligence Without Representation," modified version reprinted in John Haugeland (editor), *Mind Design II: Philosophy, Psychology, Artificial Intelligence* (Cambridge, MA: MIT Press, 1997); and *Cambrian Intelligence: The Early History of the New AI* (Cambridge, MA: MIT Press, 1999). For his claim about representations and Creatures, see "Intelligence Without Representation," pages 404–406.
39. On questioning the conceptual integrity, see Egan, "In Defense of Narrow Mindedness." On individuals as the largest units, see Segal, "Defence of a Reasonable Individualism," page 492. And on problems with wide computation interpretations, see Gabriel Segal, review of R.A. Wilson, "Cartesian Psychology and Physical Minds," *British Journal for the Philosophy of Science*, 48 (1997), pages 151–156.

8: The Embedded Mind and Cognition

1. On cognition as the filling, see Susan Hurley, *Consciousness in Action* (New York: Oxford University Press, 1998), page 401.
2. George Miller, "Informavores," in F. Machlup and U. Mansfield (editors), *The Study of Information: Interdisciplinary Messages* (New York: Wiley, 1984).
3. Ulric Neisser, "Memory: What Are the Important Questions?" reprinted in his *Memory Observed: Remembering in Natural Contexts* (San Francisco: W.H. Freeman, 1981).
4. For Neisser's views, see the work cited above as well as "The Ecological Approach to Perception and Memory," *New Trends in Experimental and Clinical*

Psychiatry, 4 (1988), pages 153–166; and "Remembering as Doing," *Behavioral and Brain Sciences*, 19 (1996), pages 203–204.
5. *An Essay Concerning Human Understanding* (Oxford: Clarendon Press, 1979), Book II, Chapter x, section 2, page 150.
6. Asher Koriat, and Morris Goldsmith, "Memory Metaphors and the Real-Life / Laboratory Controversy: Correspondence versus Storehouse Conceptions of Memory," *Behavioral and Brain Sciences*, 19 (1996), pages 167–188. See also H.L. Roediger, "Memory Metaphors in Cognitive Psychology," *Memory and Cognition*, 8 (1980), pages 231–246.
7. See Neisser, "Remembering as Doing"; Frederic C. Bartlett, *Remembering* (Cambridge: Cambridge University Press, 1932); and Bradd Shore, *Culture in Mind* (New York: Oxford University Press, 1996). The quotation is from page 201 of *Remembering*, and it is in Part II of that work, especially pages 300 forward, that Bartlett broaches the relationship between individual and societal remembering.
8. Merlin Donald, *The Origins of the Modern Mind: Three Stages in the Evolution of Culture and Cognition* (Cambridge, MA: Harvard University Press, 1991); and *A Mind So Rare: The Evolution of Human Consciousness* (New York: Norton, 2001). Donald discusses the external memory field in Chapter 8 of *Origins*.
9. Mark Rowlands, *The Body in Mind* (New York: Cambridge University Press, 1999), Chapter 6.
10. Susan Savage-Rumbaugh and Roger Lewin, *Kanzi: The Ape at the Brink of the Human Mind* (New York: Wiley, 1994); and Sue Savage-Rumbaugh, Stuart Shanker, and T.J. Taylor, *Apes, Language, and the Human Mind* (New York: Oxford University Press, 1998). See also Irene Pepperberg, "Some Cognitive Capacities of an African Grey Parrot (*Psittacus erithacus*)," *Advances in the Study of Behavior*, 19 (1990), pages 357–409; and *The Alex Studies: Cognitive and Communicative Abilities of Grey Parrots* (Cambridge, MA: Harvard University Press, 1999).
11. Michael Cole, *Cultural Psychology: A Once and Future Discipline* (Cambridge, MA: Harvard University Press, 1996); Paul Connerton, *How Societies Remember* (New York: Cambridge University Press, 1989).
12. On the concept of number, see Karen Wynn, "Children's Understanding of Counting," *Cognition*, 36 (1990), pages 155–193; and "Origins of Numerical Knowledge," *Mathematical Cognition*, 1 (1995) pages 35–60. On the concept of physical objects, see Elizabeth Spelke, "Principles of Object Perception," *Cognitive Science*, 14 (1990), pages 29–56; and Rene Baillargeon, "The Object Concept Revisited: New Directions in the Study of Infants' Physical Knowledge," in C. Granrud (editor), *Perception and Cognition in Infancy* (Hillsdale, NJ: Erlbaum, 1993).
13. For biology, Frank C. Keil, *Concepts, Kinds and Cognitive Development* (Cambridge, MA: MIT Press, 1989). For social relationships, Lawrence Hirschfeld, *Race in the Making: Cognition, Culture, and the Child's Construction of Human Kinds* (Cambridge, MA: MIT Press, 1996). For minds, Simon Baron-Cohen, *Mindblindness: An Essay on Autism and Theory of Mind* (Cambridge, MA: MIT Press, 1995). Much of the work here was pioneered by Susan Carey in her *Conceptual Change in Childhood* (Cambridge, MA: MIT Press, 1985).

14. Alexander Luria, *The Making of Mind* (Cambridge, MA: Harvard University Press, 1979), at page 43. See also Lev Vygotsky, *Thought and Language* (Cambridge, MA: MIT Press, 1962); and *Mind and Society: The Development of Higher Psychological Processes* (Cambridge, MA: Harvard University Press, 1978).
15. *Mind and Society*, page 30.
16. On literacy and augmentation, see David Olson, "Literacy," *The MIT Encyclopedia of the Cognitive Sciences*, pages 481–482. See also his *The World on Paper* (New York: Cambridge University Press, 1994); Michael Cole, *Cultural Psychology*; and Merlin Donald, *Origins of the Modern Mind*.
17. See Frank C. Keil and Robert A. Wilson (editors), *Explanation and Cognition* (Cambridge, MA: MIT Press, 2000); R.A. Wilson and F.C. Keil, "The Shadows and Shallows of Explanation," modified version reprinted in *Explanation and Cognition*; Leonid Rozenblit and Frank C. Keil, "The Misunderstood Limits of Folk Science: An Illusion of Explanatory Depth," *Cognitive Science*, 92 (2002), pages 1–42; and Donna R. Lutz and Frank C. Keil, "Early Understanding of the Division of Cognitive Labor," *Child Development*, 73 (2002), pages 1073–1084.
18. On modes of construal, see Frank C. Keil, "The Growth of Causal Understandings of Natural Kinds," in Dan Sperber, David Premack, and Ann Premack (editors), *Causal Cognition: A Multidisciplinary Debate* (Oxford: Oxford University Press, 1995). On rapprochement with a Vygotskyan slant, see Patricia Greenfield, "Culture and Universals: Integrating Social and Cognitive Development," in Larry Nucci, Geoffrey Saxe, and Elliott Turel (editors), *Culture, Thought, and Development* (Mahwah, NJ: Erlbaum, 2000); and Rom Harré, "The Rediscovery of the Human Mind: The Discursive Approach," *Asian Journal of Social Psychology*, 2 (1999), pages 43–62.
19. Lev Vygotsky, "The Genesis of Higher Mental Functions," in James Wertsch (editor), *The Concept of Activity in Soviet Psychology* (Armonk, NY: M.E. Sharpe, 1981), as quoted by James Wertsch, *Vygotsky and the Social Formation of Mind* (Cambridge MA: Harvard University Press, 1985), pages 60–61.
20. See Alan Leslie, "Pretense and Representation: The Origins of 'Theory of Mind,'" *Psychological Review*, 94 (1987), pages 412–426 on the theory of mind module; and Henry Wellman, *The Child's Theory of Mind* (Cambridge, MA: MIT Press, 1990) on the theory of mind.
21. George Lakoff and Mark Johnson, *Philosophy in the Flesh: The Embodied Mind and its Challenge to Western Thought* (New York: Basic Books, 1999); Arthur Glenberg, "What Memory is For," *Behavioral and Brain Sciences*, 20 (1997), pages 1–19; and Rick Grush, "In Defense of Some Cartesian Assumptions Concerning the Brain and its Operation," *Biology and Philosophy*, 18 (2003), pages 53–93.
22. For example, see Andy Clark, *Natural-Born Cyborgs: Minds, Technologies, and the Future of Human Intelligence* (New York: Oxford University Press, 2003), which concentrates on technology, and includes these examples. See also Donald Norman, *The Invisible Computer* (Cambridge, MA: MIT Press, 1999).

9: Expanding Consciousness

1. Thomas Nagel, "What is it Like to be a Bat?," *Philosophical Review*, 83 (1974), pages 435–450; Frank Jackson, "Epiphenomenal Qualia," *Philosophical Quarterly*, 32 (1982), pages 127–136; Joseph Levine, "Materialism and Qualia: The Explanatory Gap," *Pacific Philosophical Quarterly*, 64 (1983), pages 354–361; David Chalmers, *The Conscious Mind* (New York: Oxford University Press, 1996); David M. Armstrong, *A Materialist Theory of the Mind* (London: Routledge and Kegan Paul, 1968). For a range of recent philosophical thinking about consciousness, see Quentin Smith and Aleksandar Jokic, *Consciousness: New Philosophical Perspectives* (New York: Oxford University Press, 2003).
2. On temporally extended forms of consciousness, see Merlin Donald, *A Mind So Rare: The Evolution of Human Consciousness* (New York: Norton, 2001), pages 195–204.
3. On culturally developed cognitive tools, see Hutchins, *Cognition in the Wild*; Clark, *Natural-Born Cyborgs;* and Ron McClamrock, *Existential Cognition: Computational Minds in the World* (Chicago: University of Chicago Press, 1995). On traditional seafaring navigation, David Lewis, *We the Navigators* (Honolulu: University of Hawaii Press, 1972) and *The Voyaging Stars: Secrets of Pacific Island Navigators* (New York: Norton, 1978). And for the example of the road as a cognitive resource, see John Haugeland, "Mind Embodied and Embedded," reprinted in his *Having Thought: Essays in the Metaphysics of Mind* (Cambridge, MA: Harvard University Press, 1998), at page 234.
4. For problems with the presumption of disembodied and disembedded cognition, see Greg McCulloch, *The Life of the Mind: An Essay on Phenomenological Externalism* (New York: Routledge, 2003), Chapter 7.
5. A useful recent collection of essays on nonconceptual content is York Gunther (editor), *Essays on Nonconceptual Content* (Cambridge, MA: MIT Press, 2003).
6. For Dretske's argument, see his *Naturalizing the Mind* (Cambridge, MA: MIT Press, 1995), especially page 141; and his "Phenomenal Externalism," in E. Villanueva (editor), *Perception. Philosophical Issues* 7 (Atascadero, CA: Ridgeview, 1996).
7. For Dretske's talk of phenomenal states in general, see *Naturalizing the Mind*, page 103. See also William Lycan, *Consciousness and Experience* (Cambridge, MA: MIT Press, 1996) and "The Case for Phenomenal Externalism," in James Tomberlin (editor), *Philosophical Perspectives 15: Metaphysics* (Boston: Blackwell, 2001); and Michael Tye, *Consciousness, Color, and Content* (Cambridge, MA: MIT Press, 2000).
8. For the proposals about narrow content, see Frank Jackson and Philip Pettit, "Some Content is Narrow," in John Heil and Alfred Mele (editors), *Mental Causation* (New York: Oxford University Press, 1993) and Stephen White, *The Unity of the Self* (Cambridge, MA: MIT Press, 1991), especially Chapters 1 and 2; and "Color and Notional Content," *Philosophical Topics*, 22 (1994), pages 471–503.
9. The first of these proposals can be found in Jerry Fodor, "Cognitive Science and the Twin-Earth Problem," *Notre Dame Journal of Formal Logic*, 23 (1982),

pages 98–118; and the second in David K. Lewis, "David Lewis: Reduction of Mind," in Samuel Guttenplan (editor), *A Companion to the Philosophy of Mind* (Oxford: Blackwell, 1994).
10. David Milner and Melvyn Goodale, *The Visual Brain in Action* (New York: Oxford University Press, 1998); see also their "The Visual Brain in Action," reprinted in Alva Noë and Evan Thompson (editors), *Vision and Mind: Selected Readings in the Philosophy of Perception* (Cambridge, MA: MIT Press, 2003), pages 515–529; the quotation is from page 515 of this work. The original what/where distinction was drawn by Leslie Ungeleider and Mortimer Mishkin, "Two Cortical Visual Systems," in D.J. Ingle, M.A. Goodale, and R.J.W. Mansfield (editors), *Analysis of Visual Behavior* (Cambridge, MA: MIT Press, 1982). The quotation from Dana Ballard is on page 466 of his "On the Function of Visual Representation," reprinted in *Vision and Mind*, pages 459–479.
11. J. Kevin O'Regan and Alva Noë, "What it is Like to See: a Sensorimotor Theory of Perceptual Experience," *Synthese*, 129 (2001), pages 79–103; and "A Sensorimotor Account of Vision and Visual Consciousness," *Behavioral and Brain Sciences*, 24 (2001), pages 939–1031; the quotation is from page 943 of this paper. See also Alva Noë, "Experience and the Active Mind," *Synthese*, 129 (2001), pages 41–60; "On What We See," *Pacific Philosophical Quarterly*, 83 (2002), pages 57–80; "Is Perspectival Self-Consciousness Non-Conceptual?," *Philosophical Quarterly*, 52 (2002), pages 185–194; and *Action in Perception* (Cambridge, MA: MIT Press, in press).
12. Noë, "Experience and the Active Mind." For an earlier use of this analogy, see Donald MacKay, "Ways of Looking at Perception," in W. Wathen-Dunn (editor), *Models for the Perception of Speech and Visual Form* (Cambridge, MA: MIT Press, 1967).
13. On left-right inverting lenses, see James G. Taylor, *The Behavioral Basis of Perception* (New Haven, CT: Yale University Press, 1962). On change blindness, R.A. Rensink, J.K. O'Regan, and J.J. Clark, "On the Failure to Detect Changes in Scenes Across Brief Interruptions," *Visual Cognition*, 7 (2000), pages 127–146.
14. Paul Bach-y-Rita, *Brain Mechanisms in Sensory Stimulation* (New York: Academic Press, 1972); "Tactile Vision Substitution: Past and Future," *International Journal of Neuroscience*, 19 (1983), pages 29–36; and "The Relationship Between Motor Processes and Cognition in Tactile Vision Substitution," in A.F. Sanders and W. Prinz (editors), *Cognition and Motor Processes* (Berlin: Springer, 1984).
15. Susan Hurley, *Consciousness in Action* (New York: Oxford University Press, 1998), page 407.
16. The quotation is from *Consciousness in Action*, page 406. See also pages 30–32 and 48–50 on temporal atomization in conceptions of consciousness.
17. On making sense of actual and hypothetical experiments, see *Consciousness in Action*, Chapters 7–10. See Ivo Kohler, *The Formation and Transformation of the Perceptual World* (New York: International University Press, 1964) for his inverting lens studies, and Susan Hurley and Alva Noë, "Neural Plasticity and Consciousness," *Biology and Philosophy*, 18 (2003), pages 131–168.

18. See Ned Block, "Inverted Earth," in James Tomberlin (editor), *Philosophical Perspectives 4: Action Theory and Philosophy of Mind* (Atascadero, CA: Ridgeview, 1990). Reprinted in Ned Block, Owen Flanagan, and Güven Güzeldere (editors), *The Nature of Consciousness: Philosophical Essays* (Cambridge, MA: MIT Press, 1997); and Hurley, *Consciousness in Action*, pages 304–325, for discussion.
19. On eliminativism about qualia, see Daniel C. Dennett, "Quining Qualia," reprinted in *The Nature of Consciousness*; and *Consciousness Explained* (Boston, MA: Little, Brown, 1991).
20. See Sydney Shoemaker, "Functionalism and Qualia," *Philosophical Studies*, 27 (1975), pages 291–315; and "The Inverted Spectrum," *Journal of Philosophy*, 74 (1981), pages 357–381. Both of these are reprinted in his *Identity, Cause, and Mind* (New York: Cambridge University Press, 1984). The quotation comes from "Functionalism and Qualia," as reprinted there, on page 190.
21. See also Sydney Shoemaker, "The First-Person Perspective," *Proceedings and Addresses of the American Philosophical Association*, 68 (1994), pages 7–22, reprinted in his *The First-Person Perspective and Other Essays* (New York: Cambridge University Press, 1996), where this sort of criticism is also made of several influential arguments of John Searle's.

10: Intentionality and Phenomenology

1. Terence Horgan and John Tienson, "The Intentionality of Phenomenology and the Phenomenology of Intentionality," in David J. Chalmers (editor), *Philosophy of Mind: Classical and Contemporary Readings* (New York: Oxford University Press, 2002). I thank Terry Horgan for sending me an advance copy of this paper.
2. See John Searle, "Consciousness, Explanatory Inversion, and Cognitive Science," *Behavioral and Brain Sciences*, 13 (1990), pages 585–642; and *The Rediscovery of the Mind* (Cambridge, MA: MIT Press, 1992); the quotation is from page 156 of the book. Galen Strawson, *Mental Reality* (Cambridge, MA: MIT Press, 1994), page 208.
3. Loar's four terms for this form of intentionality are drawn, respectively, from Brian Loar, "Subjective Intentionality," *Philosophical Topics* (1987), pages 89–124; "Social Content and Psychological Content," in Robert Grimm and David Merrill (editors), *Contents of Thought* (Tucson: University of Arizona Press, 1988); "Transparent Experience and the Availability of Qualia," in Quentin Smith and Aleksandar Jokic (editors), *Consciousness: New Essays* (New York: Oxford University Press, 2003); and "Phenomenal Intentionality as the Basis of Mental Content," in Martin Hahn and Bjorn Ramberg (editors), *Reflections and Replies: Essays on the Philosophy of Tyler Burge* (Cambridge, MA: MIT Press, 2003). I thank Brian Loar for sending me advance copies of the latter two papers.
4. See "The Intentionality of Phenomenology...," cited in note 1.
5. Fred Dretske, *Naturalizing the Mind* (Cambridge, MA: MIT Press, 1995); William Lycan, *Consciousness and Experience* (Cambridge, MA: MIT Press,

1996) and "The Case for Phenomenal Externalism," in James Tomberlin (editor), *Philosophical Perspectives 15: Metaphysics* (Boston: Blackwell, 2001); and Michael Tye, *Ten Problems of Consciousness* (Cambridge, MA: MIT Press, 1995), and his *Consciousness, Color, and Content* (Cambridge, MA: MIT Press, 2000).

6. The quotation is from "The Intentionality of Phenomenology...," page 521.
7. For Harman's report, see his "The Intrinsic Quality of Experience" in James Tomberlin (editor), *Philosophical Perspectives 4: Metaphysics*, (Atascadero, CA: Ridgeview, 1990), pages 31–52.
8. See his "Transparent Experience and the Availability of Qualia" and "Phenomenal Intentionality as the Basis of Mental Content." The quotation is from the former paper, page 90.
9. "Transparent Experience...," page 92.
10. Galen Strawson, *Mental Reality*, page 5. See also Horgan and Tienson, "The Intentionality of Phenomenology...," page 523.
11. "The Intentionality of Phenomenology...," page 522.
12. See "Phenomenal Intentionality...," pages 186–189, for the build up.
13. The quotation early in this paragraph is from "The Intentionality of Phenomenology...," page 524, with the argument for this following on pages 524–526.
14. See "The Intentionality of Phenomenology...", page 525 for the quotation, with the argument following on pages 528–529.
15. For Loar's argument, see "Phenomenal Intentionality...." p. 186.
16. "Phenomenal Intentionality," page 186.
17. Loar rules out appeals to "an introspective glance" on page 190.
18. Colin McGinn, *Mental Content* (New York: Blackwell, 1989), pages 58–99. See also Gabriel Segal, *A Slim Book About Narrow Content* (Cambridge, MA: MIT Press, 2000), who rests much on the significance of the same general sort of case.
19. See Martin Davies, "Externalism and Experience," in A. Clark, J. Ezquerro, and J.M. Larrazabal (editors), *Philosophy and Cognitive Science: Categories, Consciousness, and Reasoning* (Dordrecht: Kluwer Academic Publishers, 1995). Reprinted in Ned Block, Owen Flanagan, and Güven Güzeldere (editors), *The Nature of Consciousness: Philosophical Debates* (Cambridge, MA: MIT Press, 1997). See also Elizabeth Fricker, "Content, Cause, and Function," (Critical Notice of McGinn, *Mental Content*), *Philosophical Books*, 32 (1991), pages 136–144.
20. See Greg McCulloch, *The Life of the Mind: An Essay on Phenomenological Externalism* (New York: Routledge, 2003), Chapter 7, for an argument against the coherence of brain-in-vatish thought experiments.
21. For discussion of these proposals of White, Fodor, Block, and Loar, see my *Cartesian Psychology and Physical Minds* (New York: Cambridge University Press, 1995), Chapter 9. For another appeal to the phenomenal to buttress individualism about intentionality, see Stephen White, "Color and Notional Content," *Philosophical Topics*, 22 (1994), pages 471–503.

11: Group Minds in Historical Perspective

1. For discussions of the collective psychology tradition, see Robert Nye, *The Origins of Crowd Psychology: Gustav LeBon and the Crisis of Mass Democracy in the Third Republic* (Beverly Hills, CA: Sage Publications, 1975); Susanna Barrows, *Distorting Mirrors: Visions of the Crowd in Late Nineteenth-Century France* (New Haven, CT: Yale University Press, 1981); and Jaap van Ginneken, *Crowds, Psychology, and Politics 1871–1899* (New York: Cambridge University Press, 1992).
2. For a precursor to the tradition, see Charles MacKay, *Memoirs of Extraordinary Popular Delusions and the Madness of Crowds* (Boston: L.C. Page, 1932), originally published in 1841.
3. These quotes are taken, respectively, from the Preface (page v) and page 2 of *The Crowd*.
4. This sort of point has been made by Clark McPhail, *The Myth of the Madding Crowd* (New York: Aldine de Gruyter, 1991).
5. These claims are made, respectively, in the Preface (page xxi), in Book III, Chapter III, and in Book III, Chapter II, of *The Crowd*.
6. Both Barrows, *Distorting Mirrors*, Chapter 2, and Nye, *The Origins of Crowd Psychology*, make the connection to Machiavelli. See also Nye's Chapters 5–6 on the military endorsements and uses of Le Bon's work.
7. On irrationalist theories of individual psychology, see Gordon Allport, "The Historical Background of Modern Social Psychology," in G. Lindzey and E. Aronson (editors), *The Handbook of Social Psychology*, 2nd ed. (Reading, MA: Addison-Wesley, 1968).
8. For Barrows, see *Distorting Mirrors*, Chapter 5. The claims that Le Bon makes referred to here can be found on pages 9–11 of *The Crowd*.
9. On crowds as feminine and primitive, see Barrows, *Distorting Mirrors*, Chapter 2; and on fascistic appropriations, see Nye, *Origins of Crowd Psychology*, Chapter 6.
10. See Emile Durkheim, "Individual and Collective Representations," reprinted in his *Sociology and Philosophy*, Translated by D.F. Pocock (Glencoe, IL: The Free Press, 1953), originally published in 1898; and William McDougall, *The Group Mind* (New York: Putnam, 1920).
11. McDougall, *The Group Mind*; William Trotter, *Instincts of the Herd in War and Peace* (London: Fisher Unwin, 1916).
12. Edward O. Wilson, *The Insect Societies* (Cambridge, MA: Belknap Press, 1971); David Sloan Wilson, "The Group Selection Controversy: History and Current Status," *Annual Review of Ecology and Systematics*, 14 (1983), pages 159–187. For an exception to the trend of sideline, internal histories by biologists, see Gregg Mitman's excellent *The State of Nature: Ecology, Community, and American Social Thought, 1900–1950* (Chicago: University of Chicago Press, 1992), which focuses largely on the Chicago school of ecology.
13. Clements develops the idea of communities as complex organisms in his *Research Methods in Ecology* (Lincoln, NE: University Publication Company, 1905), page 5, and that of a biome in his *Plant Succession* (Washington, D.C.: Carnegie Institute, 1916). Thanks to Michael Wade for a prod about

the work of Park within the Chicago school that builds on the physiological conception of populations. See Thomas Park, "Studies in Population Physiology: The Relation of Numbers to Initial Population Growth in the Flour Beetle *Tribolium Confusum* Duval," *Ecology*, 13 (1932), pages 172–181; and "Studies in Population Physiology. II. Factors Regulating Initial Growth of *Tribolium Confusum* Populations," *Journal of Experimental Zoology*, 65 (1933), pages 17–42.

14. For general overviews, see F.E. Clements and V.E. Shelford, *Bio-Ecology* (New York: John Wiley, 1939), Chapters 1 and 2; and W.C. Allee, A.E. Emerson, O. Park, T. Park, and K.P. Schmidt, *Principles of Animal Ecology* (Philadelphia, PA: W.B. Saunders, 1949), Chapters 1–3 and 23–30.
15. See Mary Evans and Howard Evans, *William Morton Wheeler, Biologist* (Cambridge, MA: Harvard University Press, 1970) for an easy-reading biography of Wheeler that includes a year-by-year list of his publications.
16. E.O. Wilson, *The Insect Societies*, page 317. Wheeler's appeals to individuality and regeneration are made on pages 8 and 19 of "The Ant-Colony as an Organism." The longer quote in the text is from page 26 of that essay.
17. For Wheeler's nutritive, reproductive, and defensive societies, see pages 154–157 of his "Emergent Evolution and the Development of Societies," modified version reprinted in his *Essays in Philosophical Biology* (Cambridge, MA: Harvard University Press, 1939), originally published in 1926.
18. I discuss the levels of selection in some detail in Part Four of my *Genes and the Agents of Life* (New York: Cambridge University Press, 2004).
19. McDougall, *The Group Mind*, pages 9–10.
20. Wilhelm Wundt, *Outlines of Psychology*, translated by C.H. Judd. (Leipzig: Wilhelm Engelman, 3rd edition, 1907), page 355, emphasis in the original.
21. Le Bon, *The Crowd*, pages 5–6.
22. On the transformation thesis, see Clark McPhail, *The Myth of the Madding Crowd*.
23. See Allee et al., *Principles of Animal Ecology*, Chapter 24. See especially pages 420–435; the quotation is taken from page 420. For Emerson, see his "Social Coordination and the Superorganism," *American Midland Naturalist*, 21 (1939), pages 182–209; "Basic Comparisons of Human and Insect Societies," *Biological Symposia*, 8 (1942), pages 163–176; and "The Biological Basis of Social Cooperation," *Illinois Academy of Science Transactions*, 29 (1946), pages 9–18.

12: The Group Mind Hypothesis in Contemporary Biology and Social Science

1. David Sloan Wilson, "Altruism and Organism: Disentangling the Themes of Multilevel Selection Theory," *American Naturalist*, 150 (1997), supplement, pages S122–S134, at page S128. See also David Seeley, *The Wisdom of the Hive* (Cambridge, MA: Harvard University Press, 1995); and Herbert H.T. Prins, *Ecology and Behaviour of the African Buffalo* (London: Chapman and Hall, 1996).
2. For the quote from Wilson, see "Altruism and Organism...," page S131.

3. Mary Douglas, *How Institutions Think* (Syracuse, NY: Syracuse University Press, 1986). The quotations I provide are from the Preface (page x) and page 8.
4. Richard A. Shweder, "Cultural Psychology: What Is It?," in his *Thinking Through Cultures: Expeditions in Cultural Psychology* (Cambridge, MA: Harvard University Press, 1991).
5. David Sloan Wilson, *Darwin's Cathedral: Evolution, Religion, and the Nature of Society* (Chicago: University of Chicago Press, 2002), at page 33. For his earlier work that bridges the biological and social sciences, see for example his "Levels of Selection: An Alternative to Individualism in the Human Sciences," *Social Networks*, 11 (1989), pages 257–272; and his "On the Relationship between Evolutionary and Psychological Definitions of Altruism and Egoism," *Biology and Philosophy*, 7 (1991), pages 61–68.
6. On socially distributed cognition, see Edwin Hutchins, "The Technology of Team Navigation" and Aaron V. Cicourel, "The Integration of Distributed Knowledge in Collaborative Medical Diagnosis," both in J. Galegher, R. Kraut, and C. Egido (editors), *Intellectual Teamwork: Social and Technological Foundations of Cooperative Work* (Hillsdale, NJ: Erlbaum, 1990). On distributed cognition in science, see Ronald Giere, "Scientific Cognition as Distributed Cognition," in Peter Carruthers, Stephen Stich, and Michael Siegal (editors), *The Cognitive Basis of Science* (New York: Cambridge University Press, 2002).
7. For views of holism as an *a priori* constraint, see Daniel C. Dennett, *The Intentional Stance* (Cambridge, MA: MIT Press, 1987); and Donald Davidson, *Inquiries into Truth and Interpretation* (Oxford: Oxford University Press, 1984). This sort of appeal to homeostatic mechanisms and constraints plays a central role in the philosopher Richard Boyd's account of natural kinds that I draw on and develop in Part Two of my *Genes and the Agents of Life* (New York: Cambridge University Press, 2004).
8. William McDougall, *The Group Mind* (New York: Putnam, 1920).
9. David Sloan Wilson, "Incorporating Group Selection into the Adaptationist Program: A Case Study Involving Human Decision Making," in Jeffrey A. Simpson and Douglas T. Kendrick (editors), *Evolutionary Social Psychology* (Hillsdale, NJ: Erlbaum, 1997). The quotes are from pages 358 and 359, respectively.
10. The quotation is from "Incorporating Group Selection...," page 359.
11. Both quotations are from *Darwin's Cathedral*, page 1.
12. For Wilson's discussion of "groupthink," see "Incorporating Group Selection...", pages 363–366. See also Irving Janis, *Groupthink: Psychological Studies of Policy Decisions and Fiascoes* (Boston: Houghton Mifflin, 1982), 2nd edition.
13. Ludwig Fleck, *Genesis and Development of a Scientific Fact* (Chicago: University of Chicago Press, 1979). The quotation is taken from page 39. See also Thomas Kuhn, *The Structure of Scientific Revolutions* (Chicago: University of Chicago Press, 1962); and W. Sady, "Ludwig Fleck – Thought Collectives and Thought Styles," in W. Krajewski (editor), *Polish Philosophers of Science and Nature in the 20th Century*. Poznan Studies in the Philosophy of the Sciences and the Humanities, Vol. 74 (New York: Rodopi, 2001).

14. Kuhn's quotation is from the Foreword to Fleck's book (page x). See also Douglas, *How Institutions Think*, pages 15–16.
15. *How Institutions Think*, page 128. The short preceding quotations are from pages 2 and 11.
16. On evolutionary psychology, see John Tooby and Leda Cosmides, "The Psychological Foundations of Culture," in Jerome Barkow, Leda Cosmides, and John Tooby (editors), *The Adapted Mind: Evolutionary Psychology and the Generation of Culture* (New York: Oxford University Press, 1992). On social intelligence and the Machiavellian hypothesis, see Richard Byrne and Andrew Whiten (editors), *Machiavellian Intelligence: Social Expertise and the Evolution of Intellect in Monkeys, Apes, and Humans* (Oxford: Clarendon Press, 1988) and Andrew Whiten and Richard Byrne (editors), *Machiavellian Intelligence II: Extensions and Evaluations* (New York: Cambridge University Press, 1997). For research that departs further from an individualistic tradition, see Linda R. Caporael and R.M. Baron, "Groups as the Mind's Natural Environment," in Jeffrey A. Simpson and Douglas T. Kenrick (editors), *Evolutionary Social Psychology* (Mahwah, NJ: Erlbaum, 1997); and J.C. Turner, *Rediscovering the Social Group: A Self-Categorization Theory* (Oxford: Basil Blackwell, 1987).
17. See Part Four of my *Genes and the Agents of Life* for a sustained discussion of the relationship between individual and group selection and the development of the idea that these levels are best thought of as *entwined*.
18. Wilson discusses factors that regulate group decision processes in "Incorporating Group Selection...," at pages 372–375.
19. "Incorporating Group Selection...," pages 359–360.
20. David Sloan Wilson and Elliott Sober, "Reintroducing Group Selection to the Human Behavioral Sciences," *Behavioral and Brain Sciences*, 17 (1994), pages 585–654. See especially pages 602–605.
21. On reasoning modules, see Leda Cosmides, "The Logic of Social Exchange: Has Natural Selection Shaped How Humans Reason? Studies with the Wason Selection Task," *Cognition*, 31 (1989), pages 187–276; and Patricia Cheng and Keith Holyoak, "Pragmatic Reasoning Schemas," *Cognitive Psychology*, 17 (1985), pages 391–416.
22. See my "The Mind Beyond Itself," in Dan Sperber, *Metarepresentations: A Multidisciplinary Debate* (New York: Oxford University Press, 2000); and Elliott Sober, " "Models of Cultural Evolution," in Paul Griffiths (editor), *Trees of Life* (Cambridge, MA: MIT Press, 1991).
23. On group and cultural selection, see Robert Boyd and Peter Richerson, "Group Selection among Alternative Evolutionarily Stable Strategies," *Journal of Theoretical Biology*, 145 (1990), pages 331–342; and "Culture and Cooperation," in J.J. Mansbridge (editor), *Beyond Self-Interest* (Chicago: University of Chicago Press, 1990).

References

Allee, W.C., A.E. Emerson, O. Park, T. Park, and K.P. Schmidt, 1949, *Principles of Animal Ecology.* Philadelphia, PA: W. B. Saunders.

Allport, G.W., 1968, "The Historical Background of Modern Social Psychology," in G. Lindzey and E. Aronson (eds.), *The Handbook of Social Psychology*, 2nd edition. Reading, MA: Addison-Wesley.

Ariew, A., 1996, "Innateness and Canalization," *Philosophy of Science* 63:S19–S27.

Ariew, A., 1999, "Innateness is Canalization: In Defense of a Developmental Account of Innateness," in V.G. Hardcastle (ed.), *Where Biology Meets Psychology: Philosophical Essays.* Cambridge, MA: MIT Press, pp. 117–138.

Armstrong, D.M., 1968, *A Materialist Theory of the Mind.* London: Routledge and Kegan Paul.

Armstong, D.M., 1973, *Belief, Truth, and Knowledge.* New York: Cambridge University Press.

Bach-y-Rita P., 1972, *Brain Mechanisms in Sensory Stimulation.* New York: Academic Press.

Bach-y-Rita P., 1983, "Tactile Vision Substitution: Past and Future," *International Journal of Neuroscience* 19:29–36.

Bach-y-Rita P., 1984, "The Relationship Between Motor Processes and Cognition in Tactile Vision Substitution," in A.F. Sanders and W. Prinz (eds.), *Cognition and Motor Processes.* Berlin: Springer.

Baillargeon, R., 1993, "The Object Concept Revisited: New Directions in the Study of Infants' Physical Knowledge," in C. Granrud (ed.), *Perception and Cognition in Infancy.* Hillsdale, NJ: Erlbaum.

Baldwin, J.D., 1986, *George Herbert Mead: A Unifying Theory for Sociology.* Beverly Hills, CA: Sage Publications.

Ballard, D., 1991, "Animate Vision," *Artificial Intelligence* 48:57–86.

Ballard, D., 1996, "On the Function of Visual Representation," in K. Akins (ed.), *Perception: Vancouver Studies in Cognitive Science, Volume 5.* New York, Oxford University Press, pp. 111–131. Reprinted in A. Noë and E. Thompson (eds.),

Vision and Mind: Selected Readings in the Philosophy of Perception. Cambridge, MA: MIT Press, pp. 459–479.

Ballard, D., M.M. Hayhoe, P.K. Pook, and R.P.N. Rao, 1997, "Deictic Codes for the Embodiment of Cognition," *Behavioral and Brain Sciences* 20:723–767.

Barkow, J., L. Cosmides, and J. Tooby (eds.), *The Adapted Mind: Evolutionary Psychology and the Generation of Culture.* New York: Oxford University Press.

Baron-Cohen, S., 1995, *Mindblindness: An Essay on Autism and Theory of Mind.* Cambridge, MA: MIT Press.

Barrows, S., 1981, *Distorting Mirrors: Visions of the Crowd in Late Nineteenth-Century France.* New Haven, CT: Yale University Press.

Bartlett, F.C., 1932, *Remembering.* Cambridge: Cambridge University Press.

Bateson, P.P.G., 1991, "Are There Principles of Behavioural Development?," in P. Bateson (ed.), *The Development and Integration of Behaviour.* Cambridge: Cambridge University Press, pp. 19–39.

Block, N., 1980, Introduction to his *Readings in the Philosophy of Psychology.* Cambridge, MA: Harvard University Press.

Block, N., 1986, "An Advertisement for a Semantics for Psychology," in P. French, T. Uehling Jr., and H. Wettstein (eds.), *Midwest Studies in Philosophy*, Volume 10, Philosophy of Mind. Minneapolis, MN: University of Minnesota Press.

Block, N., 1990, "Inverted Earth," in J. Tomberlin (ed.), *Philosophical Perspectives.* Atascadero, CA: Ridgeview Publishing. Reprinted in N. Block, O. Flanagan, and G. Güzeldere (eds.), *The Nature of Consciousness: Philosophical Essays.* Cambridge, MA: MIT Press.

Block, N., O. Flanagan, and G. Güzeldere (eds.), *The Nature of Consciousness: Philosophical Essays.* Cambridge, MA: MIT Press.

Bloom, P., 2000, *How Children Learn the Meanings of Words.* Cambridge, MA: MIT Press.

Blumenthal, A., 1975, "A Reappraisal of Wilhelm Wundt," *American Psychologist* 30:1081–1088.

Blumenthal, A., 1979, "The Founding Father We Never Knew," *Contemporary Psychology* 24:547–550.

Blumenthal, A., 1998, "Leipzig, Wilhelm Wundt, and Psychology's Gilded Age," in G.A. Kimble and M. Wertheimer, *Portraits of Pioneers in Psychology*, Volume 3. Mahwah, New Jersey: Erlbaum, pp. 31–50

Boyd, R., and P. Richerson, 1990a, "Group Selection Among Alternative Evolutionarily Stable Strategies," *Journal of Theoretical Biology* 145:331–342.

Boyd, R., and P. Richerson, 1990b, "Culture and Cooperation," in J.J. Mansbridge (ed.), *Beyond Self-Interest.* Chicago: University of Chicago Press.

Boyd, R.N., 1980, "Materialism without Reductionism: What Physicalism Does Not Entail," in N. Block (ed.), *Readings in the Philosophy of Psychology.* Cambridge, MA: Harvard University Press, 1980.

Brandom, R., 1994, *Making it Explicit.* Cambridge, MA: Harvard University Press.

Brooks, R.A., 1991a, "Intelligence Without Reason," *Proceedings of the 1991 International Joint Conference on Artificial Intelligence*, pp. 569–595. Reprinted in R.A. Brooks, *Cambrian Intelligence: The Early History of the New AI.* Cambridge, MA: MIT Press, 1999.

Brooks, R.A., 1991b, "Intelligence Without Representation," *Artificial Intelligence* 47:139–160. Modified version reprinted in J. Haugeland (ed.), *Mind Design II: Philosophy, Psychology, Artificial Intelligence*. Cambridge, MA: MIT Press, 1997.
Brooks, R.A., 1999, *Cambrian Intelligence: The Early History of the New AI*. Cambridge, MA: MIT Press.
Bruner, J., 1990, *Acts of Meaning*. Cambridge, MA: MIT Press.
Burge, T., 1979, "Individualism and the Mental," in P. French, T. Uehling Jr., and H. Wettstein (eds.), *Midwest Studies in Philosophy*, Volume 4, Metaphysics. Minneapolis: University of Minnesota Press.
Burge, T., 1982, "Other Bodies" in A. Woodfield (ed.), *Thought and Object: Essays on Intentionality*. Oxford: Oxford University Press.
Burge, T., 1986a, "Individualism and Psychology," *Philosophical Review* 95:3–45.
Burge, T., 1986b, "Intellectual Norms and Foundations of Mind," *Journal of Philosophy* 83:697–720.
Burge, T., 1988, "Cartesian Error and the Objectivity of Perception," in R. Grimm and D. Merrill (eds), *Contents of Thought*. Tucson: University of Arizona Press.
Burge, T., 1989, "Individuation and Causation in Psychology," *Pacific Philosophical Quarterly* 70:303–322.
Burge, T., 1992, "Philosophy of Mind and Language: 1950–1990," *Philosophical Review* 101:3–51
Burge, T., 1993, "Mind-Body Causation and Explanatory Practice," in J. Heil and A. Mele (eds.), *Mental Causation*. New York: Oxford University Press.
Byrne, R.W., and A. Whiten (eds.), 1988, *Machiavellian Intelligence*. New York: Cambridge University Press.
Caporael, L.R., and R.M. Baron, "Groups as the Mind's Natural Environment," in J.A. Simpson and D.T. Kenrick (eds.), *Evolutionary Social Psychology*. Mahwah, NJ: Erlbaum.
Carey, S., 1985, *Conceptual Change in Childhood*. Cambridge, MA: MIT Press.
Carnap, R., 1928, *The Logical Structure of the World: Pseudoproblems in Philosophy*. Translated by R. George. Berkeley: University of California Press, 1967.
Carnap, R., 1947, *Meaning and Necessity: A Study in Semantics and Modal Logic*. Chicago: University of Chicago Press. 2nd edition, 1957.
Carpendale, J.I.M., and C. Lewis, in press, "Constructing an Understanding of Mind: The Development of Children's Social Understanding within Social Interaction," *Behavioral and Brain Sciences*.
Chalmers, D., 1996, *The Conscious Mind*. New York: Oxford University Press.
Changeux, J-P., 1985, *Neuronal Man*. New York: Pantheon Books.
Cheng, P., and K. Holyoak, 1985, "Pragmatic Reasoning Schemas," *Cognitive Psychology* 17:391–416.
Cholewiak, R., and A. Collins, 1991, "Sensory and Physiological Bases of Touch," in M.A. Heller and W. Schiff (eds.), *The Psychology of Touch*. Mahwah, NJ: Erlbaum.
Chomsky, N., 1959, "Review of B.F. Skinner's *Verbal Behavior*," *Language* 24:163–186.
Chomsky, N., 1966, *Cartesian Linguistics: A Chapter in the History of Rationalist Thought*. New York: Harper and Row.
Chomsky, N., 1975, *Reflections on Language*. New York: Pantheon.

Chomsky, N., 1986, *Knowledge of Language*. New York: Praeger.
Chomsky, N., 1991a, "Linguistics and Adjacent Fields: A Personal View," in A. Kasher (ed.), *The Chomskyan Turn*. Oxford: Basil Blackwell.
Chomsky, N., 1991b, "Linguistics and Cognitive Science: Problems and Mysteries," in A. Kasher (ed.), *The Chomskyan Turn*. Cambridge, MA: Blackwell.
Chomsky, N., 1995, "Language and Nature," *Mind* 104:1–61.
Chomsky, N., 2000, *New Horizons in the Study of Language and Mind*. New York: Cambridge University Press.
Cicourel, A.V., 1990, "The Integration of Distributed Knowledge in Collaborative Medical Diagnosis," in J. Galegher, R. Kraut, and C. Egido, (eds.), *Intellectual Teamwork: Social and Technological Foundations of Cooperative Work*. Hillsdale, NJ: Erlbaum.
Clark, A., 1989, *Microcognition: Philosophy, Cognitive Science, and Parallel Distributed Processing*. Cambridge, MA: MIT Press.
Clark, A., 1997, *Being There: Putting Brain, Body, and World Together Again*. Cambridge, MA: MIT Press.
Clark, A., 2001, "Reasons, Robots and the Extended Mind," *Mind and Language* 16:121–145.
Clark, A., 2003, *Natural-Born Cyborgs: Minds, Technologies, and the Future of Human Intelligence*. New York: Oxford University Press.
Clark, A., and D. Chalmers, 1998, "The Extended Mind," *Analysis* 58:10–23.
Clements, F.E., 1905, *Research Methods in Ecology*. Lincoln, NE: University Publication Co.
Clements, F.E., 1916, *Plant Succession*. Washington, D.C.: Carnegie Institute.
Clements, F.E., and V.E. Shelford, 1939, *Bio-Ecology*. New York: John Wiley.
Cole, M., 1996, *Cultural Psychology: A Once and Future Discipline*. Cambridge, MA: Harvard University Press.
Collier, G., H.L. Minton, and G. Reynolds, 1991, *Currents of Thought in American Social Psychology*. New York: Oxford University Press.
Comte, A., 1875, *System of Positive Polity, or Treatise on Sociology, Second Volume*. New York: Burt Franklin.
Comte, A., 1974, *The Essential Comte: Selected from the Course de philosophie positive*. Edited by S. Andreski. New York: Barnes and Noble.
Connerton, P., 1989, *How Societies Remember*. New York: Cambridge University Press.
Cosmides, L., 1989, "The Logic of Social Exchange: Has Natural Selection Shaped How Humans Reason? Studies with the Wason Selection Task," *Cognition* 31:187–276.
Cosmides, L., and J. Tooby, 1992, "Cognitive Adaptations for Social Exchange," in J. Barkow, L. Cosmides, and J. Tooby (eds.), *The Adapted Mind: Evolutionary Psychology and the Generation of Culture*. New York: Oxford University Press.
Cosmides, L., and J. Tooby, 1994, "Origins of Domain Specificity: The Evolution of Functional Organization," in L. Hirschfeld and S. Gelman (eds.), *Mapping the Mind: Domain Specificity in Cognition and Culture*. New York: Cambridge University Press.
Cosmides, L., and J. Tooby, 1995, "Foreword" to S. Baron-Cohen, *Mindblindness*. Cambridge, MA: MIT Press.

Cosmides, L., and J. Tooby, 1999, "Evolutionary Psychology," in R.A. Wilson and F.C. Keil (eds.), *The MIT Encyclopedia of the Cognitive Sciences*. Cambridge, MA: MIT Press.

Cosmides, L., and J. Tooby, 2000, "Consider the Source: The Evolution of Adaptations for Decoupling and Metarepresentations," in D. Sperber (ed.), *Metarepresentations: A Multidisciplinary Perspective*. New York: Oxford University Press.

Cowie, F., 1999, *What's Within: Nativism Reconsidered*. New York: Oxford University Press.

Crane, T., 1991, "All the Difference in the World," *Philosophical Quarterly* 41:1–25.

Craver, C., in press, "Saving Your Self: Dissociative Realization and Kind Splitting," *Philosophy of Science*.

Cummins, R.C., 1983, *The Nature of Psychological Explanation*. Cambridge, MA: MIT Press.

Cummins, R.C., 1989, *Meaning and Mental Representation*. Cambridge, MA: MIT Press.

Danziger, K., 1990, *Constructing the Subject: Historical Origins of Psychological Research*. New York: Cambridge University Press.

Davidson, D., 1970, "Mental Events." Reprinted in his *Essays on Actions and Events*. Oxford: Oxford University Press, 1980.

Davidson, D., 1974a, "Belief and the Basis of Meaning." Reprinted in his *Inquiries into Truth and Interpretation*. New York: Oxford University Press.

Davidson, D., 1974b, "Philosophy as Psychology." Reprinted in his *Essays on Actions and Events*. Oxford: Oxford University Press, 1980.

Davidson, D., 1975, "Thought and Talk." Reprinted in his *Inquiries into Truth and Interpretation*. New York: Oxford University Press.

Davidson, D., 1980, *Essays on Actions and Events*. Oxford: Oxford University Press.

Davidson, D., 1984, *Inquiries into Truth and Interpretation*. Oxford: Oxford University Press.

Davies, M., 1995, "Externalism and Experience," in A. Clark, J. Ezquerro, and J.M. Larrazabal (eds.), *Philosophy and Cognitive Science: Categories, Consciousness, and Reasoning*. Dordrecht: Kluwer Academic Publishers. Reprinted in N. Block, O. Flanagan, and G. Güzeldere (eds.), *The Nature of Consciousness: Philosophical Debates*. Cambridge, MA: MIT Press.

Dawkins, R., 1976, *The Selfish Gene*. Oxford: Oxford University Press. 2nd edition, 1989.

Dawkins, R., 1982, *The Extended Phenotype*. Oxford: Oxford University Press.

Dennett, D.C., 1987, *The Intentional Stance*. Cambridge, MA: MIT Press.

Dennett, D.C., 1971, "Intentional Systems." Reprinted in his *Brainstorms*. Cambridge, MA: MIT Press.

Dennett, D. C., 1978, "Artificial Intelligence as Philosophy and as Psychology." Reprinted in his *Brainstorms*. Cambridge, MA: MIT Press, 1978.

Dennett, D.C., 1981, "True Believers: The Intentional Strategy and Why it Works." Reprinted in his *The Intentional Stance*. Cambridge, MA: MIT Press, 1987.

Dennett, D.C., 1988, "Quining Qualia," in A. Marcel and Bisiach (editors), *Consciousness in Contemporary Science*. Oxford: Oxford University Press.

Reprinted in N. Block, O. Flanagan and G. Güzeldere (eds.), *The Nature of Consciousness*. Cambridge: MIT Press, 1997.

Dennett, D.C., 1991, *Consciousness Explained*. Boston, MA: Little, Brown.

Devitt, M., 1990, "A Narrow Representational Theory of Mind," in W.G. Lycan, (ed.), *Mind and Cognition: A Reader*. Cambridge, MA: Basil Blackwell.

Donald, M., 1991, *The Origins of the Modern Mind*. Cambridge, MA: Harvard University Press.

Donald, M., 2001, *A Mind So Rare: The Evolution of Human Consciousness*. New York: Norton.

Douglas, M., 1986, *How Institutions Think*. Syracuse, NY: Syracuse University Press.

Dretske, F.I., 1981, *Knowledge and the Flow of Information*. Cambridge, MA: MIT Press.

Dretske, F.I., 1988, *Explaining Behavior: Reasons in a World of Causes*. Cambridge, MA: MIT Press.

Dretske, F.I., 1995, *Naturalizing the Mind*. Cambridge, MA: MIT Press.

Dretske, F.I., 1996, "Phenomenal Externalism," in E. Villanueva (ed.), *Perception*. Philosophical Issues 7. Atascadero, CA: Ridgeview.

Durkheim, E., 1895, *Rules of Sociological Method*. New York: Free Press, 1938.

Durkheim, E., 1897, *Suicide: A Study in Sociology*. New York: Free Press, 1951.

Durkheim, E., 1898, "Individual and Collective Representations." Reprinted in his *Sociology and Philosophy*. Translated by D.F. Pocock. Glencoe, IL: The Free Press, 1953.

Ebbs, G., 1998, *Rule-Following and Realism*. Cambridge, MA: Harvard University Press.

Edelman, G., 1987, *Neural Darwinism: The Theory of Neuronal Group Selection*. New York: Basic Books.

Egan, F., 1991, "Must Psychology be Individualistic?," *Philosophical Review* 100:179–203.

Egan, F., 1992, "Individualism, Computation, and Perceptual Content," *Mind* 101:443–459.

Egan, F., 1995, "Computation and Content," *Philosophical Review* 104:181–203.

Egan, F., 1999, "In Defense of Narrow Mindedness," *Mind and Language* 14:177–194.

Emerson, A.E., 1939, "Social Coordination and the Superorganism," *American Midland Naturalist* 21:182–209.

Emerson, A.E., 1942, "Basic Comparisons of Human and Insect Societies," *Biological Symposia* 8:163–176.

Emerson, A.E., 1946, "The Biological Basis of Social Cooperation," *Illinois Academy of Science Transactions* 29:9–18.

Evans, M.A., and H.E. Evans, 1970, *William Morton Wheeler, Biologist*. Cambridge, MA: Harvard University Press.

Fechner, G.T., 1860, *Elemente der Psychophysik*. Leipzig: Breitkopf and Hartel.

Flanagan, O., 1992, *Consciousness Reconsidered*. Cambridge, MA: MIT Press.

Flanagan, O., 2002, *The Problem of the Soul: Two Visions of Mind and How to Reconcile Them*. New York: Basic Books.

Fleck, L., 1935, *Genesis and Development of a Scientific Fact*. Chicago: University of Chicago Press, 1979.

Fodor, J.A., 1974, "Special Sciences." Reprinted in his *Representations*. Brighton, Sussex: Harvester Press, 1981a.

Fodor, J.A., 1975, *The Language of Thought*. Cambridge, MA: Harvard University Press.

Fodor, J.A., 1980, "Methodological Solipsism Considered as a Research Strategy in Cognitive Psychology," *Behavioral and Brain Sciences* 3:63–73. Reprinted in his, *Representations*. Sussex: Harvester Press, 1981a.

Fodor, J.A., 1981a, *Representations*. Sussex: Harvester Press.

Fodor, J.A., 1981b, "The Present Status of the Nativism Debate," in his *Representations*. Sussex: Harvester Press, 1981.

Fodor, J.A., 1982, "Cognitive Science and the Twin-Earth Problem," *Notre Dame Journal of Formal Logic* 23:98–118.

Fodor, J.A., 1983, *The Modularity of Mind*. Cambridge, MA: MIT Press.

Fodor, J.A., 1987, *Psychosemantics: The Problem of Meaning the Philosophy of Mind*. Cambridge, MA: MIT Press.

Fodor, J.A., 1994, *The Elm and the Expert*. Cambridge, MA: MIT Press.

Fodor, J.A., 1998, *Concepts: Where Cognitive Science Went Wrong*. New York: Oxford University Press.

Fodor, J.A., 2000, *The Mind Doesn't Work That Way*. Cambridge, MA: MIT Press.

Foucault, M., 1970, *The Order of Things: An Archaeology of the Human Sciences*. New York: Vintage Books, 1994 edition.

Foucault, M., 1975, *Discipline and Punish: The Birth of the Prison*. New York: Vintage Books, 1995 edition.

Foucault, M., 1978, *The History of Sexuality: Volume 1, An Introduction*. New York: Vintage Books, 1990 edition.

Frege, G., 1892, "On Sense and Reference." Reprinted in P. Geach and M. Black (eds.), *Translations from the Philosophical Writings of Gottlob Frege*. Oxford: Blackwell. 2nd edition, 1960.

Fricker, E., 1991, "Content, Cause, and Function," (Critical Notice of McGinn, *Mental Content*), *Philosophical Books* 32:136–144.

Galton, F., 1869, *Hereditary Genius: An Inquiry into its Laws and Consequences*. London: Macmillan. 2nd edition, 1892.

Galton, F., 1883, *Inquiries into Human Faculty and its Development*. London: Macmillan.

Garber, D., J. Henry, L. Joy, and A. Gabbey, "New Doctrines of Body and its Powers, Place, and Space," in D. Garber and M. Ayers (eds.), *The Cambridge History of Seventeenth-Century Philosophy*. New York: Cambridge University Press.

Garfield, J., (ed.), 1987, *Modularity in Knowledge Representation and Natural-Language Understanding*. Cambridge, MA: MIT Press.

Garfield, J., C.C. Peterson, and T. Perry, 2001, "Social Cognition, Language Acquisition and the Development of the Theory of Mind," *Mind and Language* 16:494–541.

Geertz, C., 1973, *The Interpretation of Cultures*. New York: Basic Books.

Giere, R., 2002, "Scientific Cognition as Distributed Cognition," in P. Carruthers, S. Stich, and M. Siegal (eds.), *The Cognitive Basis of Science*. New York: Cambridge University Press.

Glenberg, A.M., 1997, "What Memory is For," *Behavioral and Brain Sciences* 20: 1–19.
Godfrey-Smith, P., 1996, *Complexity and the Function of Mind in Nature*. New York: Cambridge University Press.
Goodwin, B., 1993, *How the Leopard Got Its Spots*. New York: Simon and Schuster.
Gopnik, A., 1993, "How We Know Our Minds: The Illusion of First-Person Knowledge of Intentionality," *Behavioral and Brain Sciences* 16:1–14.
Gould, S.J., 2002, *The Structure of Evolutionary Theory*. Cambridge, MA: Belknap Press.
Greenfield, P., 2000, "Culture and Universals: Integrating Social and Cognitive Development," in L. Nucci, G. Saxe, and E. Turel (eds.), *Culture, Thought, and Development*. Mahwah, NJ: Lawrence Erlbaum.
Grice, H.P., 1957, "Meaning," *Philosophical Review* 66:377–388.
Griffiths, P.E., 2002, "What is Innateness?," *The Monist* 85:70–85.
Griffiths, P.E., and R. Gray, 1994, "Developmental Systems and Evolutionary Explanation," *Journal of Philosophy*, 91:277–304.
Griffiths, P.E., and R.D. Gray, 2001, "Darwinism and Developmental Systems," in S. Oyama, P.E. Griffiths, and R.D. Gray (eds.), *Cycles of Contingency: Developmental Systems and Evolution*. Cambridge, MA: MIT Press, pp. 195–218.
Grush, R., 2003, "In Defense of Some Cartesian Assumptions Concerning the Brain and its Operation," *Biology and Philosophy* 18:53–93.
Gunther, Y.H., 2003 (ed.), *Essays on Nonconceptual Content*. Cambridge, MA: MIT Press.
Hahn, M., and B. Ramberg (eds.), 2003, *Reflections and Replies: Essays on the Philosophy of Tyler Burge*. Cambridge, MA: MIT Press, 2003.
Harman, G., 1990, "The Intrinsic Quality of Experience" in James Tomberlin (ed.), *Philosophical Perspectives 4: Metaphysics*, Atascadero, CA: Ridgeview, pages 31–52.
Harré, R., 1979, *Social Being*. Oxford: Blackwell.
Harré, R., 1999, "The Rediscovery of the Human Mind: The Discursive Approach," *Asian Journal of Social Psychology* 2:43–62.
Hatfield, G., 1990, *The Natural and the Normative: Theories of Spatial Perception from Kant to Helmholtz*. Cambridge, MA: MIT Press.
Hatfield, G., 2002, "Psychology, Philosophy, and Cognitive Science: Reflections on the History and Philosophy of Experimental Psychology," *Mind and Language* 17:207–232.
Haugeland J., 1995, "Mind Embodied and Embedded." Reprinted in his *Having Thought: Essays in the Metaphysics of Mind*. Cambridge, MA: Harvard University Press, 1998.
Heil, J., and A. Mele (eds.), *Mental Causation*. New York: Oxford University Press.
Hinton, G., J. McClelland, and D. Rumelhart, 1986, "Distributed Representations," in D. Rumelhart, J. McClelland and the PDP Research Group (eds.), *Parallel Distributed Processing: Explorations in the Microstructure of Cognition, Volume 1: Foundations*. Cambridge, MA: MIT Press.
Hirschfeld, L., 1996, *Race in the Making: Cognition, Culture, and the Child's Construction of Human Kinds*. Cambridge, MA: MIT Press.
Hirschfeld, L. and S. Gelman (eds.), 1994, *Mapping the Mind: Domain-Specificity in Cognition and Culture*. New York: Cambridge University Press.

Hobson, R.P., 1999, "The Grounding of Symbols: A Social-Developmental Account," in P. Mitchell and K.J. Riggs (eds.), *Children's Reasoning and the Mind*. New York: Psychology Press, pp. 11–35.

Horgan, T., 1982, "Supervenience and Microphysics," *Pacific Philosophical Quarterly* 63:29–43.

Horgan, T., 1993, "From Supervenience to Superdupervenience: Meeting the Demands of a Material World," *Mind* 102:555–86.

Horgan, T., and J. Tienson, 2002, "The Intentionality of Phenomenology and the Phenomenology of Intentionality," in D.J. Chalmers (ed.), *Philosophy of Mind: Classical and Contemporary Readings*. New York: Oxford University Press.

Horst, S., 1996, *Symbols, Computation, and Intentionality: A Critique of the Computational Theory of Mind*. Berkeley: University of California Press.

Hurley, S., 1998, *Consciousness in Action*. Cambridge, MA: Harvard University Press.

Hurley, S., 2001, "Perception and Action: Alternative Views," *Synthese* 129:3–40.

Hurley, S. and A. Noë, 2003, "Neural Plasticity and Consciousness," *Biology and Philosophy* 18:131–168.

Hutchins, E., 1990, "The Technology of Team Navigation," in J. Galegher, R. Kraut, and C. Egido, (eds.), *Intellectual Teamwork: Social and Technological Foundations of Cooperative Work*. Hillsdale, NJ: Erlbaum.

Hutchins, E., 1995, *Cognition in the Wild*. Cambridge, MA: MIT Press.

Jablonka, E., and R. Lamb, 1995, *Epigenetic Inheritance and Evolution: The Lamarckian Dimension*. Oxford: Oxford University Press.

Jackendoff, R., 1991, "The Problem of Reality," Reprinted in his *Languages of the Mind*. Cambridge, MA: MIT Press, 1992.

Jackendoff, R. J., 2002, *Foundations of Language: Brain, Meaning, Grammar, Evolution*. New York: Oxford University Press.

Jackson, F., 1982, "Epiphenomenal Qualia," *Philosophical Quarterly* 32:127–136.

Jackson, F., and P. Pettit, 1993, "Some Content is Narrow," in J. Heil and A. Mele (eds.), *Mental Causation*. New York: Oxford University Press.

James, W., 1890, *The Principles of Psychology*. New York: Dover, 1950 edition.

Janis, I., 1982, *Groupthink: Psychological Studies of Policy Decisions and Fiascoes*. Boston: Houghton Mifflin. 2nd edition.

Jeannerod, M., 1994, "The Representing Brain: Neural Correlates of Motor Intention and Imagery," *Behavioral and Brain Sciences* 17:187–245.

Karmiloff-Smith, A., 1992, *Beyond Modularity: A Developmental Perspective on Cognitive Science*. Cambridge, MA: MIT Press.

Keil, F.C., 1989, *Concepts, Kinds and Cognitive Development*. Cambridge, MA: MIT Press.

Keil, F.C., 1995, "The Growth of Causal Understandings of Natural Kinds," in D. Sperber, D. Premack, and A. Premack (eds.), *Causal Cognition: A Multidisciplinary Debate*. Oxford: Oxford University Press.

Keil, F.C., 1998, "Cognitive Science and the Origins of Thought and Knowledge," in W. Damon and R.M. Lerner (eds.), *Handbook of Child Psychology, Volume 1: Theoretical Models of Human Development*. New York: Wiley, 5th edition, pp. 341–413.

Keil, F.C., 1999, "Nativism," in R.A. Wilson and F.C. Keil (eds.), *The MIT Encyclopedia of the Cognitive Sciences*. Cambridge, MA: MIT Press, pp. 583–586.

Keil, F.C., 2000, "Nurturing Nativism," *A Field Guide to the Philosophy of Mind: Book Symposia*. Fall 1999–2000. E-journal at: http://host.uniroma3.it/progetti/kant/field/cowiesymp.htm.

Keil, F.C., and R.A. Wilson (eds.), 2000, *Explanation and Cognition*. Cambridge, MA: MIT Press.

Keller, E.F., 2000, *The Century of the Gene*. Cambridge, MA: Harvard University Press.

Khalidi, M., 2001, "Innateness and Domain-Specificity," *Philosophical Studies* 105:191–210.

Khalidi, M., 2002, "Nature and Nurture in Cognition," *British Journal for the Philosophy of Science* 53:251–272.

Kim, J., 1989, "The Myth of Nonreductive Materialism." Reprinted in his 1993a.

Kim, J., 1992, "Multiple Realization and the Metaphysics of Reduction." Reprinted in his 1993a.

Kim, J., 1993a, *Supervenience and Mind*. New York: Cambridge University Press.

Kim, J., 1993b, "The Nonreductivist's Troubles with Mental Causation." Reprinted in his 1993a.

Kim, J., 1993c, "Postcripts on Supervenience" in his 1993a.

Klatzky, R., 1999, "Haptic Perception," in R.A. Wilson and F.C. Keil (eds.), *The MIT Encyclopedia of the Cognitive Sciences*. Cambridge, MA: MIT Press, 1999, pp. 359–360.

Kohler, I., 1964, *The Formation and Transformation of the Perceptual World*. New York: International University Press.

Koriat, A., and M. Goldsmith, 1996, "Memory Metaphors and the Real-Life/Laboratory Controversy: Correspondence versus Storehouse Conceptions of Memory," *Behavioral and Brain Sciences* 19:167–188.

Kripke, S., 1980, *Naming and Necessity*. Cambridge, MA: Harvard University Press.

Kuhn, T., 1979, Foreword to L. Fleck's *Genesis and Development of a Scientific Fact*. Chicago: University of Chicago Press.

Kunda, Z., 1999, *Social Cognition*. Cambridge, MA: MIT Press.

Kuper, A., 1999, *Culture: The Anthropologists' Account*. Cambridge, MA: Harvard University Press.

Lakoff, G., and M. Johnson, 1999, *Philosophy in the Flesh: The Embodied Mind and its Challenge to Western Thought*. New York: Basic Books.

Laurence, S., and E. Margolis, 2001, "The Poverty of the Stimulus Argument," *British Journal for the Philosophy of Science* 52:217–276.

Laurence, S., and E. Margolis, 2002, "Radical Concept Nativism," *Cognition* 86: 25–55.

Le Bon, G., 1895, *The Crowd: A Study of the Popular Mind*. 2nd. edition, Dunwoody, GA: Norman S. Berg, 1968.

LeDoux, J.E., 1996, *The Emotional Brain*. New York: Simon and Schuster.

LeDoux, J.E., and M. Rogan, 1999, "Emotion and the Animal Brain," in R.A. Wilson and F.C. Keil (eds.), *The MIT Encyclopedia of the Cognitive Sciences*. Cambridge, MA: MIT Press. 1999, pp. 268–270.

Leslie, A., 1987, "Pretense and Representation: The Origins of 'Theory of Mind,'" *Psychological Review* 94:412–426.
Levine, J., 1983, "Materialism and Qualia: The Explanatory Gap," *Pacific Philosophical Quarterly* 64:354–361.
Lewis, D., 1972, *We the Navigators*. Honolulu: University of Hawaii Press.
Lewis, D., 1978, *The Voyaging Stars: Secrets of Pacific Island Navigators*. New York: Norton.
Lewis, D.K., 1968, *Convention: A Philosophical Study*. Oxford: Blackwell Press.
Lewis, D.K., 1972, "Psychophysical and Theoretical Identifications," *Australasian Journal of Philosophy* 50:291–315.
Lewis, D.K., 1986, "Introduction" to his *Philosophical Papers, Volume II*. New York: Oxford University Press.
Lewis, D.K., 1994, "David Lewis: Reduction of Mind," in S. Guttenplan (ed.), *A Companion to the Philosophy of Mind*. Oxford: Blackwell.
Lewis, D.K., 1997, "Finkish Dispositions," *Philosophical Quarterly* 47:143–158.
Lewontin, R., 1991, *Biology as Ideology*. New York: Harper Collins. 1993 edition.
Lewontin, R., 2000, *It Ain't Necessarily So: The Dream of the Human Genome and Other Illusions*. New York: New York Review of Books.
Little, D., 1991, *Varieties of Social Explanation*. Boulder, CO: Westview Press.
Lloyd, E.A., 2001, "Units and Levels of Selection: An Anatomy of the Units of Selection Debates," in R.S. Singh, C.B. Krimbas, D.B. Paul, and J. Beatty (eds.), *Thinking About Evolution: Historical, Philosophical, and Political Perspectives*. New York: Cambridge University Press.
Loar, B., 1987, "Subjective Intentionality," *Philosophical Topics*, Spring 1987:89–124.
Loar, B., 1988, "Social Content and Psychological Content," in R. Grimm and D. Merrill (eds.), *Contents of Thought*. Tucson: University of Arizona Press.
Loar, B., 2003a, "Transparent Experience and the Availability of Qualia," in Q. Smith and A. Jokic (eds.), *Consciousness: New Philosophical Perspectives*. New York: Oxford University Press.
Loar, B., 2003b, "Phenomenal Intentionality as the Basis of Mental Content," in M. Hahn and B. Ramberg (eds.), *Reflections and Replies: Essays on the Philosophy of Tyler Burge*. Cambridge, MA: MIT Press.
Locke, J., 1690, *An Essay Concerning Human Understanding*. Oxford: Clarendon Press, 1979 edition.
Ludlow, P., and N. Martin (eds.), 1998, *Externalism and Self-Knowledge*. Palo Alto, CA: CSLI Publications.
Luria, A., 1979, *The Making of Mind*. Cambridge, MA: Harvard University Press.
Lutz, D.R., and F.C. Keil, 2000, "Early Understanding of the Division of Cognitive Labor," *Child Development* 73:1073–1084.
Lycan, W., 1987, *Consciousness*. Cambridge, MA: MIT Press.
Lycan, W., 1996, *Consciousness and Experience*. Cambridge, MA: MIT Press.
Lycan, W., 2001, "The Case for Phenomenal Externalism," in J. Tomberlin (ed.), *Philosophical Perspectives 15: Metaphysics*. Boston: Blackwell.
MacKay, C., 1841, *Memoirs of Extraordinary Popular Delusions and the Madness of Crowds*. Boston: L.C. Page, 1932 edition.

MacKay, D.M., 1967, "Ways of Looking at Perception," in W. Wathen-Dunn (ed.), *Models for the Perception of Speech and Visual Form.* Cambridge, MA: MIT Press.

Marcus, G., 2001, *The Algebraic Mind: Integrating Connectionism and Cognitive Science.* Cambridge, MA: MIT Press.

Marr, D., 1982, *Vision: A Computational Investigation into the Human Representation and Processing of Visual Information.* San Francisco, CA: W.H. Freeman.

Marx, K., 1867, *Capital.* New York: International Publishers, 1967.

Matthews, R., 1988, "Comments on Burge," in R. Grimm and D. Merrill (eds.), *Contents of Thought.* Tucson: University of Arizona Press.

McClamrock, R., 1995, *Existential Cognition: Computational Minds in the World.* Chicago: University of Chicago Press.

McCulloch, G., 2003, *The Life of the Mind: An Essay on Phenomenological Externalism.* New York: Routledge.

McDougall, W., 1908, *An Introduction to Social Psychology.* Boston: John W. Luce. 2nd edition, 1909.

McDougall, W., 1920, *The Group Mind.* New York: Putnam.

McDowell, J., 1994, *Mind and World.* Cambridge, MA: Harvard University Press.

McGinn, C., 1989, *Mental Content.* New York: Basil Blackwell.

McGinn, C., 1991, "Conceptual Causation: Some Elementary Reflections," *Mind* 100:573–586.

McKitrick, J., 2003, "A Case for Extrinsic Dispositions," *Australasian Journal of Philosophy* 81:155–174.

McPhail, C., 1991, *The Myth of the Madding Crowd.* New York: Aldine de Gruyter.

Mill, J.S., 1843, *A System of Logic, Ratiocinative and Inductive, Volume 1.* London: John W. Parker, 2nd edition, 1846.

Miller, G.A., 1984, "Informavores," in F. Machlup and U. Mansfield (eds.), *The Study of Information: Interdisciplinary Messages.* New York: Wiley.

Millikan, R.G., 1984, *Language, Thought, and Other Biological Categories: New Foundations for Realism.* Cambridge, MA: MIT Press.

Millikan, R.G., 1993, *White Queen Psychology and Other Essays for Alice.* Cambridge, MA: MIT Press.

Milner, A.D., and M.A. Goodale, 1998a, *The Visual Brain in Action.* New York: Oxford University Press.

Milner, A.D. and M.A. Goodale, 1998b, "The Visual Brain in Action," *Psyche: An Interdisciplinary Journal of Research on Consciousness*, Reprinted in A. Noë and E. Thompson (eds.), *Vision and Mind: Selected Readings in the Philosophy of Perception.* Cambridge, MA: MIT Press, 2003, pp. 515–529.

Mitman, G., 1992, *The State of Nature: Ecology, Community, and American Social Thought, 1900–1950.* Chicago: University of Chicago Press.

Moss, L., 2003, *What Genes Can't Do.* Cambridge, MA: MIT Press.

Murphy, G., 2001, *The Big Book of Concepts.* Cambridge, MA: MIT Press.

Nagel, T., 1974, "What is it Like to be a Bat?," *Philosophical Review* 83:435–450.

Nebergall, W.H., H. Holtzclaw, and W. Robinson, 1980, *General Chemistry.* Lexington, MA: D.C. Heath and Co., 6th edition.

Neisser, U., 1978, "Memory: What Are the Important Questions?." Reprinted in U. Neisser (ed.), *Memory Observed: Remembering in Natural Contexts.* San Francisco: W.H. Freeman, 1981.

Neisser, U., 1981 (ed.), *Memory Observed: Remembering in Natural Contexts*. San Francisco: W.H. Freeman, 1981.
Neisser, U., 1988, "The Ecological Approach to Perception and Memory," *New Trends in Experimental and Clinical Psychiatry* 4:153–166.
Neisser, U., 1996, "Remembering as Doing," *Behavioral and Brain Sciences* 19:203–204.
Newell, A., 1982, "The Knowledge Level," *Artificial Intelligence* 18:87–127.
Noë, A., 2001, "Experience and the Active Mind," *Synthese* 129:41–60.
Noë, A., 2002a, "On What We See," *Pacific Philosophical Quarterly* 83:57–80.
Noë, A., 2002b, "Is Perspectival Self-Consciousness Non-Conceptual?," *Philosophical Quarterly* 52:185–194.
Noë, A., in press, *Action in Perception*. Cambridge, MA: MIT Press.
Norman, D., 1999, *The Invisible Computer*. Cambridge, MA: MIT Press.
Nye, R.A., 1975, *The Origins of Crowd Psychology: Gustave LeBon and the Crisis of Mass Democracy in the Third Republic*. Beverly Hills, CA: Sage Publications.
O'Regan, J.K., and A. Noë, 2001a, "What it is Like to See: A Sensorimotor Theory of Perceptual Experience," *Synthese* 129:79–103.
O'Regan, J.K., and A. Noë, 2001b, "A Sensorimotor Account of Vision and Visual Consciousness," *Behavioral and Brain Sciences* 24:939–1031.
Olson, D., 1994, *The World on Paper*. New York: Cambridge University Press.
Olson, D., 1999, "Literacy," in R.A. Wilson and F.C. Keil (eds.), *The MIT Encyclopedia of the Cognitive Sciences*. Cambridge, MA: MIT Press, pp. 481–482.
Owens, J., 1993, "Content, Causation, and Psychophysical Supervenience," *Philosophy of Science* 60:242–261.
Oyama, S., 1985, *The Ontogeny of Information*. 2nd edition, Durham, NC: Duke University Press, 2000.
Oyama, S., 2000, *Evolution's Eye: A Systems View of the Biology-Culture Divide*. Durham, NC: Duke University Press.
Oyama, S., P.E. Griffiths, and R.D. Gray (eds.), 2001, *Cycles of Contingency: Developmental Systems and Evolution*. Cambridge, MA: MIT Press.
Palmer, S., 1999, *Vision Science*. Cambridge, MA: MIT Press.
Park, T., 1932, "Studies in Population Physiology: The Relation of Numbers to Initial Population Growth in the Flour Beetle *Tribolium Confusum* Duval," *Ecology* 13:172–181.
Park, T., 1933, "Studies in Population Physiology. II. Factors Regulating Initial Growth of *Tribolium Confusum* Populations," *Journal of Experimental Zoology* 65:17–42.
Pepperberg, I., 1990, "Some Cognitive Capacities of an African Grey Parrot (*Psittacus erithacus*)," *Advances in the Study of Behavior* 19:357–409.
Pepperberg, I., 1999, *The Alex Studies: Cognitive and Communicative Abilities of Grey Parrots*. Cambridge, MA: Harvard University Press.
Perner, J., 1991, *Understanding the Representational Mind*. Cambridge, MA: MIT Press.
Pettit, P., 1993, *The Common Mind: An Essay on Psychology, Society, and Politics*. New York: Oxford University Press.
Pinker, S., 1994, *The Language Instinct: How the Mind Creates Language*. New York: Morrow.

Pinker, S., 1997, *How the Mind Works*. New York: Norton.
Pinker, S., 1999, *Words and Rules: The Ingredients of Language*. New York: Harper Collins.
Pinker, S., 2001, "Four Decades of Rules and Associations, or Whatever Happened to the Past Tense Debate?," in E. Dupoux (ed.), *Language, Brain, and Cognitive Development: Essays in Honor of Jacques Mehler*. Cambridge, MA: MIT Press.
Pinker, S., 2002, *The Blank Slate: The Modern Denial of Human Nature*. New York: Viking.
Poland, J., 1994, *Physicalism: The Philosophical Foundations*. New York: Oxford University Press.
Polger, T., in press, "Neural Machinery and Realization," *Philosophy of Science*.
Popper, K., 1945, *The Open Society and its Enemies*. Princeton, NJ: Princeton University Press. 2nd edition, 1952.
Prins, H.H.T., 1996, *Ecology and Behaviour of the African Buffalo*. London: Chapman and Hall.
Prior, E., 1985, *Dispositions*. Aberdeen: Aberdeen University Press.
Putnam, H., 1960, "Minds and Machines." Reprinted in his *Mind, Language, and Reality*. New York: Cambridge University Press, 1975.
Putnam, H., 1967a, "The Mental Life of Some Machines." Reprinted in his *Mind, Language, and Reality: Philosophical Papers, Volume 2*. New York: Cambridge University Press, 1975.
Putnam, H., 1967b, "The Nature of Mental States." Reprinted in his *Mind, Language, and Reality: Philosophical Papers, Volume 2:* New York: Cambridge University Press, 1975.
Putnam, H., 1975, "The Meaning of 'Meaning,'" in K. Gunderson (ed.), *Language, Mind and Knowledge*. Minneapolis: University of Minnesota Press. Reprinted in H. Putnam, *Mind, Language, and Reality: Philosophical Papers Volume 2*. New York: Cambridge University Press.
Pylyshyn, Z., 1984, *Computation and Cognition*. Cambridge, MA: MIT Press.
Quartz, S., 2003, "Innateness and the Brain," *Biology and Philosophy* 18:13–40.
Quartz, S., and T. Sejnowksi, 1997, "The Neural Basis of Cognitive Development: A Constructivist Manifesto," *Behavioral and Brain Sciences* 20:537–596.
Reed, E.S., 1996, *Encountering the World: Toward an Ecological Psychology*. New York: Oxford University Press.
Reed, E.S., 1997, *From Soul to Mind: The Emergence of Psychology from Erasmus Darwin to William James*. New Haven, CT: Yale University Press.
Reisberg, D., 1999, "Similarity," in R.A. Wilson and F.C. Keil (eds.), *The MIT Encyclopedia of the Cognitive Sciences*. Cambridge, MA: MIT Press, pp. 763–765.
Rensink, R.A., J.K. O'Regan and J.J. Clark, 2000, "On the Failure to Detect Changes in Scenes Across Brief Interruptions," *Visual Cognition* 7:127–146.
Richards, W., 1999, "Marr, David," in R.A. Wilson and F.C. Keil (eds.), *The MIT Encyclopedia of the Cognitive Sciences*. Cambridge, MA: MIT Press, pp. 511–512.
Roediger, H.L., 1980, "Memory Metaphors in Cognitive Psychology," *Memory and Cognition* 8:231–246.
Ross, E.A., 1908, *Social Psychology: An Outline and Source Book*. New York: Macmillan.
Rowlands, M., 1999, *The Body in Mind*. New York: Cambridge University Press.

Rozenblit, L., and F. Keil, 2002, "The Misunderstood Limits of Folk Science: An Illusion of Explanatory Depth," *Cognitive Science* 92:1–42.

Rumelhart, D., and J. McClelland, 1986, "On Learning the Past Tenses of English Verbs. Implicit Rules or Parallel Distributed Processing?," in D. Rumelhart, J. McClelland, and the PDP Research Group, *Parallel Distributed Processing: Explorations in the Microstructure of Cognition, Volume 1: Foundations.* Cambridge, MA: MIT Press.

Rumelhart, D., J. McClelland, et al., 1986, *Parallel Distributed Processing, Volume 1: Foundations.* Cambridge, MA: MIT Press.

Runeson, S., 1977, "On the Possibility of 'Smart' Perceptual Mechanisms," *Scandinavian Journal of Psychology* 18:172–179.

Sady, W., 2001, "Ludwik Fleck – Thought Collectives and Thought Styles," in W. Krajewski (ed.), *Polish Philosophers of Science and Nature in the 20th Century.* Poznan Studies in the Philosophy of the Sciences and the Humanities, Volume 74. New York: Rodopi.

Salmon, N., 1981, *Reference and Essence.* Princeton, NJ: Princeton University Press.

Samuels, R., 2002, "Nativism in Cognitive Science," *Mind and Language* 17:233–265.

Savage-Rumbaugh, S., and R. Lewin, 1994, *Kanzi: The Ape at the Brink of the Human Mind.* New York: Wiley and Sons.

Savage-Rumbaugh, S., S.G. Shanker, and T.J. Taylor, 1998, *Apes, Language, and the Human Mind.* New York: Oxford University Press.

Schank, R., and P. Abelson, 1977, *Scripts, Plans, Goals, and Understanding.* Hillsdale, NJ: Lawrence Erlbaum.

Schiller, J., 1968, "Physiology's Struggle for Independence in the First Half of the Nineteenth Century," *History of Science* 7:64–89.

Searle, J.R., 1990, "Consciousness, Explanatory Inversion, and Cognitive Science," *Behavioral and Brain Sciences* 13:586–642.

Searle, J.R., 1992, *The Rediscovery of the Mind.* Cambridge, MA: MIT Press.

Seeley, D., 1995, *The Wisdom of the Hive: The Social Physiology of Honey Bee Colonies.* Cambridge, MA: Harvard University Press.

Segal, G., 1989, "Seeing What is Not There," *Philosophical Review* 98:189–214.

Segal, G., 1991, "Defence of a Reasonable Individualism," *Mind* 100:485–494.

Segal, G., 1997, Review of R.A. Wilson, "Cartesian Psychology and Physical Minds," *British Journal for the Philosophy of Science* 48:151–156.

Segal, G., 2000, *A Slim Book About Narrow Content.* Cambridge, MA: MIT Press.

Sellars, W., 1956, "Empiricism and the Philosophy of Mind," *Minnesota Studies in the Philosophy of Science, Volume 1.* Minneapolis: University of Minnesota Press, pages 253–329.

Shapiro, L., 1993, "Content, Kinds, and Individualism in Marr's Theory of Vision," *Philosophical Review* 102:489–513.

Shapiro, L., 1997, "A Clearer Vision," *Philosophy of Science* 64:131–153.

Shapiro, L., 2000, "Multiple Realizations," *Journal of Philosophy* 97:635–654.

Shapiro, L., 2004, *The Mind Incarnate.* Cambridge, MA: MIT Press.

Shoemaker, S., 1975, "Functionalism and Qualia," *Philosophical Studies* 27:291–315. Reprinted in his *Identity, Cause, and Mind.* New York: Cambridge University Press, 1984.

Shoemaker, S., 1981a, "Some Varieties of Functionalism," Reprinted in his *Identity, Cause, and Mind*. New York: Cambridge University Press, 1984.

Shoemaker, S., 1981b, "The Inverted Spectrum," *Journal of Philosophy* 74:357–381. Reprinted in his *Identity, Cause, and Mind*. New York: Cambridge University Press, 1984.

Shoemaker, S., 1984, *Identity, Cause, and Mind*. New York: Cambridge University Press.

Shoemaker, S., 1994, "The First-Person Perspective," *Proceedings and Addresses of the American Philosophical Association* 68:7–22. Reprinted in his *The First-Person Perspective and Other Essays*. New York: Cambridge University Press, 1996.

Shoemaker, S., 2000, "Realization and Mental Causation," Reprinted in C. Gillett and B. Loewer (eds.), *Physicalism and its Discontents*. New York: Cambridge University Press.

Shore, B., 1996, *Culture in Mind*. New York: Oxford University Press.

Shweder, R., 1991, "Cultural Psychology: What Is It?," in his *Thinking Through Cultures: Expeditions in Cultural Psychology*. Cambridge, MA: Harvard University Press.

Sidelle, A., 1989, *Necessity, Essence, and Individuation: A Defense of Conventionalism*. Ithaca, NY: Cornell University Press.

Simpson, J., and D. Kendrick (eds.), *Evolutionary Social Psychology*. Hillsdale, NJ: Erlbaum.

Skinner, B.F., 1957, *Verbal Behavior*. New York: Appleton-Century-Crofts.

Smith, Q., and A. Jokic (eds.), *Consciousness: New Philosophical Perspectives*. New York: Oxford University Press, 2003.

Smith, R., 1997, *The Norton History of the Human Sciences*. New York: W.W. Norton.

Smolensky, P., 1988, "On the Proper Treatment of Connectionism," *Behavioral and Brain Sciences* 11:1–74.

Sober, E., 1991, "Models of Cultural Evolution," in P. Griffiths (ed.), *Trees of Life*. Cambridge, MA: MIT Press.

Sober, E., and D.S. Wilson, 1998, *Unto Others: The Evolution and Psychology of Unselfish Behavior*. Cambridge, MA: Harvard University Press.

Spelke, E., 1990, "Principles of Object Perception," *Cognitive Science* 14:29–56.

Spelke, E., 1995, "Initial Knowledge: Six Suggestions," *Cognition* 50:433–447.

Spelke, E., and E. Newport, 1998, "Nativism, Empiricism, and the Development of Knowledge," in W. Damon and R.M. Lerner (eds.), *Handbook of Child Psychology, Volume 1: Theoretical Models of Human Development*. New York: Wiley, 5th edition, pp. 275–340.

Sterelny, K., 1990, *The Representational Theory of Mind: An Introduction*. New York: Blackwell.

Stich, S., 1975, "Introduction," in S. Stich (ed.), *Innateness*. Berkeley: University of California Press.

Stich, S., 1978, "Autonomous Psychology and the Belief-Desire Thesis," *Monist* 61:573–591.

Stich, S., 1983, *From Folk Psychology to Cognitive Science: The Case Against Belief*. Cambridge, MA: MIT Press.

Strawson, G., 1994, *Mental Reality*. Cambridge, MA: MIT Press.

Stryker, S., 1987, "The Vitalization of Social Interaction," *Social Psychology Quarterly* 50:83–94.
Taylor, J.G., 1962, *The Behavioral Basis of Perception*. New Haven, CT: Yale University Press.
Teller, P., 1986, "Relational Holism and Quantum Mechanics," *British Journal for the Philosophy of Science* 37:71–81.
Teller, P., 1989, "Relativity, Relational Holism, and the Bell Inequalities," in J.T. Cushing and E. McMullin (eds.), *Philosophical Consequences of Quantum Theory*. Notre Dame, IN: University of Notre Dame Press.
Tomasello, M., 1999, *The Cultural Origins of Human Cognition*. Cambridge, MA: Harvard University Press.
Tomasello, M., and H. Rakoczy, 2003, "What Makes Human Cognition Unique? From Individual to Shared to Collective Intentionality," *Mind and Language* 18:121–147.
Tooby, J., and L. Cosmides, 1992, "The Psychological Foundations of Culture," in J. Barkow, L. Cosmides, and J. Tooby (eds), *The Adapted Mind: Evolutionary Psychology and the Generation of Culture*. New York: Oxford University Press.
Trotter, W., 1916, *Instincts of the Herd in War and Peace*. London: Fisher Unwin.
Turner, J.C., 1987, *Rediscovering the Social Group: A Self-Categorization Theory*. Oxford: Basil Blackwell.
Tye, M., 1995, *Ten Problems of Consciousness*. Cambridge, MA: MIT Press.
Tye, M., 2000, *Consciousness, Color, and Content*. Cambridge, MA: MIT Press.
Ullman, S., 1979, *The Interpretation of Visual Motion*. Cambridge, MA: MIT Press.
Ungeleider, L.G., and M. Mishkin, 1982, "Two Cortical Visual Systems," in D.J. Ingle, M.A. Goodale, and R.J.W. Mansfield (eds.), *Analysis of Visual Behavior*. Cambridge, MA: MIT Press.
van Gelder, T., 1999, "Dynamic Approaches to Cognition," in R.A. Wilson and F.C. Keil (eds.), *The MIT Encyclopedia of the Cognitive Sciences*. Cambridge, MA: MIT Press.
van Gelder, T., and R. Port (eds.), 1995, *Mind as Motion*. Cambridge, MA: MIT Press, pp. 244–246.
van Ginneken, J., 1992, *Crowds, Psychology, and Politics 1871–1899*. New York: Cambridge University Press.
van Gulick, R., 1989, "Metaphysical Arguments for Internalism and Why They Don't Work," in S. Silvers (ed.), *Rerepresentation*. Dordrecht, The Netherlands: Kluwer.
Vygotsky, L., 1962, *Thought and Language*. Cambridge, MA: MIT Press.
Vygotsky, L., 1978, *Mind and Society: The Development of Higher Psychological Processes*. Cambridge, MA: Harvard University Press.
Vygotsky, L., 1981, "The Genesis of Higher Mental Functions," in James Wertsch (ed.), *The Concept of Activity in Soviet Psychology*. Armonk, NY: M.E. Sharpe.
Walsh, D.M., 1998, "Wide Content Individualism," *Mind* 107:625–651.
Walsh, D.M., 1999, "Alternative Individualism," *Philosophy of Science* 66:628–648.
Watkins, J.W.N., 1957, "Historical Explanation in the Social Sciences," *British Journal for the Philosophy of Science* 8:104–117.
Webster, G., and B. Goodwin, 1996, *Form and Transformation: Generative and Relational Principles in Biology*. New York: Cambridge University Press.

Wellman, H., 1990, *The Child's Theory of Mind*. Cambridge, MA: MIT Press.
Wertsch, J., 1985, *Vygotsky and the Social Formation of Mind*. Cambridge, MA: Harvard University Press.
Wheeler, W.M., 1910, *Ants: Their Structure, Development, and Behavior*. New York: Columbia University Press.
Wheeler, W.M., 1911, "The Ant-Colony as an Organism." Reprinted in his *Essays in Philosophical Biology*. Cambridge, MA: Harvard University Press, 1939.
Wheeler, W.M., 1920, "The Termitadoxa, or Biology and Society." Reprinted in his *Essays in Philosophical Biology*. Cambridge, MA: Harvard University Press, 1939.
Wheeler, W.M., 1923, *Social Life Among the Insects*. New York: Harcourt Brace.
Wheeler, W.M., 1926, "Emergent Evolution and the Development of Societies." Modified version reprinted in his *Essays in Philosophical Biology*. Cambridge, MA: Harvard University Press, 1939.
Wheeler, W.M., 1928, *The Social Insects*. New York: Harcourt, Brace, and Co.
Wheeler, W.M., 1939, *Essays in Philosophical Biology*. Cambridge, MA: Harvard University Press.
White, S., 1982, "Partial Character and the Language of Thought," *Pacific Philosophical Quarterly* 63:347–365. Reprinted in his *The Unity of the Self*. Cambridge, MA: MIT Press, 1991.
White, S., 1991, *The Unity of the Self*. Cambridge, MA: MIT Press.
White, S., 1994, "Color and Notional Content," *Philosophical Topics* 22:471–503.
Whiten, A., and R.W. Byrne (eds.), 1997, *Machiavellian Intelligence II: Extensions and Evaluations*. New York: Cambridge University Press.
Williams, G.C., 1966, *Adaptation and Natural Selection*. Princeton, NJ: Princeton University Press.
Wilson, D.S., 1975, "A Theory of Group Selection," *Proceedings of the National Academy of Sciences USA* 72:143–146.
Wilson, D.S., 1989, "Levels of Selection: An Alternative to Individualism in the Human Sciences," *Social Networks* 11:259–272.
Wilson, D.S., 1991, "On the Relationship Between Evolutionary and Psychological Definitions of Altruism and Egoism," *Biology and Philosophy* 7:61–68.
Wilson, D.S., 1983, "The Group Selection Controversy: History and Current Status," *Annual Review of Ecology and Systematics* 14:159–187.
Wilson, D.S., 1997a, "Altruism and Organism: Disentangling the Themes of Multilevel Selection Theory," *American Naturalist* 150 (supp.):S122-S134.
Wilson, D.S., 1997b, "Incorporating Group Selection into the Adaptationist Program: A Case Study Involving Human Decision Making," in J. Simpson and D. Kendrick (eds.), *Evolutionary Social Psychology*. Hillsdale, NJ: Erlbaum.
Wilson, D.S., 2002, *Darwin's Cathedral: Evolution, Religion, and the Nature of Society*. Chicago: University of Chicago Press.
Wilson, D.S., and E. Sober, 1994, "Reintroducing Group Selection to the Human Behavioral Sciences," *Behavioral and Brain Sciences* 17:585–654.
Wilson, E.O., 1971, *The Insect Societies*. Cambridge, MA: Belknap Press.
Wilson, E.O., 1975, *Sociobiology: The New Synthesis*. Cambridge, MA: Harvard University Press. 2000 anniversary edition.

Wilson, R.A., 1995, *Cartesian Psychology and Physical Minds: Individualism and the Sciences of the Mind.* New York: Cambridge University Press.

Wilson, R.A., 1999, "The Individual in Biology and Psychology," in V. Hardcastle (ed.), *Biology Meets Psychology: Philosophical Essays.* Cambridge, MA: MIT Press.

Wilson, R.A., 2000a, "The Mind Beyond Itself," in D. Sperber (ed.), *Metarepresentations: A Multidisciplinary Perspective.* New York: Oxford University Press, pp. 31–52.

Wilson, R.A., 2000b, "Some Problems for 'Alternative Individualism,'" *Philosophy of Science* 67:671–679.

Wilson, R.A., 2002, "Locke's Primary Qualities," *Journal of the History of Philosophy* 40:201–228.

Wilson, R.A., 2003, "Individualism" in S. Stich and T.A. Warfield (eds.), *The Blackwell Companion to Philosophy of Mind.* New York: Blackwell, pp. 256–287.

Wilson, R.A., 2004, *Genes and the Agents of Life.* New York: Cambridge University Press.

Wilson, R.A., in press a, "What Computations (Still, Still) Can't Do: Jerry Fodor on Computation and Modularity," *Canadian Journal of Philosophy.*

Wilson, R.A., in press b, "Realization: Metaphysics, Mind, and Science," *Philosophy of Science.*

Wilson, R.A., and F.C. Keil, 1998, "The Shadows and Shallows of Explanation," *Minds and Machines* 8:137–159. Modified version reprinted in F.C. Keil and R.A. Wilson, *Explanation and Cognition.* Cambridge, MA: MIT Press, 2000.

Wilson, R.A., and F.C. Keil (eds.), 1999, *The MIT Encyclopedia of the Cognitive Sciences.* Cambridge, MA: MIT Press.

Wimsatt, W.C., 1999, "Generativity, Entrenchment, Evolution, and Innateness: Philosophy, Evolutionary Biology, and Conceptual Foundations of Science," in V.G. Hardcastle (ed.), *Where Biology Meets Psychology: Philosophical Essays.* Cambridge, MA: MIT Press, pp. 139–179.

Wundt, W., 1862, *Beiträge zur Theorie der Sinneswahrnehmung.* Leipzig/Heidelberg: C.F. Winter.

Wundt, W., 1863, *Vorlesungen über die Menschen- und Tierseele.* Leipzig: Voss.

Wundt, W., 1874, *Grundzüge der physiologischen Psychologie.* Leipzig: W. Engelmann.

Wundt, W., 1900–1920, *Völkerpsychologie: Eine Untersuchung der Entwicklungsgesetzen, Mythus und Sitte.* 10 volumes. Leipzig: Engelmann.

Wundt, W., 1907, *Outlines of Psychology.* Translated by C.H. Judd. Leipzig: Wilhelm Engelman, 3rd edition.

Wynn, K., 1990, "Children's Understanding of Counting," *Cognition* 36:155–193.

Wynn, K., 1995, "Origins of Numerical Knowledge," *Mathematical Cognition* 1:35–60.

Yablo, S., 1992, "Mental Causation," *Philosophical Review* 101:245–280.

Index

007 Principle, 113, 176

Abelson, Peter, 317
adaptive decision-making units, 295
Akins, Kathleen, 324
Alex, 195
Allee, W.C., 284, 332
Allport, Gordon, 312, 331
altruism, 6–7, 276, 277, 303
Amsterdam (McEwan), 208
anatomy, 32
Andreski, S., 312
animal communities, 26, 275, 283–4
animate vision, 176–7, 179, 233
anthropology, 5, 19–22, 28, 266, 273; cognitive, 13
Ariew, Andre, 315
Aristotle, 83
Armstrong, David, 214, 320
Aronson, E., 312, 331
artificial intelligence, 8, 55, 88, 92, 95, 243
associationism, 199
associative learning, 55
attention, 215, 217, 221, 223, 251
autonomy, 144, 186, 209
awareness, processes of, 215–21, 225, 294; attention as, 215, 217, 221, 223, 251; embeddedness/embodiment and, 220, 221; higher cognition as, 186–7, 215, 217, 221, 223, 224; introspection as, 215, 217, 221, 223, 294; scaffolding for, 217, 218–19, 221; spatiality of, 218; temporality of, 217–18, 220, 221–2; TESEE and, 217, 220, 223, 225, 232
Ayers, Michael, 310

Bach-y-Rita, Paul, 236, 328
background conditions, 109–11, 116, 121, 126, 130–3
Baillargeon, Rene, 325
Baldwin, J.D., 312
Ballard, Dana, 176–7, 179, 233
Barkow, Jerome, 310, 334
Baron, R.M., 334
Baron-Cohen, Simon, 59, 65, 146, 313, 325
Barrows, Susanna, 272, 331
Bartlett, Frederic C., 191, 197
Bateson, Patrick, 69, 71–2
behaviorists and behaviorism, 16, 27, 41, 54–5, 57, 140, 279
beliefs, 91, 93, 141, 186, 206–7, 242, 245, 250, 289
Berkeley, George, 45
Bernard, Claude, 32
binding problem, 218

355

biological sciences, 4–7, 8–9; dispositions in, 126; group mind hypothesis and, 26, 286–7, 298; individualism in, 11–12; internal richness/external minimalism theses and, 69–72; nativism in, 17–19, 68–72; psychology and, 28–30; realization and, 106, 114–16; superorganism tradition and, 274. *See also* natural selection
Block, Ned, 237, 240, 261, 319, 320, 321, 330
Bloom, Paul, 314
Blumenthal, Arthur, 311
bodily sensations, 215, 225–6, 227, 230, 238, 242
bodily skills, 186
Boyd, Richard, 104, 319, 320, 333
Boyd, Robert, 334
Boyle, Robert, 22–3
brain-in-a-vat thought experiments, 220, 253, 256–8, 259
brain-in-the-machine, 4
brain states, 95, 100, 101
Brandom, Robert, 90
Broca, Paul, 32
Brooks, Rodney, 177, 179
Bruner, Jerome, 49
Burge, Tyler, 99, 144, 316, 317, 323; on content, 90–3, 151, 155, 172; externalism of, 78, 86–7, 89, 156, 204, 254; on Grice's program, 87; intentionality and, 244; on Marr's theory of vision, 153–4, 155; social dimensions in, 88, 89; thought experiments of, 85–7, 102, 113, 146, 151, 154, 206, 211
Byrne, Richard, 334

canalization, 69, 72
Caporael, Linda R., 334
Carey, Susan, 325
Carnap, Rudolph, 83, 316
Carpendale, J.I.M., 315
Carruthers, Peter, 333
Cartesian duplicates, 252–3, 259

causation, 46, 144; background conditions and, 121, 130–3; determination and, 79, 80; individualism and, 94, 96–8; mental states and, 94; realization and, 102, 130–3; scaffolded processes of awareness and, 219; supervenience theses and, 136–7; taxonomies and, 96–8, 156
Chalmers, David, 142, 214, 329
Changeux, Jean-Pierre, 314
Charcot, Jean-Martin, 272
chemistry, 125–6
Cheng, Patricia, 334
children, 59, 198–206
Cholewiak, R., 321
Chomsky, Noam, 67, 87, 312; I-/E-language and, 146; internal richness/external minimalism theses and, 16, 57; linguistic revolution led by, 51–2, 54; on Marr's theory of vision, 158–9, 161, 162
Cicourel, Aaron V., 333
Clark, Andy, 113, 142, 176, 317, 321, 326, 327, 330
Clements, Frederic E., 275, 278, 284, 332
clinical psychology, 29
cognition, 50–73; behaviorism and, 54–5; children and, 198–206; computational theories of, 10, 55, 88, 144–68, 174, 177–8, 180; connectionism and, 55–6, 88; constructive learning theory and, 59; continuity thesis and, 46–8, 72–3; embedded/embodied, 89, 165, 178, 184, 187–9, 210, 212; empiricism and, 15, 54–6, 72; in evolutionary psychology, 53–4, 302–3; explanatory strategies and, 66; groups and, 176; higher, 186–7, 215, 221, 223, 224; individuals and, 28; infant dependence and, 212; internal richness/external

minimalism theses and, 56–66, 67–8, 199; locational concepts of, locational externalism and, 174–9, 180, 211; nativism and, 50–4, 56–66, 67–8, 199, 200; primitivist account of, 62–5, 66; processes of awareness and, 186–7, 215, 217, 221, 223, 224; representational systems and, 184–7, 211–2; scaffolding for, 212, 218; socially distributed, 58, 175, 176, 291, 299, 302–3; social manifestation thesis and, 302–3
cognitive anthropology, 13
cognitive development, 15, 25, 52, 59, 212, 224; cultural mediation and, 200–1, 205–6; embeddedness of, 212; internal richness/external minimalism theses and, 59–60; representation and, 183, 198–206; social manifestation thesis and, 301
cognitive metaphor, 266–7, 288. *See also* group mind hypothesis
cognitive psychology, 52, 58, 183–4, 187, 188, 212–13
cognitive resources, 210, 212, 221–2, 224. *See also* scaffolding, cultural and environmental
cognitive science, 4; computational theories and, 25, 27, 88, 95, 113, 144–80; content and, 94; continuity thesis and, 45; dispositions in, 126; explanatory strategies and, 94; externalism in, 145, 148; individualism in, 9–10, 93–6, 140, 144–50, 241; individuals in, 5, 8, 41; nativism in, 14–17, 51–2; representations and, 147–79
cognitive science gesture, 144–5, 151, 174, 178
Cole, Michael, 196, 326
collective psychology, 267–8, 294, 297; in anthropology, 273; crowds in, 269–73, 278, 279, 283, 297, 300; decision making and, 297–8; irrationalism and, 272; multilevel traits and, 269, 281, 290, 291; social manifestation thesis and, 282–3; society-individual relationship in, 281; in sociology, 273
collectives, 106, 111. *See also* collective psychology; group mind hypothesis
Collier, G., 312
Collins, A., 321
Complete Audio-Visual Environment (CAVE), 234
computational theory of mind, 25, 27, 52, 95, 113; cognition and, 55, 88, 144–80; functionalism and, 161; individuals in, 5; locational concepts of, 169–70, 179; Marr and, 145, 150–74, 178–9; separatism and, 243; subsymbolic representation and, 148; wide, 113, 145, 165–72, 174, 177–8, 179–80, 301
Comte, August, 36–7, 38
conceptual analysis, 72–3
connectionists and connectionism, 16, 27, 47, 55–6, 57, 88, 148, 168, 174, 210
connection principle, 243
Connerton, Paul, 196–7
consciousness, 25, 41, 192, 214–41; externalism and, 215; group, 290; higher-order theories of, 217; minimal mindedness and, 293–4; phenomenal states and, 215–16, 225–40, 294; processes of awareness and, 215–25, 294. *See also* TESEE (temporally extended, scaffolded, embodied, embedded)
constitutivity thesis. *See* physical constitutivity thesis
constructive learning theory, 59
constructivism, 18, 199
content, 90–3, 94, 151; folk psychology and, 91–2, 94, 98, 207; narrow, 90–1, 92–3, 98, 145, 172–4, 229, 230–1; of phenomenal states, 226, 227–31, 253; phenomenology and, 231;

content (*cont.*)
　separatism and, 242; in vision theory, 155–62, 166, 172–4; wide, 92, 135, 137, 173, 253, 254
context-sensitivity, 289; dispositions and, 120, 125–8; microphysical determinism and, 120, 121–4; neural mechanisms and, 118; nonreductive materialism and, 120, 128–30; physicalism and, 120–30; realism and, 137–9; realization and, 102–3, 107–11, 117, 118–19, 120–43, 211; relational properties and, 121–4, 136–7; smallism and, 121–4
continuity thesis, 45–8, 72–3
copy theory of mind, 147
corpuscularianism, 22–3, 33–4, 97, 122
Cosmides, Leda, 20, 53–4, 61, 146–7, 334
Cowie, Fiona, 67, 68, 312, 315
Crane, Tim, 309, 318
Craver, Carl, 319
crowds, 269–73, 278, 283, 297, 300
cultural mediation, 196, 200–1, 205–6
cultural psychology, 287
cultural selection, 305–6
Cummins, Robert, 167, 319
Cushing, James T., 320

Damon, W., 312
Danziger, Kurt, 40, 43, 44
Darwin, Charles, 6, 53
Davidson, Donald, 89, 320, 333
Davies, Martin, 256
Dawkins, Richard, 6, 11
decision making, 98, 206, 286, 295–6, 297–8, 304–5
decomposition, 65, 114, 148, 208; analytical, 32, 95; constitutive, 127, 143, 179; entity-bounded realization and, 127; functionalism and, 112; homuncular, 111; psychological states and, 105
Dennett, Daniel, 89, 319, 329, 333

Descarte, René, 23, 33, 46, 220
descriptive theories of reference, 82–3, 85, 86–7
desires, 91, 186, 206, 245, 250
determination, 103, 104, 118; causal, 79, 80; dispositions and, 44; metaphysical, 79–80, 82, 122, 195; microphysical, 120, 121–4; representation and, 52; supervenience and, 78, 80; visual experience and
developmental biology, 69–71
developmental psychology. *See* cognitive development
developmental systems theory, 70–1
Devitt, Michael, 309
Dewey, John, 35
dispositional properties, 115, 125–8, 301
dispositions, psychological, 44, 120, 190, 300–1. *See also* embedded abilities
divisions of labor: cognitive, 204–5, 291; group, 284; social, 86, 89, 204, 255, 291
Donald, Merlin, 192, 195, 197, 218, 326
doppelgänger thought experiments, 82, 155, 156, 227, 229–30, 231. *See also* Twin Earth thought experiments
Douglas, Mary, 26, 287, 298, 299, 311
Dretske, Fred, 216, 226–32, 237, 238, 239, 240, 245, 255, 322
dualists and dualism, 48, 94
DuBois-Reymond, Emil, 32
Dupoux, Emmanuel, 314
Durkheim, Emile, 13, 38, 39, 267, 270, 273, 287, 299
dynamic cycles of perception/action theory, 235–8
dynamic systems theory, 27, 210

Ebbinghaus, Hermann, 30, 45, 189
Ebbs, Gary, 316, 321
ecology, 69, 71–2, 114, 266, 268, 274–5

economics, 5, 13, 28, 116
Edelman, Gerald, 314
Egan, Frances, 155, 158–62, 163, 164, 168, 309, 318, 328
E-language, 146
elusive I problem, 218
embedded abilities, 44. *See also* dispositions, psychological
embeddedness, 307; cognition and, 89, 165, 178, 184, 187–9, 210, 212; of processes of awareness, 220, 221; representation and, 148; of tactile experience, 234; of visual experience, 249
embodiment, 184, 210; of processes of awareness, 220, 221; representation and, 148; in tactile experience, 234; of visual experience, 249
Emerson, Alfred, 284
empiricists and empiricism, 15, 33–4, 45–8, 54–6, 57–8, 72, 83, 225
encoding, 147–50, 169, 183, 188, 222
environment, 89, 115, 117, 147, 167, 176, 282, 300; causal determination and, 79; child development and, 198, 201; memory and, 192–5; reactive representational systems and, 185
epiphobia, 104
ethologists, 20, 58, 279
eugenics movement, 42
Evans, Howard, 332
Evans, Mary, 332
evil demon hypothesis, 220
evolutionary biology, 5, 11–12, 26, 96, 295. *See also* natural selection
evolutionary psychology, 53–4, 57, 266, 302–3
explanations, development of, 202–5
explanatory strategies: decomposition and, 95, 105, 111, 112, 114, 127, 143, 179, 208; imaginative evocation as, 248, 252, 259; inexpressibility of, 173; integrative synthesis as, 32, 112–13, 114, 116, 121, 127, 143, 179, 211; introspection, 27, 34–5, 37, 38, 40–1, 140, 245, 251, 261; pragmatic separatism as, 243; Ramsey-Lewis sentences as, 140–1; taxonomies, 10, 13, 17, 81–2, 96–8, 166
externalism, 10, 25–6, 69, 130, 212–13; in cognitive psychology, 188, 212–13; in cognitive science, 145, 148; consciousness and, 215; determination and, 79–82; folk psychology and, 25, 208, 224; global, 216, 225, 237, 239, 240–1; individualism and, 77–9; locational, 174–9, 180, 191–2, 211, 217, 223, 236, 241; memory and, 189–98, 224; normativity and, 89–90; phenomenology of mental states and, 88; physical, physicalism and, 99, 128; realization and, 107; social, 86–7; social manifestation thesis and, 301; taxonomy and, 81–2, 98, 174–80; vision theory and, 154–5, 157–8, 160, 178. *See also* individualism
external memory fields, 192
external minimalism thesis, 46, 47; behaviorism and, 57; biological sciences and, 69–72; cognition and, 56–66, 67–8, 199; cognitive development and, 59–60; connectionism and, 57; empiricism and, 57–8, 72; nativism and, 56–66, 199; neurosciences and, 59; statement of, 15–22, 56

false belief task performance, 60
Fechner, Gustav, 31
fitness, 115, 126, 141
Flanagan, Owen, 214, 310
Fleck, Ludwig, 26, 287, 298–9, 311
fMRI (functional magnetic resonance imaging), 189

Fodor, Jerry, 52–3, 94, 96, 97, 144, 145, 150, 260, 309, 312, 319, 324, 327; epiphobia and, 104; internal richness/external minimalism theses and, 16, 57; methodological solipsism and, 10
folk psychological systems, 113, 126, 143
folk psychology, 42, 85, 206–10, 212, 232; bare-bones, 206–8, 211; cognitive science and, 94; embeddedness of, 212; externalism and, 25, 208, 224; full-blown, 207, 290; functionalism and, 161; individualism and, 94; intentionality and, 91–2; narrow/wide content and, 91–2, 94, 98, 207; normativity and, 89; representations and, 91, 183, 208–9; scaffolding and, 219; smallism and, 208; social manifestation thesis and, 301; taxonomy and, 98; wide realizations and, 114, 206
Foucault, Michel, 25, 30
Fournial, Henry, 270
fragile sciences, as terminology, 4, 8–9
Frege, Gottlieb, 82, 83
French, Peter, 322
Fricker, Elizabeth, 256, 330
functional analysis, 105
functionalism, 92, 93, 94–5, 101, 144; computational theories and, 161; decomposition and, 112; homuncular, 111, 112, 129–30; physical constitutivity thesis and, 105, 111

Galegher, J., 333
Galileo, 23
Galton, Francis, 41–4
Galtonian paradigm, 42, 44
Garber, Daniel, 310
Garfield, Jay, 60, 313
Gassendi, Pierre, 23
Geertz, Clifford, 20–1

Gelman, Susan, 61, 313, 314
generative entrenchment, 69, 72
genes, 6–7, 11–12, 18, 106
genetics, 11, 28
geology, 96
Gestalt psychology, 27
Gibson, James J., 171, 176
Giere, Ronald, 333
Gillett, Carl, 319
Glenberg, Arthur, 210
global externalism, 216, 225, 237, 239, 240–1
Godfrey-Smith, Peter, 315
Goffman, Irving, 36
Goldsmith, Morris, 190
Goodale, Mel, 233
Goodwin, Brian, 70, 316
Gopnik, Alison, 60, 61, 67
Gould, Stephen Jay, 309
Granrud, C., 325
Gray, Russell, 70
Greenfield, Patricia, 326
Grice, Paul, 87
Griffiths, Paul, 69, 316, 334
Grimm, Robert, 317, 322, 324, 329
group mind hypothesis: in anthropology, 266; in biological sciences, 26, 286–7, 298; collective psychology tradition and, 267–74, 278, 279, 297; in cultural psychology, 287; decision making and, 286, 295–6, 297–8, 304–5; in evolutionary biology, group-only traits and, 269, 279, 281, 283, 290, 292; mindedness and, multilevel traits and, 269, 281, 290, 291; natural selection theory and, 266, 277–8, 286, 303–6; religion and, 288, 296–7; social manifestation and, 280–5, 291, 292, 296, 297, 298, 303–4; in social sciences, 26, 266, 267, 287–8; statement of, 267; superorganism tradition and, 268–9, 274–80, 297. *See also* social manifestation thesis
groups, 11–12, 26; cognition and, 176; as individuals, 8, 266–7; minimal

mindedness and, 290–4; selection and, 277–8, 286, 295–6, 302–6; sociology and, 37, 287
groupthink, 298
Grush, Rick, 210, 322
Gunther, York, 327
Guttenplan, Samuel, 319, 328

Hahn, Martin, 329
Hardcastle, V., 310, 315
Harman, Gilbert, 245, 248
Harré, Rom, 312, 326
Hatfield, Gary, 311, 313
Haugeland, John, 90, 322, 327
Heil, John, 318, 327
Heller, M.A., 321
Helmholtz, Hermann von, 31, 32
Hering, Eward, 31
heritability of traits, 42–3, 44, 69. *See also* dispositions, psychological
higher-order theories of consciousness (HOT), 217
Hildreth, Ellen, 151
Hinton, Geoff, 322
Hirschfeld, Lawrence, 61, 313, 314, 325
historical properties, 98
Hitler, Adolph, 273
Hobbes, Thomas, 23
Hobson, Peter, 60
holism, 293
Holyoak, Keith, 334
Horgan, Terence: global internalism and, 240; inseparability thesis and, 244, 245, 246, 250–1, 260, 261; methodology of, 248, 252–3; phenomenal intentionality and, 252–3, 255, 257, 258–9, 260; separatism and, 242; supervenience and, 102, 133–4, 135, 137
horizontal modules, 236
Horst, Steven, 322
human agents. *See* individuals
human sciences, 8–9
human sciences, as terminology, 8–9
Hume, David, 45, 46, 218

Hurley, Susan, 184, 216, 235–8, 240, 249
Hutchins, Ed, 58, 113, 175–6, 178, 179, 327, 333

identity, personal, 3–4
I-language, 146
imaginative evocation, 248, 252, 259
individualism, 9–14, 25, 77–99; alternative, 136–7; in biological sciences, 11–12; causation and, 94, 96–8; cognitive psychology and, 187; in cognitive science, 9–10, 93–6, 140, 144–50, 241; determination and, 79–82; in evolutionary biology, 11–12; explanatory strategies and, 10–14; externalism and, 77–9; folk psychology and, 94; functionalism and, 93, 94–5; nativism and, 17, 18, 21–2; phenomenal intentionality thesis and, 255–60; phenomenal states and, 225–6; physicalism and, 81, 93–4; in psychology, 9–11, 28–9, 41, 49, 77, 211; smallism and, 23–4; social aspects of, 87–90; social manifestation thesis and, 300–1; in social sciences, 12–14; taxonomy and, 81–2, 96–8; wide content and, 135, 137. *See also* externalism
individuals, 3, 24–5, 31, 212–13, 265, 289–90; in anthropology, 5; in cognitive science, 5, 8; crowds and, 270–1, 273, 283, 300; disciplining of, 30; in economics, 5; in evolutionary biology, 5; groups as, 8, 266–7; human agents as, 4–6, 7, 8; memory and, 198; natural selection and, 6–7, 11–12; other entities as, 7–8; in philosophy, 39; in physiology, 39; in psychology, 28–9, 39, 40–4, 48–9; in sociology, 39; subjectivity and, 121, 141–3
informational semantics, 150
Ingle, D.J., 328
inner perception. *See* introspection
Innocent, The (McEwan), 208

insects, social, 26, 266, 268, 274, 275–7, 278, 283, 284
inseparability thesis, 244–52, 258–9, 260; grain of determinateness and, 245–8, 259; modal intensity and, 245, 248–50, 259; quantificational range and, 245, 250–1, 258–9
integrative cognitive networks, 296, 297
integrative synthesis, 32, 112–13, 114, 116, 143, 179, 211; social manifestation thesis and, 301; wide realization and, 121, 127
intentionality, 89, 90–2, 240; inseparability thesis and, 244–52, 258–9, 260; meaning and, 87; of memory, 191; of mental states, 88, 226; phenomenal intentionality thesis and, 244, 252–60; phenomenology and, 25–6, 88, 252–60; of representation, 157, 162; separatism and, 242–3, 246; wide, 94. *See also* inseparability thesis
interactionism, 18; radical, 71; symbolic, 36
internal richness thesis, 46, 47, 48; behaviorism and, 57; biological sciences and, 69–72; cognition and, 56–66, 67–8, 199; cognitive development and, 59–60; connectionism and, 57; empiricism and, 57–8, 72; evolutionary psychology and, 57; nativism and, 56–66, 199; neurosciences and, 59; statement of, 15–22, 56
interpretationism, 20–1
intrinsic properties, 13, 22, 24, 78, 80–2, 96–7, 113, 115, 122–4, 125, 126, 129, 136–7, 172, 204, 211
introspection: as explanatory strategy, 27, 34–5, 37, 38, 40–1, 140, 245, 251, 261; as process of awareness, 215, 217, 221, 223, 294
Inverted Earth thought experiments, 237, 240
irrationalism, 272

Jablonka, Eva, 70
Jackendoff, Ray, 145–6, 314
Jackson, Frank, 214, 230
James, William, 28, 31, 33, 35–6
Janis, Irving, 298
Jeannerod, Marc, 321
Johnson, Mark, 210
Jokic, Aleksandar, 327, 329

Kant, Immanuel, 35
Kanzi, 195
Karmiloff-Smith, Annette, 58
Kasher, Asa, 321
Keil, Frank, 61, 67–8, 202–3, 204, 205, 312, 315, 322, 323, 325, 326
Keller, Evelyn Fox, 316
Kendrick, D., 311, 333, 334
Khalidi, Muhammad Ali, 60–2, 313
Kim, Jaegwon, 101, 104, 105, 118, 129, 310, 319
Kimble, G.A., 311
Klatzky, Roberta, 321
Kohler, I., 236
Koriat, Asher, 190
Krajewski, W., 333
Kripke, Saul, 83
Kuhn, Thomas, 298–9, 333
Kunda, Ziva, 312
Kuper, Adam, 19

Lakoff, George, 210
Lamb, Marion, 70
Landau, Barbara, 61
language, 146; acquisition of, 15, 16, 20, 28, 51–2, 56, 60, 67, 68; content and, 92–3; meaning and, 82–7; mediational/scaffolding qualities of, 200–1, 212, 219; memory and, 192; natural kinds and, 84–5, 86, 173; social aspects of, 86–8
language of thought hypothesis, 52, 147, 150
latent learning, 55
Laurence, Steve, 52, 57
learning, 15, 28, 47, 55, 57, 198, 300; associative, 55; latent, 55; skill, 55

Le Bon, Gustav, 270–3, 282–3, 294
LeDoux, Joseph, 321
left-right inversion prism experiments, 236–8
Leibniz, Gottfried, 46
Leslie, Alan, 59, 65, 326
Levine, Joseph, 214
Lewis, C., 315
Lewis, David, 87, 104, 124, 310, 320, 321, 327, 328
Lewontin, Richard, 18, 70, 71
Lindzey, G., 312, 331
Little, Daniel, 310
Loar, Brian, 324; global internalism and, 240; methodology of, 248–9; phenomenal intentionality and, 244, 251, 252, 253–4, 255, 256, 257, 258, 260, 261
Locke, John, 23, 45, 46, 47–8, 190, 191
locus of control, 185, 197–8, 289; external, 185, 209; internal, 186, 187–8, 209
logical positivism, 28, 34, 98, 298
Lolita (Nabokov), 208
Lorenz, Konrad, 279
Lotka-Volterra equations, 114
Ludlow, Peter, 317
Luria, Alexander, 200
Lutz, Donna, 204, 326
Lycan, William, 216, 227, 240, 245, 309, 319

Machiavelli, Niccolò, 272
Machiavellian hypothesis, 302
MacKay, Charles, 331
MacKay, Donald, 328
Mansbridge, J.J., 334
Marcus, Gary, 314
Margendie, François, 32
Margolis, Eric, 52, 57
Marr, David, 145, 150–74, 178–9
Marx, Karl, 13
materialism, 120, 128–30, 225. *See also* physicalism
Matthews, Robert, 155
McClamrock, Ron, 311, 327

McClelland, James, 56
McCulloch, Greg, 327, 330
McDougall, William, 38, 267, 273, 274, 280–1, 294–5
McDowell, John, 90
McEwan, Ian, 208
McGinn, Colin, 255–6, 318
McKitrick, Jennifer, 126
McPhail, Clark, 283, 331
Mead, George Herbert, 36
mechanical philosophy. *See* corpuscularianism
medical sciences, 28
Mele, Alfred, 327
Meltzoff, Andrew, 61
memory, 25, 28, 189–98, 224, 232; collective, 197; correspondence metaphor of, 190–1; cultural mediation and, 196, 201; embeddedness of, 212; enactive model of, 191–2; environment and, 300; representation and, 183, 187; semantic, 192; social, 196–7; social manifestation thesis and, 301; storehouse model of, 190, 191; symbols and, 212
mental states, 10, 13, 25, 82, 101; causation and, 94; computer metaphors and, 95; localizations and, 166; methodological solipsism and, 77, 81–2; normativity and, 89–90; phenomenology of, 88; physical constitutivity thesis and, 107; supervenience theories and, 78; taxonomies of, 166. *See also* psychological states
mental unity, law of, 271, 283
Merrill, David, 322, 324, 329
metaphysical determination, 79–80, 82, 122, 195
metaphysical sufficiency thesis, 102–4, 107–11, 113, 117–18, 128, 129, 131, 134, 135, 137. *See also* physical constitutivity thesis
metarepresentations, 187–8, 208, 224
methodological solipsism, 10, 17, 77, 81–2, 144

methodologies. *See* explanatory strategies
microphysical determination, 120, 121–4
microphysicalism, relational, 124
Mill, John Stuart, 320
Miller, George, 184
Millikan, Ruth Garrett, 322
Milner, David, 233
mimetic skills, 187
mind-world constancy, 113, 164
Mirror Earth thought experiments, 237–8
Mishkin, Mortimer, 328
Mitchell, P., 314
Mitman, Gregg, 331
modularity, 5, 52–3, 57, 58, 59, 60, 199, 204, 302, 305; ToMM and, 65, 206; vertical and horizontal, 235–6; in vision theory, 153, 161, 171
Morgan, C. Lloyd, 279
Moss, Lenny, 316
Müller, Johannes, 32
multiagent systems. *See* group mind hypothesis
multiple spatial channels theory, 169
Murphy, Greg, 314
Mussolini, Benito, 273
Myth of the Given, 89

Nabokov, Vladimir, 208
Nagel, Thomas, 214
narrow function theory, 260
nativism, 25; in anthropology, 19–22; in behavioral ecology, 69; in biological sciences, 17–19, 68–72; cognition and, 50–4, 56–66, 67–8, 199, 200; in cognitive science, 14–17, 51–2; conceptual analysis and, 72–3; continuity thesis and, 45–8, 72–3; empiricism and, 54–6, 72; heritability and, 69; individualism and, 17, 18, 21–2; internal richness/external minimalism theses and, 56–66, 199; lineage of, 51–4; smallism

and, 23–4; social manifestation thesis and, 300; in social sciences, 19–22
natural kinds, 84–5, 86, 173
natural selection: altruism and, 6–7, 276, 277, 303; fitness and, 115, 126, 141; group mind hypothesis and, 266, 277–8, 286, 303–6; individuals in, 6–7, 11–12; selfish genes and, 6, 266–7, 302; social manifestation thesis and, 302–6; superorganism tradition and, 277–8
Nebergall, W. H., 320
Neisser, Ulric, 189, 191
neo-behaviorism. *See* behaviorists and behaviorism
neural correlates, 100, 118
neural mechanisms, 100, 118
neuronal selection theory, 59
neurophysiology, 29
neurosciences, 59, 88, 100, 118, 214
Newell, Alan, 323
Newton, Isaac, 22
nociceptive system. *See* pain
Noë, Alva, 216, 233, 236, 249, 328
Norman, Donald, 326
normativity, 45, 89–90
Nucci, Larry, 326
Nye, Robert, 331

Olson, David, 326
O'Regan, J. Kevin, 216, 233, 236, 249
Owens, Joseph, 318
Oyama, Susan, 70

pain, 107–9, 132, 138, 140, 141, 238, 242, 289; consciousness and, 215; inseparability thesis and, 244; as phenomenal state, 215, 216, 225–6, 227–30, 294; TESEE and, 240
Palmer, Stephen, 322
Park, Thomas, 275
particularism, 122–4
Pepperberg, Irene, 195
perceptual physiology, 31
perceptual psychology, 52

perceptual systems, 15, 54, 55, 113, 164, 169, 232, 235–8; multiple spatial channels theory and, 169
Perner, Joseph, 60, 67
Peterson, Candida, 60
PET (positron emission tomography), 189
Pettit, Philip, 89, 230
phenomenal intentionality thesis, 244, 252–60
phenomenal states, 215–6, 225–40, 294; bodily sensations as, 215, 225–6, 227, 230, 238, 242; content and, 226, 227–31, 253; individualism and, 225–6; inseparability thesis and, 244–52, 258–9, 260; pain as, 215, 225–6, 227–30, 238, 242; separatism and, 242–3, 246; TESEE and, 225, 232–40; visual experience as, 215, 216, 225, 226, 232–5, 238
phenomenology, 231; intentionality and, 25–6, 88, 240, 252–60. *See also* inseparability thesis
philosophy, 30, 31–6, 39, 101–2, 118
phrenologists, 32
physical constitutivity thesis, 104–7, 111–14, 116, 117–18, 121, 128, 133–4, 135, 137; physicalism, 128; token-token identity theory and, 128; total realization and, 130. *See also* metaphysical sufficiency thesis
physicalism, 48, 101, 118, 125, 144; conceivability of zombies and, 257; context-sensitivity and, 120–30; externalism and, 99, 128; individualism and, 81, 93–4; realization and, 103, 104, 107, 120–30; supervenience thesis and, 122. *See also* materialism
physics, 124
physiology, 31–4, 36, 37, 39
Piaget, Jean, 199
picture theory of mind. *See* copy theory of mind
Pierce, Charles Sanders, 35

Pinker, Steven, 20, 56, 314
plant communities, 26, 274–5, 283–4
Plato, 45
Poggio, Tomas, 151
Poland, Jeffrey, 319
Polger, Thomas, 319, 323
Popper, Karl, 13, 53, 310
population ecology, 114
Port, Robert, 322
positron-emission tomography (PET), 28
predator-prey systems, 114–15
primary qualities, 23
primitivism, 62–5, 66
Prince, The (Machiavelli), 272
Prins, Herbert, 287
Prinz, W., 328
Prior, Elizabeth, 320
problem solving, 58, 193–5
processes of awareness. *See* awareness, processes of
properties: dispositional, 115, 125–8, 301; historical, 98; intrinsic, 13, 22, 24, 78, 80–2, 96–7, 113, 115, 122–4, 125, 126, 129, 136–7, 172, 204, 211; relational, 24, 97, 98, 114, 121–4, 125, 127–8, 130, 136–7; subvenient, 80–1
psychiatry, 28
psychological primitives, 62
psychological states: computational theories of, 167–8; decomposition and, 105; methodological solipsism and, 10, 77. *See also* mental states
psychology, 27–30; collective psychology tradition and, 273; computational, 161–2; departments of, 27–8; disciplining of, 8, 30–1; explanatory strategies of, 28, 34–5, 36, 40–1; individualism in, 9–11, 28–9, 41, 49, 77, 211; individuals and, 28–9, 39, 48–9; philosophy and, 31–6, 39; physiology and, 31–6, 39; psychophysics and, 29, 31; realization and, 100, 118; social sciences and, 28–30, 36–40

psychophysics, 29, 31
psychosemantics, 150
Putnam, Hilary, 101, 144; divisions of labor and, 89; externalism of, 78, 86–7, 204, 254; intentionality and, 244; natural kinds and, 82, 83–5; thought experiments of, 90–3, 102, 113, 146, 206, 211
Pylyshyn, Zenon, 323

Quartz, Steve, 59
Quetelet, Adolphe, 42

radical interactionism, 71
Ramberg, Bjorn, 329
Ramsey-Lewis method, 140–1
rational choice theory, 13, 28
rationalists and rationalism, 33, 45–8, 72
rational morphology, 70
realism, 121, 132, 137–9
realization, 25, 32, 95, 97, 100–19; background conditions and, 109–11, 116, 121, 130–3; in biological sciences, 106, 114–16; compositional view of, 128–9; context-sensitive, 102–3, 107–11, 117, 118–19, 120–43, 211, 289; core, 107–11, 118, 126, 128, 130, 131, 137–8, 289; dispositions and, 120, 125–8; entity-bounded, 112, 117, 121, 127, 140–1; folk psychological states and, 114, 206; individualism and, 102, 105–6, 117; metaphysical sufficiency and, 102–4, 107–11, 113, 117–18, 128, 129, 131, 134, 135, 137; in neuroscience, 100, 118; in philosophy, 101–2, 118; physical constitutivity thesis and, 104–7, 111–14, 116, 117–18, 121, 128, 130, 133–4, 135, 137; in psychology, 100, 118; social manifestation thesis and, 301; in social sciences, 106, 116; standard view of, 102, 103–7, 117–18, 130–7, 145, 289; total, 108–11, 112, 128, 130, 131, 137–9, 166, 208; wide,

112–16, 117–18, 127–8, 130, 139–42, 165, 167, 206, 208;
reasoning experiments, 53
reductionism, 22, 102, 280
Reed, Edward S., 311, 324
reference, theories of, 82–3, 85, 86–7
reflexes, 186
relational microphysicalism, 124
relational properties, 24, 97, 98, 114, 121–4, 125, 127–8, 130, 136–7
religion, 3–4, 288, 296–7
Rensink, R.A., 328
representation, 92, 145–6, 147–79, 210; cognitive psychology and, 183–4, 198–206; collective, 39, 273; connectionism and, 148, 210; determination and, 52; distributed, 148; enactive systems of, 185, 186, 188–9; encoding and, 147–50, 169, 183, 188, 222; exploitative, 145, 162–72, 176–7, 178, 183–4, 209, 210, 222, 223; folk psychology and, 91, 183, 208–9; intentionality of, 157, 162; memory and, 183, 187; metarepresentations and, 187–8, 208, 224; phenomenal state theories of, 226, 245; primitive, 154, 155, 157, 174; reactive systems of, 185, 186, 188; subsymbolic computation and, 148; symbolic systems of, 186, 187, 188, 211–12, 218–19
research strategies. See explanatory strategies
Richards, Whitman, 322
Richerson, Peter, 334
Riggs, K.J., 314
rigidity assumption, 163
robotics, 177
Roediger, H.L., 325
Rogan, Michael, 321
Ross, Edward A., 37–8
Rossi, Pasquale, 270
Rowlands, Mark, 171, 192, 197, 311, 323
Rozenblit, Leonid, 202–3, 326

rule-following, 89
Rumelhart, David, 56, 322
Runeson, Sverker, 323
Rush Hour game, 193–5
Russell, Bertrand, 82

Sady, W., 333
Salmon, Nathan, 317
Samuels, Richard, 60, 62–5, 66
Sanders, A.F., 328
Savage-Rumbaugh, Sue, 195
scaffolding, cultural and environmental, 217, 218–19, 221. *See also* cognitive resources
Schank, Roger, 317
schema theory, 13, 28
Schiller, J., 311
Schopenhauer, Arthur, 35
Searle, John, 214, 243, 322, 329
Seeley, David, 287
Segal, Gabriel, 155–8, 163, 172, 173, 317, 324, 330
Sejnowski, Terry, 59
selfish gene, 6, 266–7, 302
self-knowledge, 215, 216, 294
self-observation. *See* introspection
Sellars, Wilfrid, 89
sensorimotor contingency theory, 216, 233–5, 236, 238
sensory experiences. *See* phenomenal states
sensory systems, 15, 54, 55
separatism, 242–3, 246
Shapiro, Lawrence, 161, 163, 319, 323
Shelford, V.E., 284, 332
Shoemaker, Sydney, 104, 107–8, 239–40, 319, 329
Shore, Bradd, 312–13, 325
Shweder, Richard, 287
Sidelle, Alan, 317
Sighele, Scipio, 270
Silvers, Stuart, 318
Simpson, J., 311, 333, 334
simulationists, 60
single-cell recording, 28
situated cognition. *See* embeddedness

skepticism, 46, 225
skill learning, 55
Skinner, B.F., 54
smallism, 22–4, 34, 212, 310; context sensitivity and, 121–4, 129; folk psychology and, 208; realization and, 105–7, 121–4, 129; relational properties and, 24; taxonomy and, 97
Smith, Quentin, 327, 329
Smith, Roger, 8, 312
Smolensky, Paul, 322
Sober, Elliott, 11, 305, 309, 334
social actions, 131, 139, 141, 299
social cognition, 36
social interactions, 201, 212, 219
sociality and asociality, 41, 88, 89, 266, 275–8, 302
social manifestation thesis, 281, 291, 292, 296, 297, 298, 299–302; cognition and, 302–3; cognitive development and, 301; externalism and, 301; folk psychology and, 301; group mind hypothesis and, 280–5, 291, 292, 296, 297, 298, 303–4; group selection and, 302–6; individualism and, 300–1; integrative synthesis and, 301; memory and, 301; nativism and, 300; realization and, 301. *See also* group mind hypothesis
social psychology, 26, 29, 36–8, 40
social sciences, 4–7, 8, 38; collective psychology tradition and, 273, 274; disciplinization of, 8; dispositions in, 126; group mind hypothesis and, 26, 266, 267, 287–8, 298–9; individualism in, 12–14; nativism in, 19–22; psychology and, 28–30, 36–40; realization and, 106, 116
sociobiology, 22
sociology, 26, 36–40, 273, 287
spatiality, 3, 123; multiple spatial channels theory and, 169; processes of awareness and, 218; of tactile experience, 234, 235; of visual experience, 169, 170

368   Index

species, 7, 11, 96, 114, 275, 286
Spelke, Elizabeth, 312, 315, 325
Spencer, Herbert, 53, 280
Sperber, Dan, 320, 326, 334
Spinoza, Benedict, 46
Stalnaker, Robert, 77
Sterelny, Kim, 322
Stich, Stephen, 61, 94, 144, 145, 317
Strawson, Galen, 242–61
Stryker, Sheldon, 312
subjectivity, 121, 141–3
subvenience theses, 80, 104
subvenient properties, 80–1
sufficiency thesis. *See* metaphysical sufficiency thesis
suggestions and suggestibility, 38–9, 272
superorganism tradition; "Chicago school" and, 26, 266, 268–9, 274–80, 284, 297; group-only traits and, 269, 279, 281, 283, 290, 292; social insects and, 26, 266, 268, 274, 275–7, 278, 283, 284; social manifestation thesis and, 283–4
supervenience theses, 10, 13, 24, 77, 78, 80, 82, 93–4, 96–7, 100, 102, 104, 118; causation and, 136–7; determination and, 78; explanation and, 204; physicalism and, 122; realization and, 105; regional, 133–4, 135, 137
symbolic interactionism, 36

tactile experience, 234, 235
tactile visual sensory systems (TVSSs), 234
Taine, Hippolyte, 270
Tarde, Gabriel, 270
taxonomies, 10, 13, 17, 81–2; causation and, 96–8, 156; externalism and, 81–2, 98, 174–80; memory and, 190; of mental states, 166
Taylor, J.G., 236–7, 328
teleosemantics, 150
Teller, Paul, 122, 320, 321
temporality, 3; of perceptions/actions, 236; processes of awareness and,

217–18, 220, 221–2; of tactile experience, 234; in vision theory, 153, 157, 158, 161, 170, 171
TESEE (temporally extended, scaffolded, embodied, embedded), 25, 215–16; global externalism and, 240–1; phenomenal states and, 225, 232–40; processes of awareness and, 217, 220, 223, 225, 232; sensory experiences and, 235
theory of mind module (ToMM), 65, 206
theory-theorists, 60
Thompson, Evan, 328
thought collective *(Denkkollektiv)*, 26, 287, 298–9
thought experiments, 102, 146, 151, 206, 211, 239–40; brain-in-a-vat, 220, 253, 256–8, 259; *doppelgängers* in, 155, 156, 229–30, 231; inseparability thesis and, 248–50; Inverted Earth, 237, 240; Mirror Earth, 237–8; Twin Earth, 82, 83–4, 90–3, 113, 172–3, 228, 252, 257
Tienson, John: inseparability thesis and, 244, 245, 246, 250–1, 260, 261; methodology of, 248, 252–3; phenomenal intentionality and, 252–3, 255, 257, 258–9, 260; separatism and, 242
Tinbergen, Nikko, 279
token-token identity theories, 99, 128–9
Tomasello, Michael, 60
Tomberlin, James, 330
Tooby, John, 20, 53–4, 61, 146–7, 334
transformation thesis, 283
Trotter, William, 274
Turing machines, 101
Turner, J.C., 334
Twin Earth thought experiments, 82, 83–4, 113, 172–3, 228, 252, 257 content and, 90–3
Tye, Michael, 216, 227, 240, 245

Ullman, Shimon, 158, 163
Ungeleider, Leslie, 328

van Gelder, Tim, 322
van Ginneken, Jaap, 331
van Gulick, Robert, 318
vertical modules, 235–6
virtual reality experiences, 234
vision, theory of, 145, 150–74, 178–9; algorithmic level in, 152, 153, 154, 160–1, 164; computational level in, 152–3, 159–61, 164, 179; externalism and, 154–5, 157–8, 160; implementational level in, 152; integrative synthesis and, 179; modular dimension of, 153, 161, 171; temporal dimension of, 153, 157, 158, 161, 170, 171
visual experience, 215, 225, 226, 232–8, 249; content and, 226, 231; metaphysical determination of, sensorimotor contingency theory of, 216, 233–5, 236, 238; spatiality and, 169, 170
Vygotsky, Lev, 60, 196, 200–2, 205–6

Wade, Michael, 269, 331
Walsh, Denis, 133, 134–7, 318
Wason, Peter, 53
Wathen-Dunn, W., 328
Watkins, John, 13, 310
Webster, Gerry, 316

Wellman, Henry, 326
Wertsch, James, 326
Wheeler, William Morton, 268, 275–7, 278
White, Stephen, 230, 260
Whiten, Andrew, 334
Williams, George, 11, 309
Wilson, David Sloan, 291, 309, 311; group mind hypothesis and, 26, 286–7, 295–8; group minds and, 303–5; group selection and, 11, 26, 274; religion and, 288
Wilson, Edward O., 20, 274, 276
Wilson, Robert A., 309, 310–11, 313, 314, 315, 316, 317, 318, 319, 320, 321, 322, 326, 330, 332, 333, 334
Wimsatt, William, 315
Woodfield, Andrew, 317
Wundt, Wilhelm, 31, 33, 34–6, 38, 40–1, 251, 261, 273, 282
Wundtian paradigm, 43
Wynn, Karen, 315, 325

Yablo, Stephen, 319

Zola, Emile, 270
zombies, 249, 257
zoology, 9